AF566955

ADHESION OF DUST AND POWDER

ADHESION OF DUST AND POWDER

Anatolii D. Zimon
Institute of Physical Chemistry
Academy of Sciences of the USSR
Moscow, USSR

Translated from Russian

Translation Editor
Morton Corn
Department of Occupational Health
Graduate School of Public Health
University of Pittsburgh
Pittsburgh, Pennsylvania

PLENUM PRESS • NEW YORK–LONDON • 1969

Anatolii Davydovich Zimon was born in Moscow in 1924. In 1952 he was graduated from the Moscow Chemical Engineering Institute, and since 1959 he has done research work under the direction of Academician B. V. Deryagin, the foremost Soviet scientist in the field of dispersed systems and a Fellow of the Institute of Physical Chemistry of the Academy of Sciences. Zimon specializes in the physical chemistry of surface phenomena and particle/surface adhesion.

АНАТОЛИЙ ДАВЫДОВИЧ ЗИМОН

АДГЕЗИЯ ПЫЛИ И ПОРОШКОВ

ADGEZIYA PYLI I POROSHKOV

Library of Congress Catalog Card Number 69-12547

A Division of Plenum Publishing Corporation
227 West 17th Street, New York, N.Y. 10011

United Kingdom edition published by Plenum Press, London
A Division of Plenum Publishing Company, Ltd.
Donington House, 30 Norfolk Street, London W.C.2, England

Printed in the United States of America

Foreword

This monograph is the first systematic treatment of the subject of particle adhesion (particles to surfaces) and autohesion (particles to particles). The author first attacks the subject from the theoretical standpoint of the physical and chemical factors giving rise to adhesion under ideal conditions of particle shape and surface characteristics, electrical forces, and substrate properties. The cases of adhesion in air and liquid media are differentiated, but the common factors are very nicely discussed. Following the establishment of the theoretical bases of the subject and a discussion of widely utilized experimental methods to obtain adhesion data for particles, the author proceeds to review available data for a variety of situations of practical importance where particle adhesion is intrinsic to the phenomenon. These topics include adhesion of particles to paint and varnish coatings, adhesion and removal of particles in an air flow and in a flow of water, adhesion of particles in air-cleaning equipment, particle adhesion in electrophotography, and soil erosion and silt transport.

The level of treatment of the subject should appeal to the beginner as well as to the advanced scientific worker. In all cases, Zimon starts at the beginning and derives fundamental equations. The coverage of the vast literature on the subject will surprise even the expert reader, as will the ubiquitousness of particle adhesion and its fundamental importance to so many aspects of our technological environment. Finally, the reader will admire the skill with which the author has unified the apparently diverse aspects of particle adhesion.

Pittsburgh, Pennsylvania
July 1969

Morton Corn

Preface

Whereas the results of recent investigations into the adhesion of films and gluing processes were generalized in earlier monographs [1-5], current information relating to the adhesion of dust particles and powders is spread out over a number of articles published in various specialized journals, or alternatively forms merely a component part of specific fundamental treatises. Thus, Deryagin and Krotova's monograph [1] was mainly devoted to the adhesion of films, setting out the theory of interactions between solid bodies and considering the relation between adhesion and friction, while Fuks [6, 7] considered certain questions relating to the sticking of particles in an air flow. There have been a number of theoretical and experimental treatments relating to the adhesion of particles in a liquid medium (Deryagin, Fuks, and Buzach [8]). In these, adhesion was studied as a function of the properties of the liquid bounding the contiguous solid particles and the thickness of the liquid layer, using simulation techniques to model the interaction of the particles. In our own investigations, we have developed and perfected methods of determining forces of adhesion and have made an attempt at analyzing the causes responsible for this phenomenon; we have determined the relationship between the forces of adhesion and the properties of the contiguous bodies and surrounding medium, and we have considered the conditions required for removing particles under the influence of air flows, water flows, an electric field, and so forth.

The number of books and papers devoted to the study of particle adhesion alone is very limited; at the same time, there are many publications in which this phenomenon is considered in conjunction with others.

Adhesion is widely used in industry and agriculture. For example, this phenomenon plays a considerable part in the filtration and separation of dry materials, the cleaning of surfaces, and spraying, in electrophotography, in the treatment of plants with pesticides, etc. Adhesion plays no small part in processes taking place in nature. In the absence of adhesion, dust settling on the ground would be continuously returned to the atmosphere by air currents and its concentration would reach vast proportions. On the other hand, soil-erosion processes are largely due to the failure of the soil particles to interact sufficiently with each other.

Adhesive properties sometimes have a decisive effect on the choice of methods and conditions for the preparation, preservation, application, and transportation for powdered materials. These properties must never be neglected when constructing and preparing the working parts of mechanisms intended for working with powders.

Thus, the existence of a certain number of publications of a theoretical and experimental nature on the adhesion of dusts and powders, on the one hand, and the wide use of this phenomenon in various fields of production, on the other, constitute a reasonable basis for writing a monograph. The author was presented with the by-no-means easy problem of analyzing, classifying, and systematizing disconnected and sometimes contradictory information regarding the adhesion of particles; he nevertheless hopes that this work will in some measure aid the reader in understanding the phenomenon of adhesion and guide subsequent investigations in the direction of a deeper and more systematic understanding of the problem.

The author wishes to thank Professor B. V. Deryagin, who promoted the scientific aspect and guided the direction of the investigations, Professors G. I. Fuks, N. A. Fuks, and I. S. Adamovich and V. E. Titov, N. N. Fursov, and B. I. Myakov, and other colleagues who read the monograph in manuscript form and offered valuable suggestions.

The author would also be grateful to any readers who wish to make comments.

A. Zimon

Contents

Chapter I

Fundamental Concepts of the Adhesion of Particles

§1. Adhesion of Particles

The Concept of the Adhesion of Particles. Usually in defining the concept of "adhesion" (sticking) one understands the adhesion of liquids or films to a solid surface [9]. The interaction of particles with a plane surface is often known as "adherence," and the interaction of particles among themselves as "conglomerance." This terminology is neither universal nor really suitable. For the sake of uniformity in terminology, and also in view of the fact that processes taking place when films, dust particles, and powders come into contact with a solid surface are all of an analogous nature [2], we shall call the interaction of particles with a solid surface adhesion and the interaction of particles among themselves autohesion. Adhesion and autohesion are due to analogous causes.

In order to distinguish the adhesion of paints to a substrate from the sticking of dust to paint coatings, we propose calling the latter the "dust-holding capacity of paint coatings" [10].

Cohesion is a phenomenon akin to adhesion and autohesion. By "cohesion" we understand the attraction between the molecules within the bounds of a single body.

Adhesion is often treated as the molecular coupling between two different contiguous bodies (phases) [9]. This definition is to some extent valid for the adhesion of films and paint coatings, since, in such cases, the effect of the surrounding medium may be neglected; however, it cannot reflect the whole complexity of the processes taking place during the adhesion of particles to a solid surface.

In an air (or gaseous) medium, microscopic particles adhere to a solid surface not only as a result of molecular forces, but also under the influence of capillary forces in the liquid condensing in the space between the contiguous particles, the double electric layer formed in the zone of contact, and also Coulomb interaction and other like causes. Coulomb forces arise between charged particles and may considerably exceed the molecular forces. This fact is used in particular for holding pesticide particles sprayed in an electric field on the leaves of plants.

In the presence of oil pollution or an adhesive film, the sticking of particles is due to the "tackiness" of the substrate. In this case either the adhesive interaction at the solid—oil interface or the cohesive forces of the viscous film on the solid substrate are overcome when the particles are detached (see Chapter III).

The adhesive forces of microscopic particles in a liquid medium are made up of the molecular attraction of contiguous bodies and the repulsive forces of the thin layer of liquid in the contact zone (see Chapter IV).

Hence, the adhesion of dust particles and powders constitutes the interaction of microscopic particles with a solid surface due to forces which depend both on the properties of the bodies in contact and on those of the surrounding medium.

In this monograph we shall consider the adhesion of solid microscopic particles with solid substrates in gaseous and liquid media. We must therefore pay some attention to the concept of "microscopic particles," i.e., estimate their sizes, which are related both to the conditions required for the very existence of the particles and to the properties of the contiguous bodies and surrounding medium. Since an estimate of particle size is only of interest for present purposes in connection with the possibility of revealing adhesive properties, the main criterion must be the adhesive forces required to ensure the retention of the particles on the surface. The minimum particle size is in general limited by the meaning of the word "micro." By analogy with the minimum sizes of colloidal particles, the lower limit in the present case is taken as a quantity of the order of 10^{-7} cm. The upper limit of particle size is hard to set unequivocally; it may, for example, increase on changing the external medium or conditions of contact

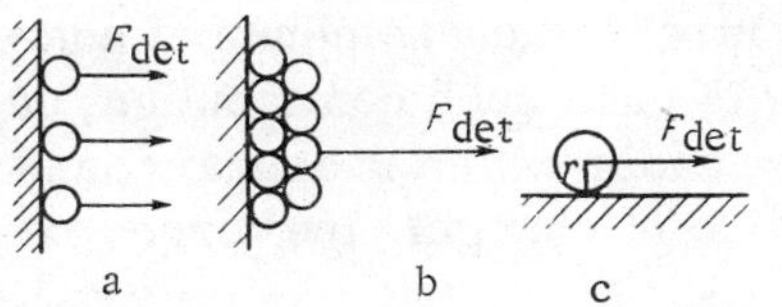

Fig. I.1. Possible cases of the detachment of particles: detachment of a monolayer (a) and a thicker layer (b) by a normal force, and of an individual particle by a tangential force (c).

for the same type of contiguous bodies. Thus, the maximum size of the particles capable of being held on some paint coatings may be 10^{-2} cm (100 μ) [10, 11]. However, if the surface is contaminated with oil or carries an adhesive film, the upper limit of particle size may be much greater.

Macroscopic particles are those for which the forces of interaction with a plane surface are smaller than their weight.

Classification of Adhesion. Adhesion is classified by reference to a variety of indices, which may include: the properties of the medium, the forces determining adhesion, the number of particles sticking, the change in adhesion in the course of detachment, and the direction of the detaching forces.

The forces of adhesion are to a great extent determined by the properties of the surrounding medium. Thus, 90% of glass particles 40-60 μ in diameter adhering to steel surfaces may be detached by a force of $2 \cdot 10^{-1}$ dyn in air or $8 \cdot 10^{-4}$ dyn in water. Thus, in a liquid medium the adhesion of particles is much smaller than in air [11]. In view of this it is practically convenient and theoretically justifiable to distinguish adhesion in air (gas) and liquid media [11].

In the presence of water vapor (or the vapor of any other liquid) in the surrounding air, drops of vapor may condense between the contiguous bodies (the particle and the plane surface). If the depth of the layer of water (or other liquid) in the gap between the bodies in contact exceeds the height of the meniscus formed on wetting these bodies, the ashesion may be regarded as of the liquid type.

Adhesion is caused by a variety of forces, differing in their nature. Under certain conditions individual contributions to the forces of adhesion may prevail over others. Thus, in an air humidity of over 70% the increase observed in the adhesion of microscopic particles is due to capillary forces. If the particles are highly charged, the value of the Coulomb forces exceeds that of other contributions to adhesion.

In order to discover how to control the phenomenon of adhesion, we must make a detailed study of each such contribution, determine its nature and the conditions under which it appears, and also establish what factors are capable of changing the forces responsible for adhesion.

Adhesion may also be classified by reference to the number of interacting particles. If there are only a few particles, then these frequently have no contact with each other and form a so-called monolayer on the surface (Fig. I.1a). If there are many particles, then these form a layer consisting of one or several rows of particles in contact with each other, i.e., connected by autohesive interaction (Fig. I.1b).

In the case of the adhesion of a monolayer (Fig. I.1a), the detaching force acts on each particle, and if $F_{det} > F_{ad}$ (the latter being the adhesive force) the adhering particles will be detached from the surface. When whole layers adhere to the surface (Fig. I.1b), the force acts on all the particles forming the layer or layers. The strength of this layer depends not only on its adhesion to the surface, but also on the autohesion of the particles themselves. If $F_{ad} < F_{det} < F_{aut}$ (the latter being the force of autohesion), adhesive detachment will occur; if $F_{ad} > F_{det} > F_{aut}$, autohesive detachment will take place. If $F_{ad} \sim F_{aut}$, there may be mixed adhesive—autohesive detachment.*

A classification of adhesion as a function of the changes in the interaction between contiguous bodies in the course of detachment was given by Deryagin [1] on the basis of the analogy between adhesion and friction. (Friction prevents the tangential motion of particles and adhesion prevents the motion of particles in a direction perpendicular to the surface on which they have been deposited.)

By analogy with friction, we may distinguish static and kinetic adhesion. Static adhesion is characterized by a resistance to the onset of detachment, and kinetic adhesion by the interaction between the particles and the surface in the course of detachment.

* In view of the fact that we shall subsequently only consider the adhesion of a layer of particles in air, the rheological properties of the adhering layer need not be taken into consideration.

In order to detach particles from a surface, the main problem is to overcome the force of static adhesion, since the kinetic adhesion is always smaller than this. This was pointed out by Fuks [12], who indicated that static friction was measured by the force directed tangentially to the substrate.

The detachment of dust particles (static adhesion) depends on the magnitude and direction of the force applied to the particle. If the forces are applied normally to the dusty surface (Fig. I.1a), then for the particles to be detached we must have $F_{det} > F_{ad}$. For a tangentially directed force (Fig. I, 1c) the moment of the forces, i.e., $M_{det} = F_{det} \cdot r$ (r = particle radius) is operative. The first stage in the detachment process will in this case be the rolling or sliding of the particle, i.e., friction as well as adhesion will have to be overcome.

§ 2. Estimating the Value of the Adhesion

Direct Methods. Let us first consider methods of estimating the adhesion of a monolayer of particles. Adhesion is characterized by the force which arises on contact between particles and various surfaces. For particles having roughly the same diameters, the forces of adhesion will not necessarily be exactly the same under similar conditions [13], i.e., a particular detaching force will only remove a certain proportion of these particles, and not all. Noting this, Buzach* introduced the concept of the adhesion number [14]. According to Buzach, if we deposit particles on a surface by sedimentation in a liquid, the adhesion number (γ_F) equals the ratio of the number of particles (N) remaining after rotation of the plate to the number (N_0) originally deposited on the plate from the liquid.

Subsequently, the concept of the adhesion number [13, 15] was extended, and at the present time we understand it to mean the ratio of the number of particles (N) remaining after the application of a given force to the number (N_0) originally attached to the test

* We note, by the way, that Buzach determined the rolling angle of all the particles with respect to a vertical plate and deduced the adhesion number from this, rather than the distribution of the particles with respect to adhesive forces.

surface. The adhesion number is often expressed in percents:

$$\gamma_F = \frac{N}{N_0} \cdot 100$$

Adhesion may also be characterized by the number α_F (in %), equal to the ratio of the number of particles detached under the influence of a specific force to the original number:

$$\alpha_F = 100 - \gamma_F$$

The relation between the adhesion number and the forces holding singly dispersed dust particles on a surface is generally characterized by integral curves of adhesive force [13] (Fig. I.2). On the integral curves the force of detachment may be expressed either in absolute units or in units of g.* Fuks [12] called the force of detachment expressed in units of g, i.e., the ratio of the detaching force to the weight of the particles, the *coefficient of adhesion*.

Sometimes the force of detachment is expressed indirectly, for example, in the form of the number of rotations of a centrifuge, the frequency of the oscillations of a vibrating plate, the angle of rotation of a dusty surface, etc. (see Chapter II for more details).

Adhesion under different conditions may be compared both by reference to the force of detachment (or retention) of an equal number of particles, and also by reference to the adhesion number (i.e., the number of particles remaining) for the same value of the detaching force

In estimating adhesion by reference to the detaching force, we recognize the existence of a "minimum" force (F_{min}), under which the first few (e.g., [15], 2%) of the particles are removed, and a maximum force (F_{max}), under which the majority of the particles are removed (so that, for example, only 2% remain [16]). However, the concept of F_{min} is rather indefinite, while the detachment of the last few particles, especially in air, is quite difficult; as a rule, the integral curves of adhesion forces degenerate into a straight line with decreasing adhesion number. In addition to this, there are certain cases in which the valves of F_{min} and F_{max} of a

*g is the gravitational acceleration, equal to 980 cm/sec^2.

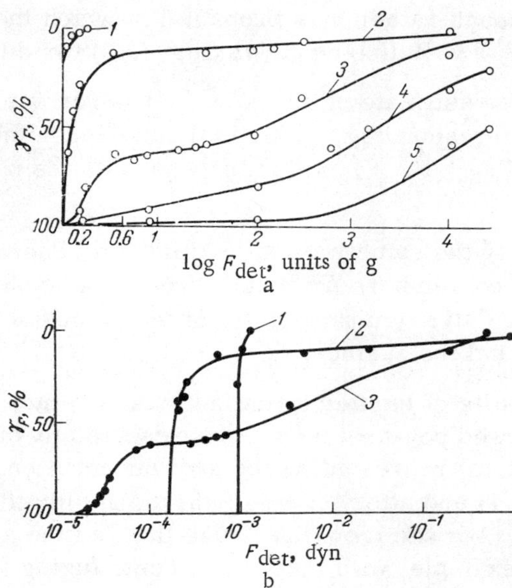

Fig. I.2. Integral adhesion curves for spherical glass particles of different diameters adhering to a steel surface of the 13th class of finish in units of g (a) and in absolute measure (b). 1) d_p = 80-100; 2) 40-60; 3) 20-30; 4) 10-20; 5) 5-10 μ.

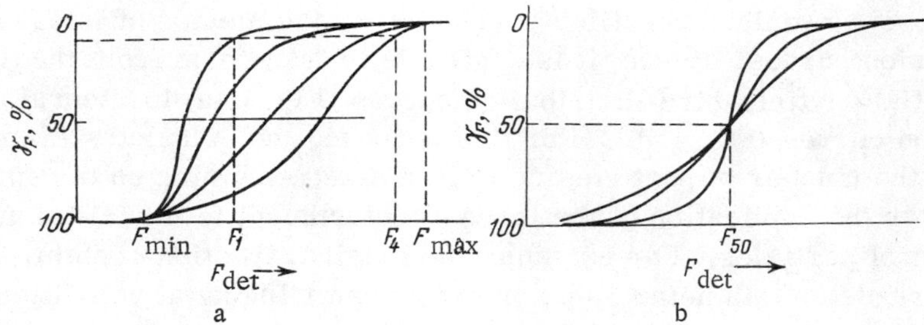

Fig. I.3. Families (a, b) of integral adhesive-force curves.

family of integral curves (Fig. I.3a), all come together at equal adhesion numbers; this applies to both F_{min} (according to Fuks, at γ_F = 98%) and F_{max} (according to Kordecki and Orr [16] at γ_F = 2%). However, F_1 is certainly much smaller than F_4, i.e., in the first case the majority of the particles are more easily detached than in the fourth. It is clear that here it is better to estimate adhesion [13] by reference to the force of detachment for an adhesion number of 50% (see Fig. I.3a). It should be noted that there may be cases, although rather less probable, in which these forces of detachment (F_{50}) will also be equal (Fig. I.3b) for different curves.

Thus, any estimate of adhesion by reference to the forces of detachment corresponding to identical adhesion numbers (i.e., by reference to F_{min}, F_{max}, or F_{50}) incorporates the possibility of error.

In exactly the same way, an estimate of adhesion by reference to adhesion numbers for equal forces of detachment may only be used as a relative characteristic of the forces acting between the particles and the surface.

The results of earlier investigations [16] into the detachment of poly-dispersed powders may be represented by differential curves (Fig. I.4a) representing the size distribution of the particles before (curve 1) and after (curves 2-5) the application of the detaching force. We see from the figure that, as the applied force increases, for example, with increasing centrifuging velocity of the dusty surface, the number of adhering particles diminishes, the larger ones tending to be removed; the maxima on the differential curves move in the direction of smaller particles.

However, despite the possible errors and inaccuracies, it is more convenient to estimate adhesion by the method of integral adhesion curves. Hence, it is desirable to transform from the differential particle-size-distribution curves (Fig. I,4a) to integral adhesion curves (Fig. I.4b). For this purpose, one calculates the ratio of the number of particles of a given size remaining on the surface after the application of the force of detachment to the initial number of particles. For example, the original fraction contains 34 particles of diameter 50 μ, and after centrifuging at velocities of 1500, 5850, 9600, and 13,050 rpm there remain, respectively, 25, 15, 3, and 1 particles. On dividing the number of particles remaining by the initial number, we obtain the adhesion numbers, which are; respectively, equal to 73.5, 44, 9, and 3% (curve 4, Fig. I.4b). In an analogous way the adhesion numbers may be calculated for particles of other sizes and the integral curves plotted.

The forces of adhesion acting on the particles may be expressed by the following equation:

$$\Sigma F_{ad} = \int_{F_0}^{F_{100}} x f(x)\, dx \tag{I.1}$$

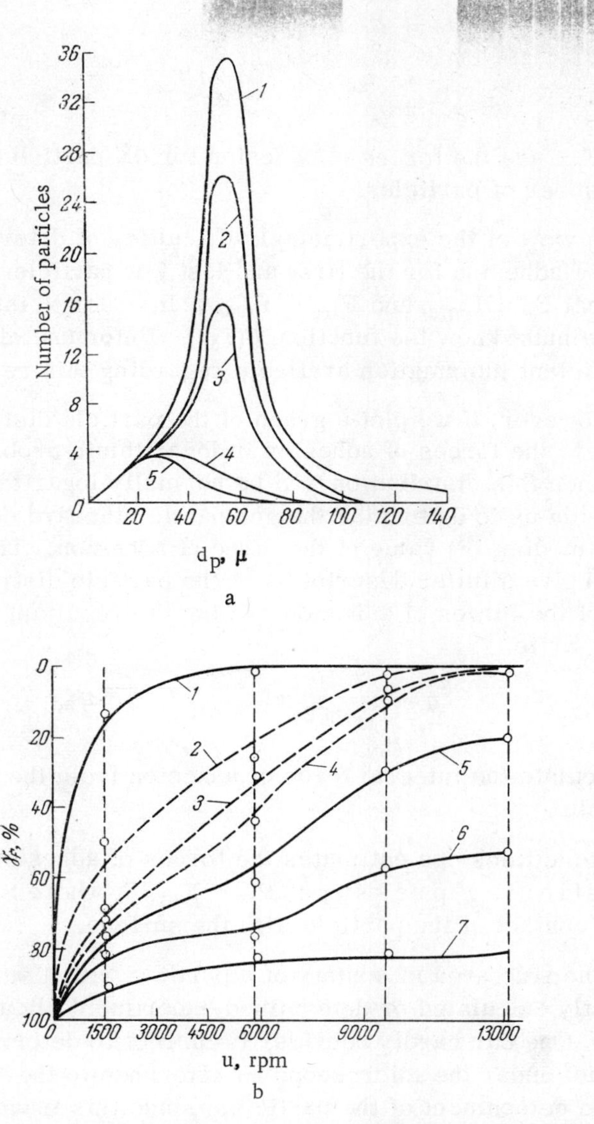

Fig. I.4. a) Differential particle-size-distribution curves. 1) Initial; 2-5) after centrifuging at u = 1500, 5850, 9600, and 13,050 rpm. b) Integral curves characterizing the adhesion of particles: 1) d_p = 100; 2) 90; 3) 60; 4) 50; 5) 40; 6) 30; 7) 20 μ.

with

$$x = F_{ad}; \qquad f(x) = f(F_{ad}) = \frac{dN}{N}$$

where F_{ad} are the forces of adhesion for dN particles, and N is the total number of particles.

In view of the experimental difficulties of determining the forces of adhesion for the first and last few particles, we may consider that $F_0 = F_{min}$ and $F_{100} = F_{max}$. In order to integrate Eq. (I.1), we must know the function $f(F_{ad})$. Unfortunately, there is not yet sufficient information available regarding this relationship.

However, if we plot a graph of the particle distribution with respect to the forces of adhesion in logarithmic probability coordinates, then this distribution will be normally logarithmic, which will enable us to determine the geometric standard deviation (σ), and the median ($\overline{F}$) value of the force of adhesion. The parameters σ and $\overline{F}$ give a fuller description of the particle distribution in terms of the forces of adhesion. Using the resultant parameters, we may write

$$\frac{dN}{N} = \frac{0.43}{F_{ad} \cdot \sigma \sqrt{2\pi}} e^{-\frac{(\log F_{ad} - \log \overline{F})^2}{2\sigma^2}} dF_{ad}$$

and calculate the integral force of adhesion from the resultant relationship.

Sometimes one estimates the forces of adhesion by reference to the *sticking pressure* $P_S = F_{ad}/S$, where S is the true area of contact of the particle with the surface.

The true area of contact of a particle with a surface cannot be exactly calculated or determined experimentally under real conditions. One can hardly consider it correct to determine the area of contact under the microscope by reference to the "trace" left after the detachment of the particles, since this trace may be established in a tacky, or at any rate inelastic, substrate, contradicting the very concept of a sticking pressure. For adhesion in an air medium, the elastic deformation of the zone of contact will be different at different points. The mean integral value of the sticking pressure may be expressed in the form

$$\bar{P}_S = \frac{1}{S} \int\limits_0^{P_{S,\max}} dP \qquad \text{(I.2)}$$

The maximum pressure of a spherical particle is determined by the formula [17]

$$P_{S,\max} = 0.578 \sqrt[3]{\frac{F_{\text{ad}}}{\eta_c^2 \cdot r^2}} \qquad \text{(I.3)}$$

where η_c is the elastic constant of the materials of the bodies in contact, and r is the radius of the sphere.

The quantity η_c is equal to

$$\eta_c = \frac{1 - \mu_1^2}{E_1} + \frac{1 - \mu_2^2}{E_2} \qquad \text{(I.4)}$$

where μ_1, μ_2 and E_1, E_2 are the Poisson ratio and Young's modulus of the particle and the surface, respectively.

Even if it were possible to measure the area of contact between a particle and the surface, the mean value of the sticking pressure would not be an objective characteristic of adhesion, owing to the nonuniformity of the pressure distribution in the contact zone [see formulas (I.2)-(I.4)].

Let us now consider methods of estimating the adhesion of a full layer of powder; these differ from those employed in estimating the adhesion of a monolayer. We must first note some inaccuracies in the conventional methods described by Cremer, Buzach, and other authors.

Cremer, Conrad, and Kraus [18] proposed expressing the adhesive force (F_l) of a layer of powder in terms of the area occupied by this layer on the surface of the plate and the dimensions of the particles

$$F_l = a \frac{S_l}{d_p} \qquad \text{(I.5)}$$

where a is a coefficient having the dimensions of surface tension, dyn/cm, S_l is the area of the plate on which the layer of powder lies, and d_p is the diameter of the particles.

The coefficient a should not depend on the dimensions of the particles; it is determined solely by the properties of the bodies in contact. Hence, this coefficient may be used for calculating the adhesive force of a powder layer formed by particles of different diameters to substrates of the same type under the same conditions. However, later investigations [19, 20, 21] showed that a did, in fact, depend on the particle size, and this cast doubt on the validity of Cremer's calculations.

Chicherin et al. determined the force required to detach a layer of dust from a glass cloth skeletal filter by a centrifugal method in the presence of a pressure simultaneously acting on the adhering layer [22]:

$$F_c = F_{hr} + F_l \tag{I.6}$$

where F_c is the centrifugal force detaching the powder (per unit surface), F_{hr} is the pressure arising from the hydraulic resistance of the dust layer associated with the blowing of air through it, and F_l is the adhesive force of the dust layer.

In the opinion of the authors of [22], F_l is the force corresponding to the autohesion of individual particles in the layer,i.e., the specific strength of the powder layer. However, this assertion does not entirely agree with the facts. For $F_{ad} > F_{aut}$, we in fact have $F_l = F_{aut}$, but if $F_{ad} < F_{aut}$, we have $F_l = F_{ad}$, and then the detachment of the particle layer is of the adhesive type.

Buzach [14] proposed calculating the adhesive force of a powder layer from the equation

$$F_l = N\pi r^2 P_s \tag{I.7}$$

where N is the original number of particles, r is the particle radius, and P_s is the specific sticking force (or sticking pressure).

It follows from (I.7) that the adhesive force of a powder layer is proportional to the middle section of the particle and not the area of actual contact between the particle and the surface. In view of the fact that the middle section is usually tens or even hundreds of times greater than the actual contact area of the particles, the values of specific sticking force determined from Eq. (I.7) are low in comparison with experimental values. Hence, the specific sticking force is a relative quantity characterizing the adhesion of a layer of powder; it is not equal to the true adhesive force of the powder layer. Other authors [23, 24] have also calculated the adhesive force of a layer by reference to the specific sticking force.

In order to secure a more accurate estimate of the adhesive force we must base our calculations on the number of contacts between the particles and the surface rather than the cross-sectional area of the particles [21]. Then the adhesion of a particle layer (F_l) may be expressed by the equation

$$F_l = F_{ad} \cdot N \tag{I.8}$$

After determining F_l experimentally, and knowing N, we may use (I.8) to calculate F_{ad} and compare the result with the measured adhesive force.

This method firstly eliminates the inaccuracies associated with the indeterminacy of the area of contact (it should be noted that the area of true contact between the particle and the surface has still not been defined), and, secondly, it enables us to compare the adhesive force of a complete layer of powder with the adhesive force of a monolayer, i.e., to compare the two cases of adhesion.

The following data were obtained by experimentally studying the detachment of layers of spherical glass particles on inclining a surface (a steel surface worked to a Class 9 finish) and by calculating the adhesive forces of individual close-packed particles on the basis of these experimental results, using Eq. (I.8):

d_p, μ	60-90	40-60	20-30	10-20
F_l (referred to 1 cm^2), dyn	1.1	21.7	208	370
F_{ad} (refered to one particle), dyn	$6.7 \cdot 10^{-5}$	$4.8 \cdot 10^{-4}$	$1.2 \cdot 10^{-3}$	$0.7 \cdot 10^{-3}$

On comparing the adhesive force of the particles F_{ad} calculated from Eq. (I.8) with experimental data relating to the detachment of a monolayer, it is easy to see* that F_{ad} corresponds to the force exerted by the most weakly held particles of the monolayer, i.e., to the initial part of the integral adhesion curves. Hence, when

*This comparison is only possible in connection with the adhesion-type detachment of the particle layer.

a powder layer is detached from a surface by inclining the latter it is the mean adhesive force of the easily removed particles which is measured. Sliding off, these particles remove the remaining particles on the avalanche principle. If the adhesive force between the layer and the substrate is greater than the autohesion in the layer, then detachment takes place with respect to the weakest autohesive interparticle bonds.

Indirect Methods. The Mackrle brothers [25, 26], studying the adhesion of mineral particles in the presence of $Al(OH)_3$ to the surface of filter grains in the filtration of water, estimated the adhesion by reference to the Ma adhesion-similarity criterion. For particles composed of a material with a density close to that of water, the authors neglected gravitational and inertial forces. On such an assumption it may be considered that a particle situated in a moving flow is acted upon at the surface of a filter grain by a van der Waals attraction and a resistive force due to the viscosity of the medium (the authors also neglected the disjoining pressure of the thin layer of liquid, although this was not absolutely justified, see §20). Then a particle of suspended matter will be acted upon by an attractive force (see §4)

$$F_{ad} = \frac{Ad_p}{12H^2} \qquad (I.9)$$

and an opposing force

$$F_{det} = 3\pi\eta d_p v_H \qquad (I.10)$$

Here, A is a constant, d_p is the diameter of a suspended particle, H is the gap between the particles and the surface, η is the viscosity of the liquid, and v_N is the velocity of the particle in a direction normal to the surface.

If $F_{ad} = F_{det}$, then from Eqs. (I.9) and (I.10),

$$3\pi\eta v_N = \frac{A}{12H^2} \qquad (I.11)$$

and for all similar systems,

$$\frac{A}{v_N \eta H^2} = \text{const} = \text{Ma} \qquad (I.12)$$

Thus for such systems the quantity Ma should be the same.

However, it is insufficient for the similarity of two filtration processes simply that the adhesion-similarity criteria should be the same; it is also essential to consider the mechanical similarity criteria of the moving liquid, i.e.,

$$f(\mathrm{Ma}, \mathrm{Ho}, \mathrm{Fr}, \mathrm{Eu}, \mathrm{Re}) = 0 \tag{I.13}$$

where Ho is the homochronic criterion, Fr the Froude number, Eu the Euler number, and Re the Reynolds number.

For stationary filtration under isothermal conditions,

$$\mathrm{Ho} = 0, \ \mathrm{Fr} = 0, \text{ and } \mathrm{Eu} = f(\mathrm{Re})$$

Then Eq. (I.13) simplifies considerably and takes the form

$$\mathrm{Ma} = f(\mathrm{Re}) \tag{I.14}$$

By analyzing experimental data relating to the filtration of $Al(OH)_3$ and $Fe(OH)_3$ particles 1-25 μ in diameter, first subjected to coagulation, an equation in the following form was obtained:

$$\mathrm{Ma} = \mathrm{Re}^{2.16} \tag{I.15}$$

From Eq. (I.15) we may determine the proportion of particles held as a function of the height of the filter charge for any rate of filtration of the suspension in question.

It should be remembered, however, that Eq. (I.15) is only valid for specific particle and filter-grain sizes. Equation (I.15) has been obtained on the basis of a number of important limitations (namely, that the density of the particle matter equals the density of water and the disjoining pressure of the liquid is not taken into account), and this narrows the prospects of calculating forces of adhesion by the similarity method.

A change in the properties of the suspension, in particular the turbidity of the water flow [27], and also the rheology of cohesive dispersed systems, are due to forces of autohesion. However, other still insufficiently understood factors also affect these processes. Hence, Kurgaev's attempt to associate autohesion with the rate of compaction of the residues of certain suspensions cannot be considered successful or his calculations of the forces of autohesion reliable [29].

Another method (theoretically better based) has been proposed for calculating the forces of autohesion from the limiting shear stress of the suspension (P_{sh}), taking the number of coagulation bonds per unit area into consideration [30, 31]. Yakhnin and Taubman [31] related P_{sh} to the properties of the medium

$$P_{sh} = k \frac{F_{ant}}{d_p^2 \rho (V_s - V_N)} - P_e \tag{I.16}$$

where k is a coefficient depending on the shape of the particles and also the proportion of structural particles in the chain, F_{aut} is the force of autohesion, ρ is the density of the particles, V_s is the volume of the system formed by the particles, V_N is a constant characterizing the volume of the particles, allowing for their packing, and P_e is an empirical correction taking account of the inhomogeneity of the packing and the unrealized strength.

However, owing to the indeterminacy of the empirical corrections, even this method of calculation is only of theoretical value and cannot be used in practical calculations of the quantity F_{aut}.

§ 3. Adhesion and Friction

Detachment of Particles. The force acting perpendicularly to the dust-laden surface on an adhering particle determines the value of the static adhesion. If this force is directed tangentially to the surface, then the static friction is measured when the particle becomes detached. Under practical conditions the detaching force may be directed at an oblique angle to the dust-laden surface.

The conditions under which the detachment of the particle is possible are expressed in the following manner:

$$F_{det} \geqslant \beta F_{ad} + (1 - \beta) F_{fr} \tag{I.17}$$

where F_{det}, F_{ad}, and F_{fr} are, respectively, the forces of detachment, static adhesion, and friction, while β is the proportion of the forces of detachment used in overcoming the forces of adhesion.*

For a force of detachment directed perpendicularly to the surface $\beta = 1$ and $F_{det} \geq F_{ad}$; if the force is directed tangentially,

*The quantity β may be indirectly determined from the adhesion number.

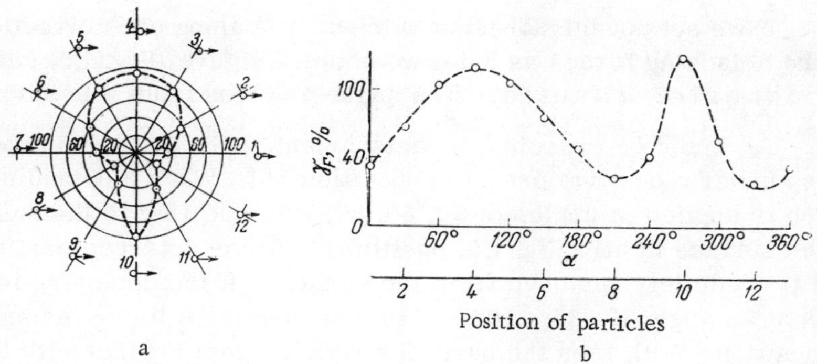

Fig. I.5. Adhesion rosette (a) and its development (b) obtained for the adhesion of spherical glass particles 20 to 30 μ in diameter to a steel surface of the 13th class of finish for a detaching force of $4.1 \cdot 10^{-3}$ dyn.

then $\beta = 0$ and $F_{det} \geq F_{fr}$. The friction of the adhering particles (in the absence of an external load) is due to adhesion: $F_{fr} = \mu F_{ad}$ (where μ is the coefficient of friction). Then for $\beta = 0$, $F_{det} \geq F_{ad}$.

The relation between the adhesion numbers determined by the tangential and normal detachment forces is expressed by empirical formulas obtained as a result of an analysis of experimental data [13]:

$$\gamma_{\alpha/N} = \frac{\gamma_\alpha}{\gamma_N} = 1 - 0.6\,(\cos\alpha)^3 \tag{I.18}$$

$$\gamma_{\alpha/T} = \frac{\gamma_\alpha}{\gamma_T} = 2.5 - 1.5\,(\cos\alpha)^3 \tag{I.19}$$

where γ_N, γ_T, and γ_α are the adhesion numbers for a given detachment force acting normally (N), tangentially (T), or at an angle of α to the dust-laden surface (α is the angle between the direction of the force of detachment and the plane).

Equations (I.18) and (I.19) are obtained with angles of $180° \geq \alpha \geq 0°$ for the adhesion of spherical glass particles to steel surfaces [13]; by using these equations and the adhesion numbers for the normal or tangential force of detachment, we may determine the adhesion numbers $\gamma_{\alpha/N}$ and $\gamma_{\alpha/T}$ for the case in which the detaching force is directed at an angle α ($\alpha \neq 0$ and $\alpha \neq 90°$) to the dust-laden surface.

If we set out the adhesion number (γ_F) along each direction of the detaching forces as axis, we obtain a figure (Fig. I.5) called the adhesion rosette. The angle α varies from 0 to 360°.

As would be expected, no detachment of the particles takes place under a positive pressure (position 10). When a detaching force is applied at angles of 30, 60, 90, 120, and 150° to the dust-laden surface in air (Fig. I.5, positions 2-6) the detached particles are immediately removed from the surface. If the detaching force acts at an angle of 180, 210, or 240° (the particles lie on the surface in positions 7-9), then the particles retain direct contact with the surface as they move. Here the force of interaction with the surface equals the kinetic sticking force. When particles adhering to the underside of the horizontal plane are detached (positions 11, 12, and 1) the centrifugal force is aided by gravity, which breaks the contact between the particles and the surface.

In an air medium $\gamma_N > \gamma_T$. Fuks [12] observed an analogous relationship in the adhesion of quartz particles 0.8-15μ in diameter in water and in solutions of certain electrolytes. In this case, $\gamma_N = (1.2\text{-}2.0)\gamma_T$, depending on the electrolyte concentration. After the particles have been detached by a force directly tangential to the dust-laden surface they execute a complex motion: rolling, sliding, and jumping.

The frictional force in air equals [32]

$$F_{fr} = \tau S \tag{I.20}$$

where S is the area of true contact and τ is the shear resistance of the adhesive bonds.

If, however, the contact is made in a liquid medium, the force of static friction will be [33]

$$F_{fr} = n_1 S \tau_M + n_2 S \tau_H \tag{I.21}$$

where n_1 and n_2 are the parts of the nominal area of contact on which microscopic shear (τ_M) between surfaces separated by a monomolecular layer or layers and shear in the actual boundary layer (τ_H), respectively, take place.

Since the area of contact is proportional to the normal force (F_N), we obtain an equation known as the Amonton law of friction [34]:

$$F_{fr} = \mu F_N \qquad (I.22)$$

However, the Amonton law has not yet been verified for the friction of individual particles. It is hard to measure the frictional force associated with the sliding of microscopic particles experimentally and to ensure true sliding without simultaneous rolling of the particles. It is therefore more convenient to consider the friction and adhesion of monolayers and layers of powder rather than individual particles.

Detachment of a Layer of Particles. In removing a layer of particles, the particles slip or slide along the surface. Under the influence of autohesive forces the particles in the layer form a continuous mass, which prevents them from rolling or being detached.

In earlier investigations on the friction associated with the sliding of a layer of powder, particles of relatively large dimensions (0.1-2.5 mm in diameter) and irregular shapes were used. The particles were separated into fractions by a sifting method [34]. Later [21], spherical glass particles less than 100 μ in diameter were studied after preliminary separation into constant-size fractions by an air-fractioning procedure. Steel surfaces of the 9th class of surface finish were given continuous powder layers over an area of 3 cm^2 (all the experimental data were referred to 1 cm^2 of the test surface).

By varying the number of powder layers, i.e., the mass of the powder, the force with which the layer of powder presses on the surface may also be varied:

$$F_N = mg \cos\alpha$$

where m is the mass of the powder and α is the slope of the surface.

The force of detachment for the layer

$$F_{det} = mg \sin\alpha \qquad (I.23)$$

determines the friction of the powder layer as this slips off the inclined surface. If Amonton's law given in formula (I.22) were true for the detachment of the powder layer, then the straight line characterizing the relation between $mg \sin\alpha$ and $mg \cos\alpha$ would have to pass through the origin of coordinates. Analysis of experimental

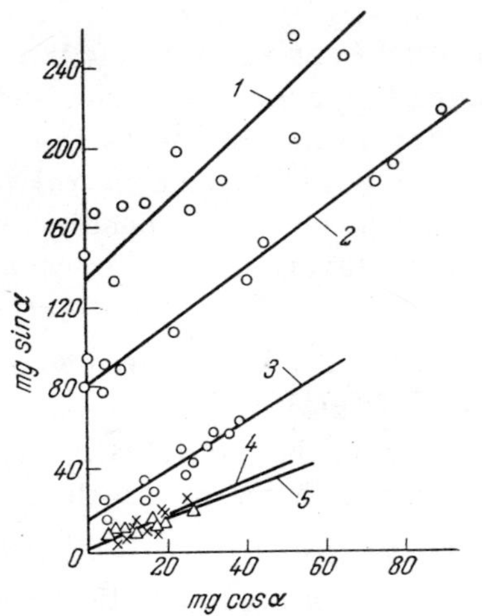

Fig. I.6. Forces of adhesion and friction in a powder layer as functions of the particle size: 1) d_p = 10-20; 2) 20-30; 3) 40-60; 4) 60-90; 5) 100-150 μ.

data (Fig. I.6) shows that this kind of relationship is only valid for relatively large particles (more than 100 μ in diameter), i.e., for particles such that the force of interaction with the surface is smaller than their weight [13] (macroscopic bodies).

The detachment of microscopic bodies on rotating the dust-laden surface will be resisted by both friction and adhesive forces, and this is not allowed for in the Amonton law [21].

The Deryagin Two-Term Friction Law. In the generalized law of friction proposed by Deryagin [35], both the external load and the adhesive interaction are taken into account:

$$F_{fr} = \mu (F_N + F_0) \quad (I.24)$$

$$F_0 = P_S \cdot S \quad (I.24a)$$

where F_{fr} is the force of static friction, F_N is the normal force, F_0 is the force of interaction between the surfaces in contact, μ is the coefficient of friction, S is the area of true contact, and P_S is the sticking pressure acting on the area S.

Strictly speaking, the quantity F_0 is not equal to the adhesive force, since, under the influence of the normal force F_N, the zone of contact will be flattened and the adhesive interaction will increase. This flattening may not occur if the actual and nominal contact areas between the two bodies are equal. This situation is brought about by the fusing of the two bodies [36] (Wood's alloy), by using material capable of plastic deformation [37], and also in the adhesion and friction of steel surfaces separated by a lubricating layer [38]. In the latter case the straight line obtained (Fig. I.7), which represents a linear relationship between the friction of the surfaces and the normal force, does not pass through the origin

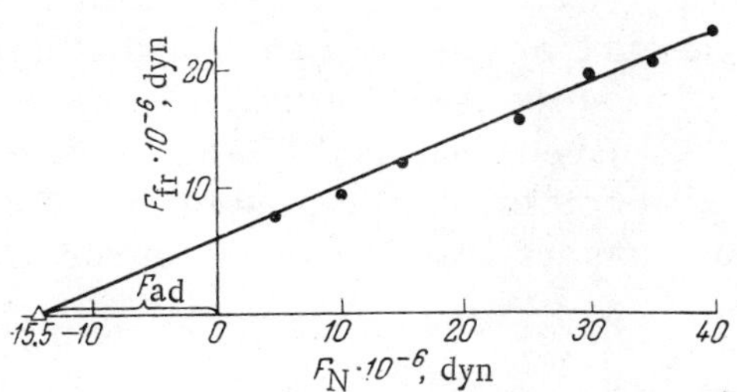

Fig. I.7. Friction of steel surfaces worked to the 14th class of surface finish as a function of the load in the presence of a layer of lubricant (oleic acid) 0.01 μ thick between the surfaces.

of coordinates. In the absence of load, i.e., when $F_N = 0$, the friction between the surfaces is due to the adhesion F_{ad}, which corresponds to the intercept on the x axis made by the straight line $F_{fr} = f(F_N)$.

The forces of adhesion may either be calculated in the manner indicated or measured directly, for example, by the detachment of the steel surfaces from one another [39].

The forces of adhesion to steel surfaces of the 14th class of surface finish obtained experimentally* (measured with a triboadhesiometer) and derived by extrapolation of $F_{fr} = f(F_N)$ straight lines are compared below:

Lubricant	Oleic acid	Industrial lubricant	Vaseline oil
$F_{ad} \cdot 10^{-6}$, dyn			
Measured	15.7	19.6	17.6
Extrapolated	15.2	21.8	18.6

The investigations of Akhmatov experimentally confirmed that the second term in Eq. (I.24) of the Deryagin friction law was due to the forces of adhesion.

Remembering that under conditions of complete lubrication the asperities on the steel surfaces are leveled out, we may consider that the actual area of contact of the steel plates is equal to

*Lubricant layer 0.01 μ thick; plate area 2.7 cm^2.

the nominal area determined by the dimensions of the sample. This offers the possibility of calculating the specific force of adhesion from (I.24a). It must be emphasized once again that Eqs. (I.24) and (I.24a) are valid in practice if one of the bodies in contact is made of ductile (easily deformed) material [36-38]. When the actual and nominal contact areas differ sharply, the forces of adhesion only increase the actual contact area of the contiguous bodies.

On the basis of the two-term law (I.24) we may express the equilibrium of the forces at the instant at which the powder layer is detached from the inclined surface, as follows [21]:

$$mg \sin \alpha = \mu mg \cos \alpha + F_l \tag{I.25}$$

where m is the mass of all the particles (layer of powder) in g, μ is the coefficient of friction, and F_l is the adhesive force of the powder layer in dynes.

Then the intercepts cut off by straight lines 1, 2, and 3 on the y axis of Fig. I.6 will, according to Eq. (I.25), be equal to the forces of adhesion of the powder layer F_l.

For $F_N \gg F_0$ [see (I.24)] the two-term law passes into the Amonton law. The condition may be satisfied either by increasing the load, i.e., F_N (for microscopic particles this is difficult to do, since only the particles' own weight affects their sticking), or by reducing or eliminating the forces of adhesion. Such an elimination of the forces of adhesion may be achieved if we measure the friction in a liquid medium where, owing to the disjoining pressure, the molecular forces of interaction between the surfaces may, in fact, not occur [40]. We see from (I.25) that estimates of adhesion based on the force of detachment ($mg \sin \alpha$) are not exact, since, as the angle α changes so does the pressure of the powder on the surface ($mg \cos \alpha$) which, in turn, affects the particle—surface interaction.

§ 4. Theory of the Molecular Interaction and Adhesion of Particles

Molecular interaction is characterized by van der Waals forces appearing between the molecules at a distance equal to between one and several hundred molecular diameters. The gap between the contiguous bodies is no greater than a few molecular diameters in either the adhesion of particles (in the absence of layers of liquid between these) or the adhesion of films. Hence, molecular interaction exerts an appreciable influence on the development of adhesive forces [1-5].

The interaction energy between two molecules i and j lying at a distance of H may be calculated from the equation [41]

$$E_{i,j} = -\frac{\lambda_{i,j}}{H^6} \tag{I.26}$$

where $\lambda_{i,j}$ is a constant characterizing the orientational (${}_{o}\lambda_{i,j}$), inductive (${}_{i}\lambda_{i,j}$), and dispersive (${}_{d}\lambda_{i,j}$) interactions:

$$\lambda_{i,j} = {}_{o}\lambda_{i,j} + {}_{i}\lambda_{i,j} + {}_{d}\lambda_{i,j} \tag{I.27}$$

The minus sign in Eq. (I.26) arbitrarily denotes an attractive energy.

Calculations show that the inductive and orientational effects are insignificant and need not be considered, particularly for condensed bodies [41, 42]. Then $\lambda_{i,j} = {}_{d}\lambda_{i,j}$ and ${}_{d}\lambda_{i,j}$ may be determined from the formula

$${}_{d}\lambda_{i,j} = -\frac{3}{2} h \left(\frac{\nu_i \cdot \nu_j}{\nu_i + \nu_j} \right) \alpha_i \cdot \alpha_j \tag{I.28}$$

where h is Planck's constant, ν_i and ν_j are the frequencies of the oscillations of the interacting electron oscillators, and α_i and α_j are the polarizabilities of molecules i and j.

A characteristic feature of the dispersion (London) forces is their additivity. A molecule induces periodic dipoles in several neighboring molecules. The induced dipole is attracted to the original dipole. In view of this, the energy of attraction between the two bodies may be regarded as the sum of the energies of attraction between the corresponding pairs of molecules forming the bodies in question.

Strictly speaking, London interaction is valid for two very rarefied systems, i.e., gases. The extension of the additivity of the forces to condensed systems not constituting a simple sum of free molecules has not yet been given a firm theoretical basis.

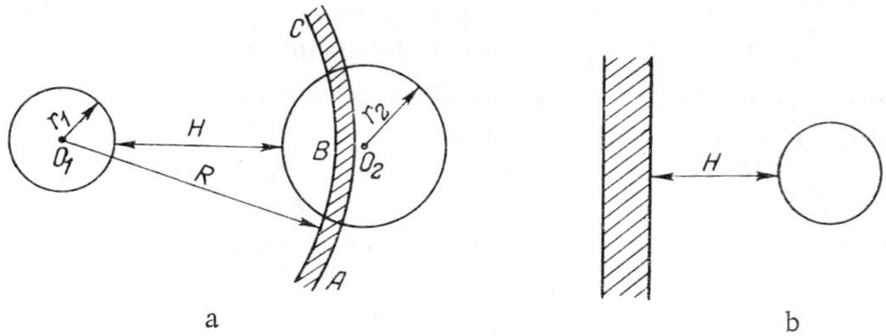

Fig. I.8. Interaction between two spherical particles (a) and between a particle and a plane (b).

However, Bradley [43] experimentally determined the force of interaction between two quartz and borate spheres; the result was close to the calculated value obtained on the principle of additive molecular interaction. Hence, we may *a priori* accept the additivity of London interaction and extend this to condensed systems, since, at the present time, there are no other methods of estimating the molecular interaction of such bodies when these are separated by a small gap.

Hamaker indicated the transformation from the molecular interaction energy to the energy of interaction in condensed systems [44].

The energy of interaction between elementary volumes of two spherical surfaces (Fig. I.8a) containing n molecules per cm^2 is, according to Eq. (I.16), equal to

$$E = -\int_{V_1} dV_1 \cdot \int_{V_2} \frac{n^2 \cdot \lambda_{i,j}}{H^6} dV_2 \qquad (I.29)$$

where $\lambda_{i,j}$ is the London constant, H is the distance between the molecules, and V_1 and V_2 are the total volumes of the two spheres.

The integration of Eq. (I.29) is cumbersome and there is no point in carrying it out completely. We need only give an idea of the integration and the way in which the power of H is reduced on passing from interaction between molecules to interaction between condensed bodies.

Let us site one molecule of the first body arbitrarily at a particular point (Fig. I.8a) and from this describe a cone with a generator R cutting off part of the medium in the second body. A dispersion interaction will appear between the first molecule and all the molecules in the elementary volume of the neighboring particle, i.e.,

$$dV_2 = S_{ABC} \cdot dH \tag{I.30}$$

$$S_{ABC} = \pi \frac{H}{R} [r_1^2 - (R - H)^2] \tag{I.31}$$

where H is the distance between the condensed bodies.

If we allow for the mutual energy of all pairs of molecules in the elementary volumes of the two bodies (dV_1 and dV_2) and integrate Eq. (I.29), we obtain

$$E = -\frac{A}{12}\left(\frac{y}{x^2 + xy + x} + \frac{y}{x^2 + xy + x + y} + 2\ln\frac{x^2 + xy + x}{x^2 + xy + x + y}\right) \tag{I.32}$$

where A is a constant equal in the present case to

$$A = \pi^2 n^2 \lambda_{i,j} \tag{I.33}$$

In general, the quantity A is a constant of the molecular interaction of the condensed bodies (van der Waals constant):

$$A = \pi^2 \sum_i \sum_j n_i \cdot n_j \cdot \lambda_{i,j} \tag{I.34}$$

where n_i and n_j are the numbers of i and j molecules in 1 cm^3.

In Eq. (I.32) the dimensions of the bodies and the gap between them are expressed in relative units, namely, $x = H/d_p$ and $y = r_2/r_1$ (Fig. I.8a).

Thus, by replacing the elementary volumes in Eq. (I.29) by their values in (I.30) and (I.31), and then integrating, we secure a simplification of the expression for the interaction energy. The power of H implied in Eq. (I.32) is in fact smaller than in Eq. (I.29).

Let us consider some cases of interaction between bodies of regular geometric form. For the interaction of two similar spheres, i.e., when $r_1 = r_2$, $y = 1$,

$$E_1(x) = -\frac{A}{12}\left(\frac{1}{x^2 + 2x} + \frac{1}{x^2 + 2x + 1} + 2\ln\frac{x^2 + 2x}{x^2 + 2x + 1}\right) \tag{I.35}$$

If one of the contiguous surfaces is plane, i.e., $y \to \infty$ and $y \gg x$ (Fig. I.8b), then we obtain from (I.32)

$$E_{\infty}(x) = -\frac{A}{12}\left(\frac{1}{x} + \frac{1}{x+1} + 2\ln\frac{x}{x+1}\right) \tag{I.36}$$

For contact between microscopic particles and a plane, surface, when $x \ll 1$,

$$E_{\infty}(x) = -\frac{A}{12x} \tag{I.37}$$

It follows from Eq. (I.37) that the interaction energy between a spherical particle and a plane is inversely proportional to the distance between them. The interaction energy between two bodies, referred to 1 cm^2, equals

$$E = -\int_H^{\infty} \frac{\pi n\lambda}{6}\cdot\frac{1}{H^3}\, n\, dH = -\frac{A}{12}\cdot\frac{1}{H^2} \tag{I.38}$$

From Eqs. (I.32) and (I.36)-(I.38) it is quite easy to obtain an expression for the force of interaction between two condensed bodies. Considering that the force of interaction equals

$$F = \frac{\partial E}{\partial H} \tag{I.39}$$

and substituting into (I.39) the expression for the energy [formulas (I.32), (I.35), and (I.37)], afterwards differentiating the resultant relations, we find for y = const:

$$F = \frac{\partial E}{\partial H} = -\frac{A}{d_p}\cdot\frac{\partial E_y(x)}{\partial x} = -\frac{A}{d_p}\cdot F_y(x) \tag{I.40}$$

For $y = 1$, the force of interaction equals

$$F = -\frac{A}{d_p} F_1(x)$$

$$F_1(x) = \frac{1}{6}\left[\frac{2(x+1)}{x^2+2x} - \frac{x+1}{(x^2+2x)^2} - \frac{2}{x+1} - \frac{1}{(x+1)^3}\right] \tag{I.41}$$

For contact between a plane and a spherical particle, when $y \to \infty$

$$F_{\infty}(x) = \frac{1}{12}\left[\frac{2}{x} - \frac{1}{x^2} - \frac{2}{x+1} - \frac{1}{(x+1)^2}\right] \tag{I.42}$$

For contact between a macroscopic particle and a plane, when $x \ll 1$,

$$F_{\infty}(x) = -\frac{1}{12x^2} \tag{I.43}$$

The formulas (I.32)-(I.43) relate the work and force of adhesion and enable us to calculate these quantities as functions of x, the ratio of the gap between the contiguous bodies and the particle diameter.

Below we give the energy and force of interaction as functions of the relative gap (A = 1):

x	0.001	0.01	0.1	1.0
Interaction of two similar particles:				
$E_1(x)$, from (I.35), erg. . . .	$4.07 \cdot 10$	3.58	$1.74 \cdot 10^{-1}$	$6.58 \cdot 10^{-4}$
$F_1(x)$, from (I.41), dyn . . .	$4.15 \cdot 10^4$	$4.00 \cdot 10^2$	2.83	$2.32 \cdot 10^{-3}$
Interaction of a spherical particle and a plane:				
$E_\infty(x)$, from (I.36), erg . . .	$8.23 \cdot 10$	7.65	$5.09 \cdot 10^{-1}$	$9.48 \cdot 10^{-3}$
$F_\infty(x)$, from (I.42), dyn. . .	$8.62 \cdot 10^4$	$8.17 \cdot 10^2$	6.88	$2.08 \cdot 10^{-2}$

It follows from the data presented that the energy and force of molecular interaction depend greatly on the gap width ($x = H/d_p$) separating the contiguous bodies. For a sphere and a plane the energy and force of interaction are greater than for two spheres. This is due to the greater distance between the two bodies resulting from the curvature of the spheres.

It follows from Eq. (I.39) that the power of H (the gap width) in the expression for the force is greater by unity than in the expression for the energy.

The energy and force of the van der Waals interaction between two gas molecules may be put in the form

$$E_{i,j} = -\frac{\lambda_{i,j}}{H^6} \tag{I.44}$$

$$F = \frac{6\lambda_{i,j}}{H^7} \tag{I.45}$$

If we considered condensed systems, then for two spheres

$$E = -\frac{Ar}{12H} \quad \text{(I.46)}$$

$$F = \frac{Ar}{12H^2} \quad \text{(I.47)}$$

for a sphere and a plane

$$E = -\frac{Ar}{6H} \quad \text{(I.48)}$$

$$F = \frac{Ar}{6H^2} \quad \text{(I.49)}$$

and for two planes,

$$E = -\frac{A}{12\,\pi H^2} \quad \text{(I.50)}$$

$$F = \frac{A}{6\,\pi H^3} \quad \text{(I.51)}$$

We must emphasize once again the difference between the interaction of individual molecules and the molecular interaction of condensed systems, depending on the extent of the gap separating them. In the first case, the interaction energy is inversely proportional to H^6 and the force to H^7; in the second case we have H and H^2, respectively for two spheres (or a sphere and a plane) and H^2 and H^3 for two planes. On the basis of Eqs. (I.46)-(I.51) we may calculate the energy and force of the interaction between two bodies as a function of the gap separating them.

The work expended in overcoming the forces of adhesion as the gap varies may clearly be expressed in the form

$$E_{ad} = \int_0^\infty F dH \quad \text{(I.52)}$$

where F is the force of interaction between the particles and the surface; this diminishes with increasing width of the gap between the contiguous bodies, falling from F_{max} to F_{min}, equal to the weight of the particles P.

Let us suppose that in the course of the detachment of particles in a liquid medium the value of F_{max} falls exponentially in accordance with the gap width. The validity of this assumption is confirmed by experimental results, as we shall see later (§20). Then,

$$F = F_{ad} \cdot e^{-kH} \tag{I.53}$$

where F_{ad} is the force measured at the instant of particle detachment [45] and k is a coefficient depending on the properties of the surface and the surrounding medium.

In order to determine the coefficient k, we take logarithms of Eq. (I.53), taking the gap as equal to $2 \cdot 10^{-5}$ cm. If the gap is wider than this, the forces of adhesion acting on the particle in water become negligible; for the removal of such particles it is only necessary to overcome their weight (P), i.e., $F_{min} = P$

$$k = -5 \cdot 10^4 \ln \frac{P}{F_{ad}} = -1.15 \cdot 10^5 \log \frac{P}{F_{ad}} \tag{I.54}$$

(we take a value of F_{ad} corresponding to a specific adhesion number). Knowing F_{ad}, we may calculate the coefficient k. For example, on painted surfaces, for 10μ particles, the value of $k = 3.7 \cdot 10^5$ (in water) [45].

Substituting the value of F from (I.53) into (I.52), we obtain, after integration,

$$E_{ad} = \int_0^\infty F_{ad} \cdot e^{-kH} dH = -F_{ad} \frac{e^{-kH}}{k} \tag{I.55}$$

Similar calculations may be made for the adhesion of particles in an air medium. Thus, for particles 10 μ in diameter in this case [13], the coefficient $k = 4.6 \cdot 10^5$. However, in an air medium the exponential relationship (I.53) is only roughly true.

Below we present the values of energy required to detach particles 10 μ in diameter from painted surfaces in air and water.*

Medium	Air	Water
Energy of detachment, erg:		
for one particle	$\approx 10^{-12}$	$\approx 10^{-13}$
for densely packed particles (referred to 1 m²)	$\approx 10^{-2}$	$\approx 10^{-3}$

*Calculations are made for the following adhesion numbers: in air, 50%; in water, 90%.

Table I.1. Values of the Constant A According to Various Authors *

Medium	Material of surfaces in contact	Distance between surfaces in contact, A°	$A \cdot 10^{12}$, erg
Vacuum [43]	Sodium borate spheres	3	4.7
	Quartz spheres	3	2.2
Vacuum and nitrogen [48]	Glass spheres and plates of Pyrex glass	Direct contact	0.1-10*
Air [51]	Mica layers	5-25	0.1-10†
Air [52, 53]	Quartz spheres and metal plates	1000	0.05
Air [49]	Glass plates	2500	1.1-30
	Quartz plates	3000	11-30
	Silvered quartz plates	8000	39-78

*Theoretical value of A only calculated for vacuum [49, 50]; the value equalled $1 \cdot 10^{-12}$ erg.

† The authors regard their data as approximate.

Thus, in order to detach particles in air we must expend energy an order of magnitude higher (for 10μ particles and given γ_F) than in a liquid medium.

In order to determine the force of molecular interaction, we must know the constant A and the gap between the contiguous bodies [see (I.44)-(I.51)].

The constant A may be calculated from Eq. (I.34) if we know the chemical composition of the bodies in contact and the surrounding medium, the refractive index, the dipole moment, and the density of the substance.* De Boer [47] calculated the dispersion—interaction constant $d\lambda_{i,j}$ and the constant A for NaCl molecules; these constants, respectively, equalled 10^{-58} erg $\cdot$ cm^6 and ~10^{-12} erg.

The value of the constant A may also be calculated from the experimentally determined values of the force of interaction between the two bodies and the width of the gap separating them [see (I.47), (I.49), (I.51)]. The results of experimental determinations of the constant A are presented in Table I.1.

*In the present case the effect of the surrounding medium may be neglected; a calculation of the constant A allowing for the properties of the surrounding medium is given on page 117.

In view of the fact that it is impossible to measure the gap between the contiguous surfaces directly, especially in the case of the adhesion of microscopic particles, one is compelled to use methods simulating adhesion interaction (see § 8). These methods enable us to determine the gap between the contiguous bodies quite accurately and hence calculate the molecular component of the forces of adhesion.

The interaction of the molecules is determined by the dispersion forces for not only nonpolar but also a considerable number of polar molecules (except for those with strong polar properties). A qualitative estimate of the character of the dispersion interaction was given by Dubinin [54]. Studying the adsorption of various molecules on silica gel, Dubinin used Eq. (I.28) to calculate the approximate values of the constant of dispersion interaction for the three typical groups

$$\equiv Si{-}OH, \quad \equiv Si{-}F, \quad \text{and} \quad \equiv Si{-}CH_3$$

of molecules in the surface under consideration and the corresponding molecules interacting with these.

Below we give some values for the relative* energy of dispersion interaction between functional groups and molecules:

	N_2	C_6H_{12}	C_6H_6	H_2O
$\equiv Si{-}F$ . .	0.62	0.56	0.55	0.61
$\equiv Si{-}CH_3$.	0.65	0.72	0.67	0.64

The interaction energy between the molecules indicated and surface hydroxyl groups is 30-45% higher than for silica gels with methylated or fluorinated surfaces.

We note that an important feature in the molecular interaction of condensed systems is not merely the presence but also the number of molecules adsorbed on the surface, which determines

*The interaction energy with the hydroxyl group is taken as the unit.

the value of the constant A [see (I.34)].

Analysis of the data presented shows that the dispersion interaction is a maximum for molecules of the same polarity (for example, $-\overset{|}{\underset{|}{Si}}-CH_3$ and C_6H_{12} and C_6H_6) and a minimum for molecules of different polarity ($-\overset{|}{\underset{|}{Si}}-CH_3$, H_2O). This phenomenon is reflected for the adhesion of solid surfaces in the empirical De Bruyne rule [5]: the maximum adhesive force occurs for surfaces of identical hydrophily and the minimum for those with different hydrophilies. This rule (see §§10 and 27) is generally valid and of considerable significance in changes of adhesive interaction.

However, the theory of London dispersion forces on which the calculations in question are based takes no account of electromagnetic lagging. This is equivalent to the assumption that the rate of propagation of electromagnetic waves is infinitely great and the distances between the molecules infinitely small compared with the absorption wavelengths λ which are characteristic of the atoms and molecules of the contiguous bodies. The London theory is valid when the gap between the surfaces in direct contact is no greater than 10 Å, i.e., when it is of the order of magnitude of the absorption wavelengths of the atoms and molecules (for $-H$; $-O$; $-CH_3$; $-OH$; $-Cl$; $-F$ this is 5-7 Å). For surfaces bordering the contact zone in air and also for cases in which there is a liquid layer between the contiguous surfaces and the gap width is greater than the absorption wavelength, the London theory cannot be used.

Lifshits [55] used quantum electrodynamics to develop a theory for the molecular interaction of condensed macroscopic bodies, allowing for electromagnetic lagging. The value of this theory lies in that the forces of interaction calculated from it agree closely with the earlier-obtained [53, 56] experimental data on the interaction of spherical glass bodies with a plane metallic surface. According to the Lifshits theory, for a small gap between the contiguous surfaces, i.e., in the case $H \leq \lambda$, the force of interaction between two similar plane surfaces equals

$$F_{\infty} = \frac{\hbar}{8\pi^2 H^3} \int_0^{\infty} \left(\frac{\varepsilon_{i.\,\varphi} - 1}{\varepsilon_{i.\,\varphi} + 1} \right)^2 d\xi \qquad \text{(I.56)}$$

where $\hbar = h/2\pi$ is Planck's constant; $\varepsilon_{i,\varphi}$ is the imaginary part of the complex dielectric constant of the substance, considered as a function of frequency, and ξ is the argument.

For large values of gap, when $H \gg \lambda$, the force of interaction between two dielectrics equals

$$F_{\infty} = \frac{\hbar c}{H^4} \cdot \frac{\pi^2}{240} \cdot \left(\frac{\varepsilon_0 - 1}{\varepsilon_0 + 1}\right)^2 \varphi(\varepsilon_0) \tag{I.57}$$

where c is the velocity of light, ε_0 is the dielectric constant of the substance, and $\varphi(\varepsilon_0)$ is a function determined graphically [53].

By comparing expressions (I.56) and (I.57) with (I.47), (I.49), and (I.51), we see that in both cases the form of the formulas for the molecular force of interaction is the same:

$$F = \frac{A}{H^n} \tag{I.58}$$

the power of H for the case of two planes being 3 in the equations derived by either method.

In order to transform to the interaction between a plane and a sphere, we use the Deryagin formula [57]

$$F = Q \int_0^{\infty} F_{\infty}\, dH \tag{I.59}$$

where F is the force of attraction between the sphere and the plane, Q is a geometrical factor, and F_{∞} is the force associated with unit area of contact between plane-parallel surfaces.

For $H \ll \lambda$,

$$F = \pi r \frac{A}{H^2} \tag{I.60}$$

and for $H \gg \lambda$,

$$F = \pi r \frac{A}{H^3} \tag{I.61}$$

Expression (I.60) is analogous to (I.49), which was obtained by Hamaker on the basis of the London dispersion theory for the interaction between a sphere and a plane. Hence, for a small value of the gap between the contiguous bodies (i.e., in the zone of direct contact between the particles and the surface) the electromagnetic lagging is not serious, and the force of intermolecular interaction may be calculated either from Eq. (I.60) or from (I.47), (I.49), and (I.51).

Below we present some values of the forces of intermolecular interaction calculated from (I.60) for a gap width equal to 10 A in the case of the particles for which the adhesive force was determined experimentally in [13]:

Radius of particles, μ	7.5	12.5
F, calculated from (I.60), dyn	$1.2 \cdot 10^{-3}$	$2.0 \cdot 10^{-3}$
F_{ad} ensuring the retention of 50% of the particles, dyn.	$6.2 \cdot 10^{-3}$	$2.15 \cdot 10^{-3}$

We see from the data presented that the molecular interaction plays an important part in establishing the adhesive forces of the particles.

§ 5. Deryagin's Theory and the Possibility of Calculating Forces of Adhesion

Deryagin developed a theory of adhesion in [57]; he established that adhesion took place under the influence of surface forces and could be considered as a thermodynamic equilibrium and reversible process, provided that the radius of curvature of both the surfaces greatly exceeded the radius of action of the surface forces.

Deryagin's theory is based on the assumption that the force of adhesion is a function of the gap width H separating the spherical surfaces of contiguous bodies $F = f(H)$. When the gap vanishes $(H \to 0)$, the adhesive force becomes equal to

$$F = -\frac{2\pi}{\sqrt{\varepsilon\varepsilon_1}} f(0) \tag{I.62}$$

where

$$f(0) = \sigma_{1.2} - \sigma_{1.3} - \sigma_{2.3}$$

ε and ε_1 being the curvatures of the contiguous bodies; $\sigma_{1,2}$ is the surface tension at the boundary between phases 1 and 2 (two solids), and $\sigma_{1,3}$ and $\sigma_{2,3}$ those at the boundaries between the solids and phase 3 surrounding them.

For the adhesion of two spheres made of the same material ($\sigma_{1,3} = \sigma_{2,3}$; $\sigma_{1,2} = \sigma_{1,1}$) with radii r_1 and r_2, Eq. (I.62) acquires the following form:*

$$F = 2\pi \frac{r_1 \cdot r_2}{r_1 + r_2} (2\sigma_{1,3} - \sigma_{1,1}) \tag{I.63}$$

If one of the surfaces is a plane ($r_1 \gg r_2$) and the quantity $\sigma_{1,1}$ may be neglected, we obtain

$$F = 4\pi\sigma r \tag{I.64}$$

It follows from Deryagin's theory that the force of adhesion depends on the curvature [formula (I.62)] of the surfaces in contact. The influence of the properties of the surfaces on the adhesion is taken into account by the free energy $f(0)$; the effect of capillary forces and the particle charges on the adhesive force, however, is not taken into consideration (see Chapter II, §§12-14 on this point).

If we take the mean value of the surface tension of the solid bodies [58] as $\sigma \approx 800$ erg/cm^2, we find from Eq. (I.64) that the adhesive forces for particles of radius 25 μ equals 25 dyn.

According to experimental data, the adhesive force for particles of radius 25 μ lies between 0.06 and 0.083 dyn [16], i.e., two or three orders lower than the calculated value.

It would appear that Fuks is right in considering that, in Eq. (I.64), r should be taken to mean not the particle radius but the radius of the submicroscopic projections by which the particle makes actual contact with the surface. However, neither the area of these projections nor the area of contact have yet been determined. In view of this, and also in view of the fact that methods of determining the surface tension of solids are still imperfect, we

* Bradley [43] obtained an analogous relationship between the adhesive force and the diameters of spherical particles (d_1 and d_2): $F = Ad_1d_2/(d_1 + d_2)$. The original premises of Bradley's derivation differed from those used in constructing Deryagin's theory. Bradley integrated the equations for the energy and force of the intermolecular interaction (see § 4) on the assumption that the distance between the bodies corresponded to the dimensions of a molecule; this limits the applications of the derivation.

cannot calculate the forces of adhesion from Eq. (I.64). Although the individual components of adhesive forces (for example, capillary and electric forces) are in some cases susceptible to calculation (see §§11-13), at the present time adhesive forces can only be determined experimentally.

Chapter II

Methods of Determining Forces of Adhesion

§ 6. Methods of Measuring Forces Associated with the Detachment of Microparticles

The adhesive force is numerically equal and opposite to the force required in order to detach particles from a surface. Therefore, if we measure the latter (i.e., the force at the instant of detachment) by varying the slope of the surface or by centrifugal, vibrational, or pulse methods, we may determine the adhesive force.

Method of Varying the Slope of the Surface. The simplest and most accessible method is that of determining the detaching force by reference to the slope of the surface α

$$F_{\text{det}} = V(\rho_1 - \rho_2)\, g \sin\alpha \qquad \text{(II.1)}$$

where V is the volume of the particle, and ρ_1, ρ_2 are the densities of the particle and liquid, respectively.

This method was first used by Buzach [14] for determining adhesive forces in liquid media. Buzach placed the "dusty" (dust-laden) plate at the bottom of a vessel, which he then rotated through a specific angle. An analogous method was used by Fuks [12, 59], who designed a special cuvette [60]. Later, the construction of the cuvette was slightly improved [45] (Fig. II.1). The cuvette is fixed to the object stand of a microscope and may be rotated together with this through a specified angle. This arrangement may be used for determining forces of adhesion in liquid media by the centrifugal method also. The method in question is applicable when the force of interaction between the particles and the

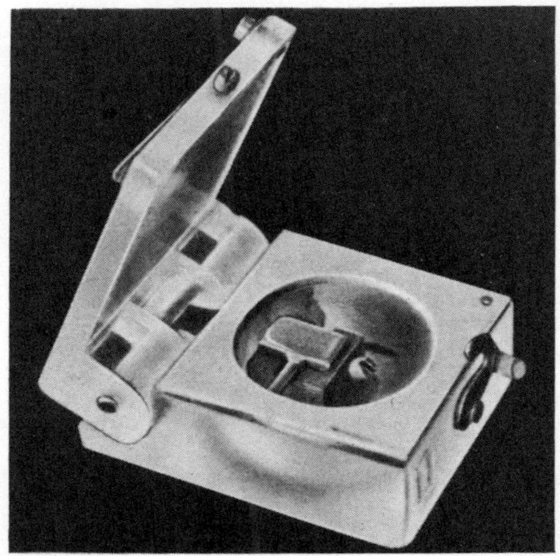

Fig. II.1. Cuvette for determining adhesive forces.

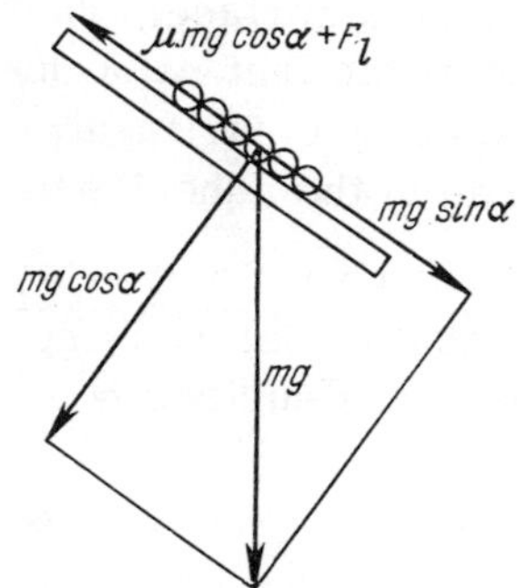

Fig. II.2. Forces acting on a layer of particles on an inclined surface.

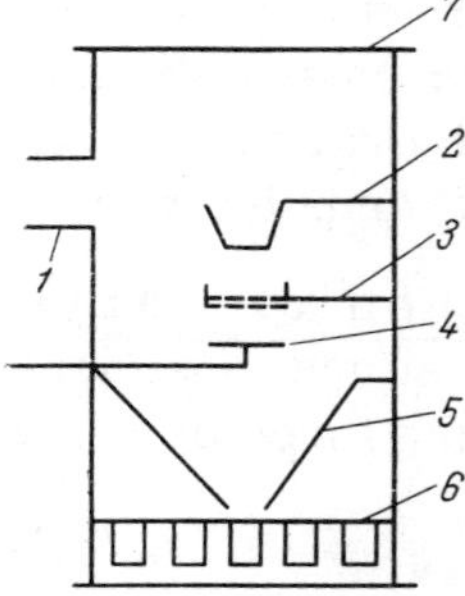

Fig. II.3. Apparatus of Patat and Schmid for determining forces of adhesion. 1) Body; 2) crucible containing the dust particles; 3) mesh; 4) plate; 5) cone; 6) space for collecting fractions of dust; 7) lid.

surface is smaller than the weight of the particles. The detaching force then acts tangentially to the dust-laden surface, i.e., essentially just one of the components of the forces of friction is determined.

The adhesive force of powder layers in air was also determined by Cremer [18] and later Batel [19], Patat and Schmid [20], Pecht [61], and also in our own investigations [21].

When the layer creeps (Fig. II.2), the detaching force may be expressed by Eq. (I.23). It is assumed that in the course of detachment there is no slipping inside the powder layer or conversion of the powder mass into a monolayer of particles. By varying the mass of the powder, we may obtain an experimental straight line which on intersecting the ordinate axis cuts off a section equal in magnitude to F_l (see Fig. I.6).

A successful practical embodiment of the method was achieved by Patat and Schmid [20] (Fig. II.3). The particles fall on the horizontal test plate 4, on rotating of which they slide and are collected in the collector 6, which has 12 sockets, so that 12 measurements may be made in one experiment. A potential of up to 1.5 kV may be applied to the plate. The body of the apparatus is airtight, so that experiments may be made in vacuum or in various vapors and gases at various temperatures.

For small particles (less than 20 μ in size) the detaching force has to be greater than that obtained by rotating the dust-laden surface; hence, the majority of such particles are held on the surface [21]. It is not surprising that in order to estimate the adhesive forces of a layer of powder, large particles are chosen (usually with a diameter of more than 60 μ). It was noted in [20, 21] that even in this case a certain number of smallish particles (60–100 μ) remained on the substrate after the detachment of the majority of the particles.

In determining the force required to detach a monolayer of particles, the accuracy of the method is determined by the error in measuring the diameter of the particles d_p and the slope α. The error of determination B (in %) equals

$$B = \frac{\Delta F_{\text{det}}}{F_{\text{det}}} \cdot 100 = \left(3\,\frac{\Delta d_{\text{p}}}{d_{\text{av}}} + \cot \alpha \cdot \Delta \alpha \right) \cdot 100 \qquad \text{(II.2)}$$

where Δd_p is the scatter in the diameters of the particles, and $\Delta\alpha$ is the error in measuring the slope.

Figure II.4 indicates the accuracy of the method. According to the results presented for $\Delta\alpha = \pm 2°$, it is best to conduct the measurements with $\alpha > 20°$ (curve 3), when the relative error is no greater than 10%. Of course, by reducing $\Delta\alpha$ we may also reduce the minimum slope. Thus, for $\Delta\alpha = \pm 1°$, a relative experimental error of 10% will occur at $\alpha = 10°$.

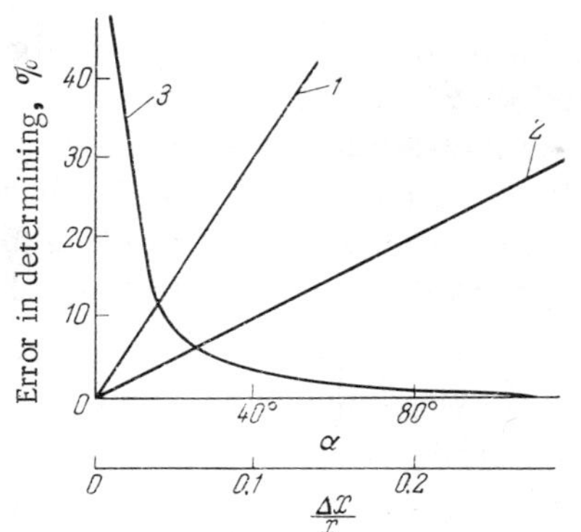

Fig. II.4. Error in determining the detaching force on rotating a dust-laden surface. 1) $3\Delta d_p/d_{av}$; 2) $\Delta m/m$; 3) $\cot\alpha \cdot \Delta\alpha$ (for $\Delta\alpha = \pm 2°$).

Figure II.4 shows the accuracy of the experiment as a function of the indeterminacy of the particle diameters in the fractions (curve 1). For $\Delta d_p/d_{av} = 0.05$, the error of the method is 15%. The slope method may therefore only be used for the detachment of particles of a single size. If the sizes of the particles in a fraction fluctuate appreciably, the accuracy of the method diminishes (for example, for $d_p = 40–60\ \mu$, $\Delta d_p/d_{av} = 0.2$ and $B = 60\%$) and in practice it becomes unsuitable for determining the adhesive forces of the particles.

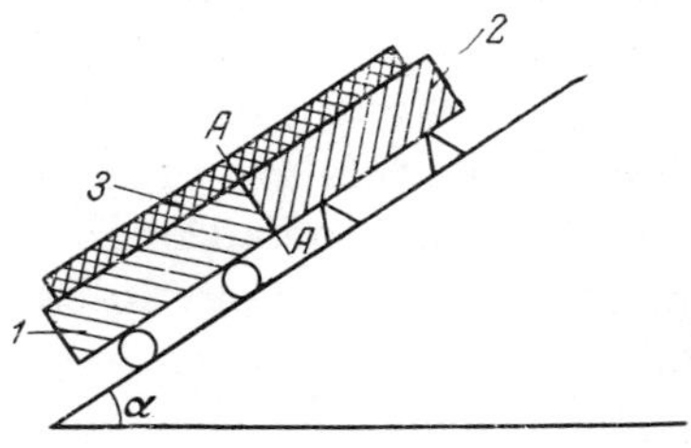

Fig. II.5. Apparatus for determining the autohesion of an adhering layer. 1) Movable part; 2) immovable part of the glass platform; 3) layer of dust.

When a layer of particles is detached, their mass is measured. In this case, the accuracy of the method may be estimated as follows:

$$B = \frac{\Delta F_{det}}{F_{det}} \cdot 100 = \left(\frac{\Delta m}{m} + \cot\alpha \cdot \Delta\alpha\right) \cdot 100 \qquad \text{(II.3)}$$

In order that the error should not exceed 10%, the ratio $\Delta m/m$ should not be smaller than 0.1 (for $\Delta\alpha = 0$). This ratio is made up of the accuracy of the weighing (Δm) and the mass of the powder (m) sticking to the surface. The mass may be increased (in order to reduce the experimental error) by increasing the dust-laden area.

The method of inclining a dust-laden surface may be used to measure the autohesive force of an adhering layer [24, 62] when $F_{ad} > F_{aut}$. The dust layer is deposited simultaneously on the movable and immovable parts of a glass platform (Fig. II.5). For a specific slope α the movable part of the platform detaches itself from the fixed part and the dust layer is broken along the section A—A.

By measuring the area of the break in the surface layer of dust and weighing the moving part of the glass platform, we may calculate the force associated with unit area of the cross section of the adhering layer from the formula

$$F_l = \frac{P \sin \alpha}{S} \quad \text{(II.4)}$$

where P is the weight of the moving part of the glass platform together with the particles adhering to it, α is the slope at which the sections are uncoupled, and S is the area of cross section of the detached part of the dust layer.

Centrifugal Method. The centrifugal method of determining the forces of adhesion is based on the detachment of particles by rotating the dust-laden surface around a horizontal or vertical axis. The value of the detaching force may be found from the formula

$$F_{det} = V(\rho_1 - \rho_2)(\bar{j} + \bar{g}) \quad \text{(II.5)}$$

with

$$j = \omega^2 x \text{ and } \omega = \frac{2\pi n}{60}$$

Here, $\bar{j}$ is the centrifugal acceleration, $\bar{g}$ is the gravitational acceleration, ω is the angular velocity of the rotation of the dust-laden surface, x is the distance of the dust-laden surface from the axis of rotation, and n is the number of rotations of the dust-laden surface in 1 min.

In order to determine the value of the detaching force acting on the particle, the vectors representing the centrifugal acceleration and the acceleration due to gravity must be added (Figs. II.6, II.7). On rotating the dust-laden surface around a horizontal axis, the force of gravity tends to assist the detachment of hanging particles (Fig. II.6c) and oppose that of particles lying on the surface (Fig. II.6b). On rotating the surface around a vertical axis, if the value of g cannot be neglected, the detaching force is directed at an angle to the surface.

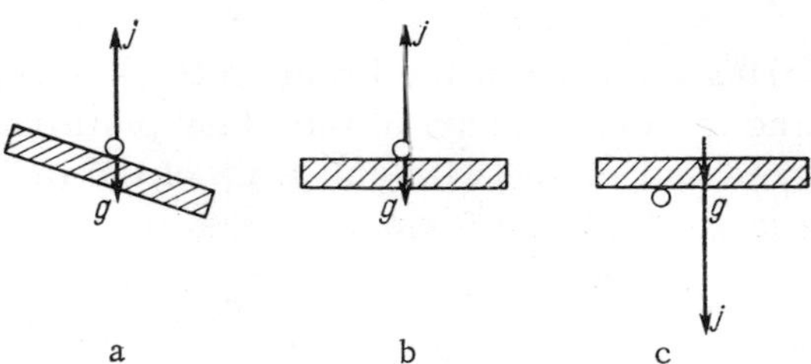

Fig. II.6. Acceleration acting on a particle on rotating the dust-laden surface around a horizontal axis. a) General case; b) particle lying on the surface; c) hanging particle.

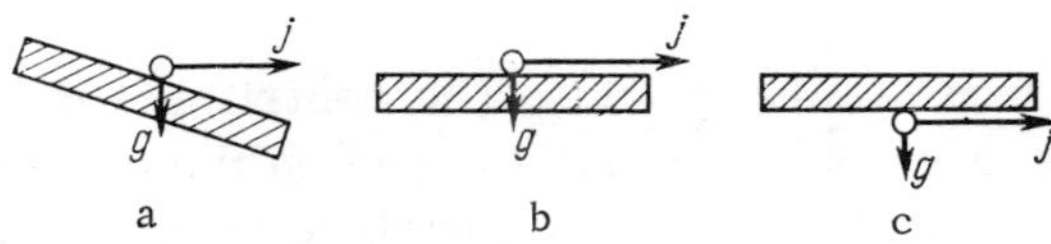

Fig. II.7. Acceleration acting on a particle on rotating the dust-laden surface around a vertical axis. a, b, c) See Fig. II.6.

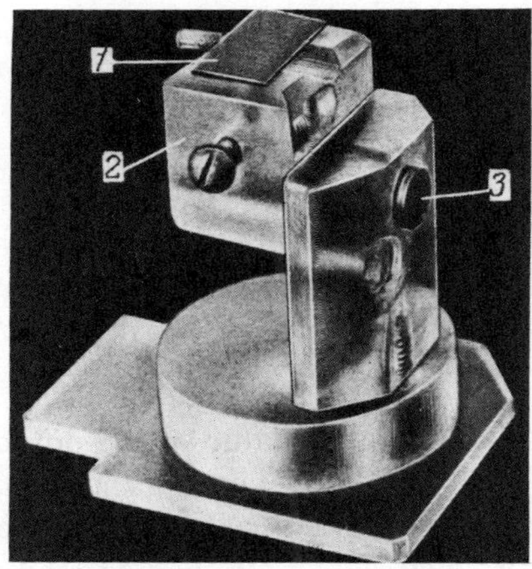

Fig. II.8. Plate holder with rotating head. 1) Test plate; 2) rotating head; 3) axis.

For an air medium, when $\rho_1 \gg \rho_2$, and for $\omega^2 x \gg g$, Eq. (II.5) is simplified:

$$F_{det} = V\rho_1\omega^2 x \qquad \text{(II.6)}$$

The centrifugal method has been widely used for determining adhesive forces both in air [10, 16, 45, 63-66] and in liquids [10, 45, 59].

Either a rotating dust-laden cylinder or a centrifuge may be used for investigations of this kind.

If detachment of the particles takes place as a result of forces directed at an angle to the dust-laden surface, then the original substrate is placed at the appropriate angle in the sockets of the centrifuge, or else a holder (Fig. II.8) with a rotating head for the plate is employed.

The accuracy of the centrifugal method is determined from the equation

$$B = \frac{\Delta F_{det}}{F_{det}} \cdot 100 = \left(3\frac{\Delta d_p}{d_{av}} + 2\frac{\Delta n}{n} + \frac{\Delta l}{l}\right) \cdot 100 \qquad \text{(II.7)}$$

where B is the error of determination in %.

The error represented by the quantity $\Delta n/n$ (curve 2, Fig. II.9) depends on the accuracy of measuring the rotations of the centrifuge. For TsLN and TsUM centrifuges, $\Delta n = 200$ rpm. In air, the detachment of the particles usually occurs at $n > 4000$ rpm.* In this case the experimental error is no greater than 10%. By increasing the accuracy of measuring the number of revolutions, the relative error $\Delta n/n$ may be greatly reduced and brought to a minimum. When determining adhesive forces in liquid media in our experiments [45], the number of revolutions was usually 200-3000 rpm. Under these conditions the number of revolutions had to be checked very carefully.

The error represented by the ratio $\Delta l/l$ (curve 3) is due to the different distances between the axis of rotation and the dust-laden surface when the latter lies at an angle to the axis of rotation of the centrifuge [$\alpha \neq 0$; $\Delta l = (L/2)\sin\alpha$, where L is the length of

*In determining adhesive forces in air, the angular velocity of the centrifuge may reach 50,000 rpm and the detaching force 10^5 g.

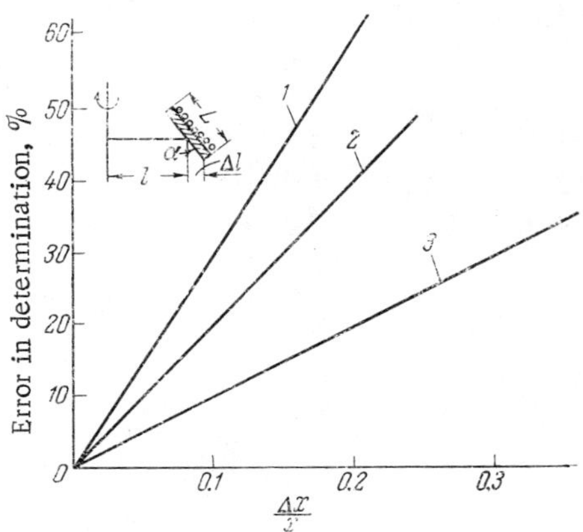

Fig. II.9. Error in determining forces of detachment by the centrifuge and vibration methods. 1) $3\Delta d_p/d_{av}$; 2) $(2\Delta n/n) \cdot (2\Delta\nu/\nu)$; 3) $(\Delta l/l)(\Delta y/y)$.

the plate]. Under our conditions, $\Delta l/l < 5\%$. The error associated with the determination of $\Delta l/l$ may be reduced if only particles lying on the axis of the plate are considered ($\Delta l \approx 0$).

The error associated with the different sizes of the particles in the fraction (curve 1), i.e., $\Delta d_p/d_{av}$, was considered earlier (page 40).

On comparing the various possible errors [see (II.7)], we may conclude that the main error in the centrifugal method is due to the different sizes of the particles in a specified fraction.

In order to avoid errors associated with the rate of rotation, we must increase (or reduce) the revolutions of the centrifuge smoothly in order to eliminate the effects of inertial forces, holding the specified number of revolutions for several seconds [16]. Further increasing the time of centrifuging has no effect on the detachment of dust particles in air. In centrifuging it is important to take proper precautions against vibrations of the body and beating of the centrifuge axis, since these effects may distort the results of measurements made on the force of detachment.

In the practical use of this method for liquid media a number of special features have to be taken into consideration. The whole space in the cylinder (or cuvette) must be filled with liquid in order to prevent the latter from moving in the course of centrifuging and thus eliminate the influence of side effects. Owing to the difficulty of hermetizing the cylinder (or cuvette) the number of revolutions of the centrifuge in the methods employed is no greater than 3000 and the detaching force no greater than 10^2–10^3 g. The time of centrifuging should be about 1 min, so that hydrodynamic factors associated with the way in which the dust-laden surface is situated in the liquid medium may be fully taken into account (see § 19).

The centrifugal method of measuring the value of the detaching force is the principal method used in determining forces of adhesion. The advantages of this method lie in its simplicity and accessibility, and also in the reliability of the results and the rapidity of the measurements. In addition to this, a variety of conditions may be created in the centrifuge test tubes (humidity, temperature, pressure, etc.), which widens the experimental potentialities of the method. However, in order to obtain the integral adhesion curve several measurements must be made with different numbers of revolutions.

It is possible to use the centrifugal method for determining the adhesion of a layer of particles. In this case, when calculating the detaching force, the quantity $m = V\rho_1$ must be understood to mean the mass of the particles adhering to unit area of the substrate.

In estimating the detaching force one must allow for the autohesion of the particles.

The centrifugal method may be used for determining the adhesion of particles in a drop of water [67]. Dust is deposited on a carefully cleaned surface by means of a very thin, smooth, glass rod. Only a very small part of the surface ($D \approx 1.6$ mm) need have the dust deposited on it. A drop of water from a calibrated capillary is placed on the dust-laden surface in such a way as to cover the whole of the dust with the drop. Plates are fixed in special holders and with the help of these the detaching forces acting on the drop and particles in directions tangential and perpendicular to the surface may be determined.

Vibration Method. The vibration method was first used for determining the adhesion of films [68]. Later, Larsen [69] determined the adhesive forces of spherical particles to fibers vibrating at a frequency of the order of tens of cycles. This method was improved and extended [13] by using sonic and ultrasonic vibrations.

The vibration method is only used for determining the adhesive force of dust in air. For this purpose either low-frequency (20-30 cps) or high-frequency (hundreds or thousands of cps) vibrations are used.

With the high-frequency system, the sound vibrations, previously amplified, set in motion a dynamic diffuser to which the dust-laden plate is attached. The frequency of the oscillations is usually no greater than 2 kcs and the value of the detaching force reaches 2500 g.

By varying the frequency of the vibrations, one may vary the detaching force over a wide range. In order to increase the range of detaching forces, one may use an ultrasonic system generating oscillations at a frequency of 10-20 kcs [13]. Here the value of the detaching force will be $(10\text{-}24) \cdot 10^4\ g$.

Thus, the frequency of the vibrations is given by the generator, while the amplitude may be measured (visually or with an oscillograph). On the basis of these data one may calculate the detaching force acting on a particle when the dust-laden surface vibrates:

$$F_{\text{det}} = m\,(j + g) \tag{II.8}$$

where

$$j = 4\pi^2\nu^2 y \cos\left(\omega t - \frac{\pi}{2}\right) \quad \text{and} \quad \omega = 2\pi\nu$$

Here, m is the mass of a dust particle, j is the acceleration of the vibrational motion, ν is the frequency, y is the amplitude of the vibrations, and t is the time.

The method under consideration enables the distribution of the particles with respect to adhesive forces to be determined in a single experiment, i.e., it yields the integral curve of adhesive forces.

It is essential to remember that the value of the detaching force calculated from formula (II.8) is maximum when $\cos(\omega t - \pi/2) = 1$. The plate vibrates together with the particles, the vibrational force pressing the particle to the surface [cos $(\omega t - \pi/2) < 0$] and the zone of contact may be distorted. If the substrate is ductile, the vibrational method is inapplicable.

The error (%) of the vibration method is determined in the following way:

$$B = \frac{\Delta F_{det}}{F_{det}} \cdot 100 = \left(3\frac{\Delta d_p}{d_{av}} + \frac{2\Delta\nu}{\nu} + \frac{\Delta y}{y}\right) \cdot 100 \quad \text{(II.9)}$$

The error $\Delta\nu/\nu$ in the determination is associated with the accuracy of measuring the vibration frequency (see Fig. II.9). Usually, the frequency of vibration is given by the generator and the vibrations are reproduced by the dust-laden surface. The error $\Delta y/y$ is due to the accuracy of measuring the amplitude of the vibrations. In our experiments, the combined error $\Delta\nu/\nu + \Delta y/y$ was never greater than 5%. Hence, as in the previous case, the main error in the method may be attributed to the unequal sizes of the particles.

Below we present some comparative data relating to the detachment of particles 40–60 μ in diameter from steel surfaces of the ninth class of finish by the vibration and centrifugal methods:

F_{det}, dynes	$2.5 \cdot 10^{-4}$	$5.7 \cdot 10^{-4}$	$4.2 \cdot 10^{-3}$	$1.7 \cdot 10^{-2}$	$5.9 \cdot 10^{-2}$	1.6
γ_F, %						
vibration method · ·	51.7	30.0	31.4	28.5	25.5	4.9
centrifugal method	60.2	–	37.3	30.1	21.2	1.2

For relatively small forces of detachment, a greater number of particles are detached in the vibrational than in the centrifugal method. The difference becomes less marked as the applied detaching force becomes greater. The detachment of firmly held particles takes place more intensively in the centrifugal than in the vibration method.

The vibration method may be accompanied by simultaneous measurement of the electric charges arising from the detachment of the particles, these being required in order to calculate the electric component of the adhesive forces (see § 11). For this purpose we may use an apparatus [70] differing from those employed earlier (of the electrometer type [11, 13, 71]), using electronic and loop oscillographs. This obviates the dependence on visual measurements and photographically records the electrical processes taking

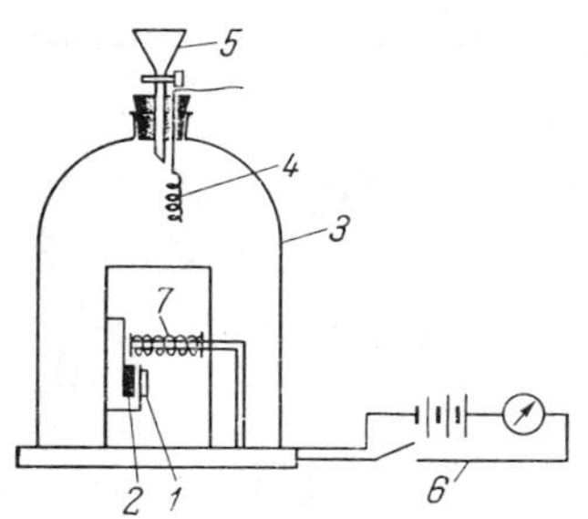

Fig. II.10. Apparatus for determining adhesive force in various gases and vapors. 1) Dust-laden surface; 2) impinging arm(striker); 3) bell-jar; 4) heating system; 5) dropping funnel; 6) electrical supply; 7) electromagnet.

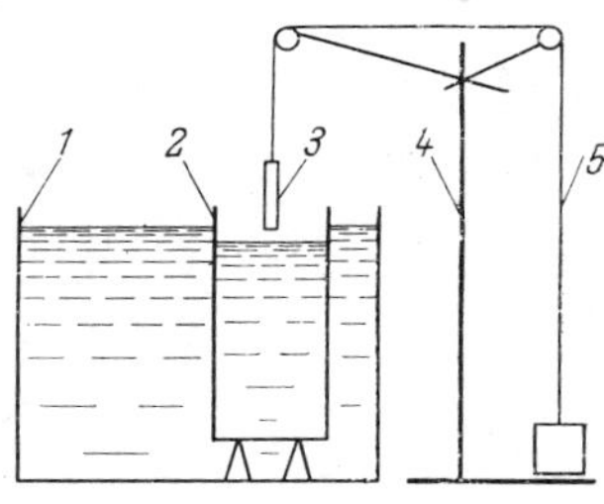

Fig. II.11. Apparatus for determining adhesive force by the immersion method. 1) Thermostat; 2) vessel containing solution; 3) cassette for samples; 4) stand; 5) thread carrying weight.

place in the zone of contact between the dust particles and substrate as functions of time. The apparatus consists of three units: I) a system for vibrating the dust-laden plates; II) an amplification (electrometric) system; and, III) a photographic unit. Both the sign and absolute magnitude of the charge arising as a result of contact or disconnection between the particles and test surfaces may be determined in this way; the variation in the charge on the particles may be determined as a function of time both visually and photographically, and the relationship between the adhesive force of the dust and the charge on breaking contact may be established by varying the value of the detaching force (the smallest value of charge capable of determination is $5 \cdot 10^{-13}$ C).

Indirect Methods of Determining Adhesion. small particles (of under 10μ diameter) stick so firmly to surfaces that forces corresponding to accelerations of the order of $(10^3-10^4)\,g$ are incapable of overcoming the adhesive forces [13]. This explained the tendency to use ultracentrifuges for detaching small particles [72, 73]. However, an attempt at using an ultracentrifuge of the UTs-2-A type (produced by Mikrotechna of Prague) was unsuccessful. On rotating a sphere in the magnetic field of this centrifuge in vacuum, the sphere with the particles attached to it became heated. The heat melted particles consisting of fusible materials (for example, polymers) [72], and this distorted the results of the measurement.

On using an ultracentrifuge with a revolving rotor the detaching force equals $(5 \cdot 10^4 \text{ to } 10^5)g$ [73]. Even under these conditions, however, less than 20% of particles of diameter 3 μ are detached in the case of such materials as gold.

In view of this, one is forced to use indirect methods for determining the adhesion of such particles by estimating the adhesive forces from experiments on the detachment of different particles under identical conditions. The special characteristic of these methods lies in the fact that the value of the detaching force is not calculated, but kept constant [74].

A variation on the vibration method is the pulse method, which we used to determine adhesion in vacuum and in the vapor of various liquids (Fig. II.10).

The dust-laden test surface was placed under a vacuum bell-jar and the required vapor concentration was created by evaporating the liquid released by a dropping funnel in the jar. The space under the jar could be filled with any gas. Detachment of the particles was effected by means of a blow from an arm 2 on the dust-laden surface 1. The elasticity of the substrate and the force of the blow from the impinging arm were the same in all the experiments; this ensured a constant value of the detaching force and made it possible to compare experimental results.

One variant of the impulse method was proposed by Deryagin and co-workers [75]. The detachment of the dust particles was effected by shooting pellets at the dust-laden surface from an air gun.

In order to detach dust particles in liquid media, the loading method may be employed [76]. The dust-laden surface is immersed in the test solution for a specified time interval (Fig. II.11). The conditions of immersion and extraction of the samples from the solutions should be exactly the same.

For removing particles from a dust-laden surface, the latter may be subjected to an air blast [77]. Here, even for particles of a single size, the value of the detaching force is not susceptible to exact calculation, while for powder of many particle sizes the calculation is still more difficult (see Chapter VI). In order to create identical conditions for detaching the particles, the rate of air flow and the direction of the flow relative to the dust-laden surface must be exactly the same.

For removing a layer adhering to a surface, the layer may be subjected to the impact of freely falling solid spheres [78]. The size of the spheres, the repetition frequency, and the height from which they fall must be kept constant during the experiments, thus ensuring constant conditions for the detachment of the layers of particles.

§ 7. Methods of Determining Forces of Interaction Between Macroscopic Bodies

The adhesion of microscopic particles may be judged from the character of the interaction between macroscopic particles and from the dependence of this on air humidity, particle size, surface properties, temperature, pressure, and other factors. Thus, the methods considered may be used for modeling the adhesion of microscopic particles. Attempts to use these methods for determining the adhesive forces of microscopic particles directly have not so far yielded the desired results. The advantage of such methods lies in the fact that they may be employed for determining adhesive forces in so-called "pure" conditions (in air, in vacuum, etc.), and hence offer the possibility of obtaining reproducible results.

In experiments intended to determine the forces of interaction between macroscopic particles one uses either two spherical particles of a diameter of the order of a few millimeters, or one such particle and a plane, or the fused ends of glass, quartz, or metal filaments (Fig. II.12). The form of the ends may be rounded (a), spherical (b), or plane (c).

The literature [43, 48, 79-84] contains descriptions of a number of pieces of apparatus in which the forces of interaction between spherical (or spherical and plane) surfaces coming into contact with one another are determined. Despite differences in structural formulation, all forms of apparatus contain the following principal units: a spring or balance to which one of the bodies coming into contact is fixed, an arrangement for fixing the other body, and a reading device.

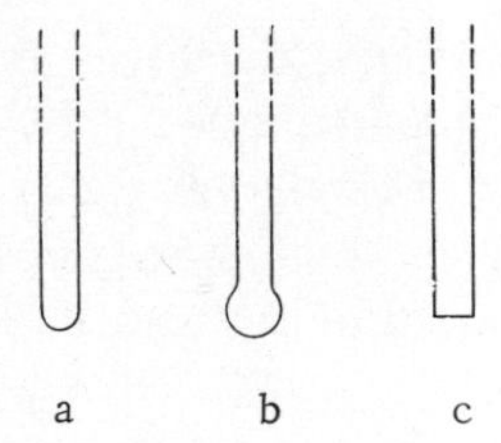

Fig. II.12. Form of the fused ends of filaments.

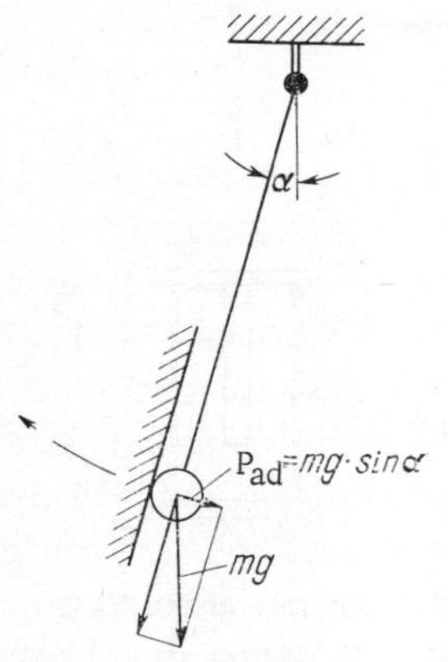

Fig. II.13. Pendulum method of determining forces of adhesion.

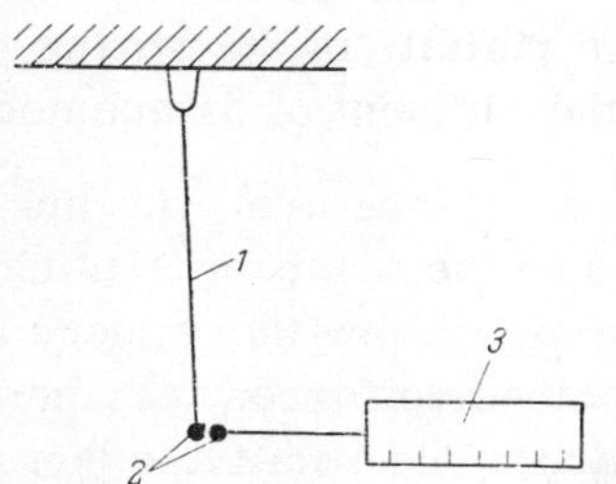

Fig. II.14. Principle of Stone's apparatus for determining the adhesive force between particles. 1) Glass fiber; 2) fused ends (or particles); 3) micrometer.

These systems are usually quite small and are frequently placed in a special hermetically sealed space in which the temperature, humidity, and pressure may be varied. There are two types of apparatus, differing in the manner of detaching the particles. In systems of the first type [48, 79, 81, 82] a vertical plate (or sphere) is brought up to a freely hanging sphere or the fused end of a filament until contact occurs (Fig. II.13). Then the plate is moved in a direction perpendicular to the area of contact. The deviation (angle α) of the suspended sphere from the vertical due to the action of the adhesive forces serves as a measure of the adhesion:

$$F_{ad} = mg \sin \alpha \qquad \text{(II.10)}$$

The angle α is either measured directly [48, 81, 82] or calculated from the deviation (measured with a micrometer) of the filament with the fused end from the original position (Fig. II.14). This method is quite accurate, since the error of the method is determined by the accuracy of measuring the angle α.

In systems of the second type, the adhesive forces are estimated by reference to the elongation of the quartz spring at the instant of particle detachment, i.e.,

$$F_{ad} = c\Delta h \qquad \text{(II.11)}$$

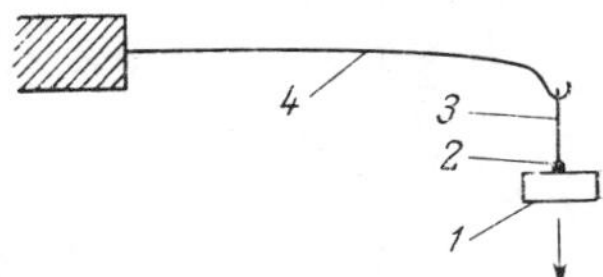

Fig. II.15. Principle of the Corn adhesiometer. 1) Substrate; 3) fiber with fused end (2); 4) arm of microbalance.

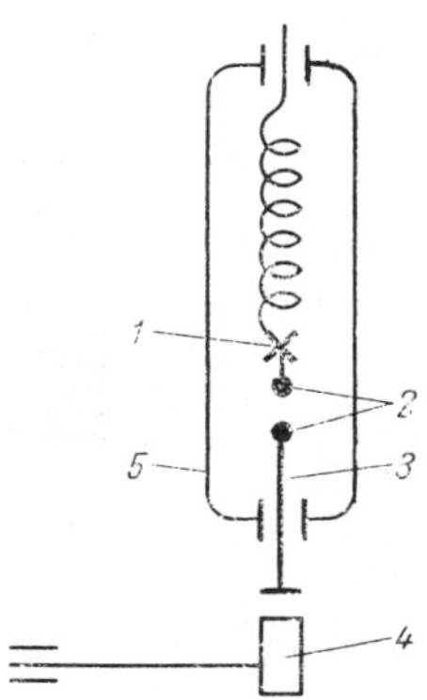

Fig. II.16. Principle of Bradley's adhesiometer. 1) Point observed with the cathetometer; 2) bodies to be brought into contact; 3) quartz fiber; 4) eccentric; 5) casing.

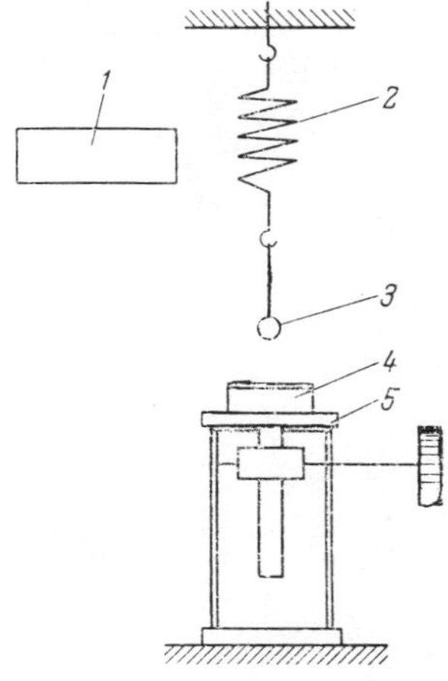

Fig. II.17. Principle of the NIIOGAS adhesiometer. 1) Microscope; 2) quartz spring; 3) glass sphere with adhesive film; 4) dust-laden plate; 5) elevating stand.

where c is a parameter characterizing the rigidity of the spring, and Δh is the elongation of the spring at the moment of detachment.

Corn [80] used a microbalance with a balance arm. In this case, the quantity Δh means the deflection of the arm; the arm carries a filament of the material under investigation with a sphere at the end (Fig. II.15). With this apparatus adhesive forces may be measured over a wide range (10^{-1}-10^{-4} dyn). The relative error of the experiment is no greater than 4-10% for particles 21-31 μ in size and increases to 24% when the particle size rises to 90 μ.

Using a spring balance, Bradley [43] was able to determine the adhesive force of quartz particles 0.4-1 mm in size (Fig. II.16). The original position of the spring is established by means of a cathetometer, and then the elongation Δh is determined at the instant of contact with the particle. The force of adhesion is calculated from formula (II.11).

The methods considered for determining the forces of interaction between particles are very cumbersome, and the apparatus required is *ad hoc* and complex. In addition to this, results obtained by different authors for the forces of interaction between the same surfaces differ. This is because the preliminary compressive force bringing the bodies into contact with one another (which determines the true contact area, and naturally affects the force of interaction between the particles) differs for different authors (although generally remaining constant for any particular set of experimental conditions), and is indeed not usually quoted.

Despite the fact that the method in question was developed for determining the forces of interaction between macroscopic particles, attempts have nevertheless been made to use spring balances for measuring the forces of detachment of adhering microscopic particles. For this purpose, an adhesive film was deposited on a glass sphere (Fig. II.17) fixed to a quartz spring and brought into contact with the dust-laden surface. On letting the stand down, the particles were detached from the substrate. In Tekenov's apparatus [83, 84] an adhesive substance was deposited on a fiber and the force of adhesion was determined from the bending of the fiber. However, these methods failed to give accurate results, since, at the instant of contact between the dust-laden surface and the film, there was an inevitable displacement of the adhering particles, which distorted the results of the measurements.

§ 8. Methods of Modeling the Adhesion of Microparticles

Adhesive forces depend not only on the properties of the contiguous bodies and the medium, but also on the gap separating the bodies. Existing methods of determining the adhesion of microscopic particles and also the majority of methods of determining the interaction between macroscopic particles, make no provision for measuring the intervening gap. Hence, in order to study the effect of the gap width on the forces of interaction, we must employ methods based on modeling or simulating adhesion. The bodies in contact may be fibers, spheres, or bodies with plane surfaces (in different combinations).

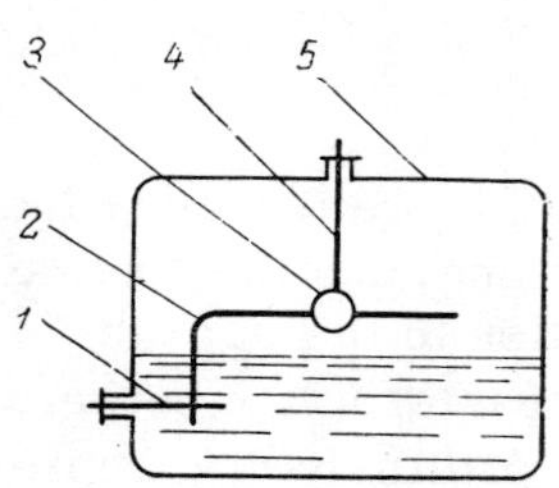

Fig. II.18. Apparatus for measuring forces of adhesion by the crossed-filament method. 1,2) Platinum filaments 300 μ in diameter; 3) mirror; 4) torsion suspension; 5) vessel.

A basis for the simulation of interaction between particles is Deryagin's [1] thermodynamic theory of interaction between surfaces, according to which (see §5)

$$F = QE\,(H) \qquad \text{(II.12)}$$

where F is the force of interaction between the two contiguous surfaces, Q is a geometric factor expressed in terms of the radius of curvature of the surfaces, and E (H) is the energy of interaction between plane surfaces separated by a plane-parallel gap of width H, calculated for unit contact area.

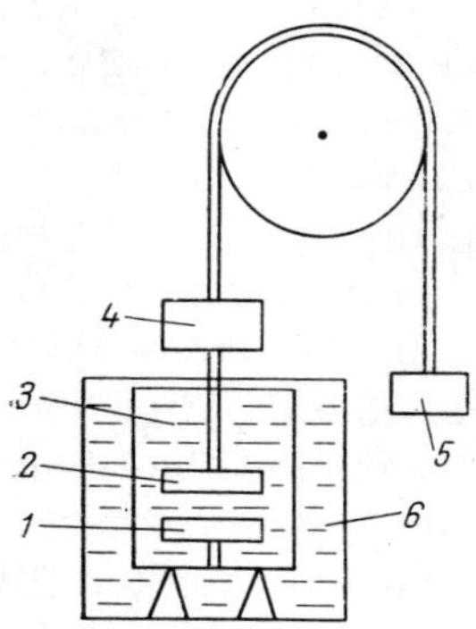

Fig. II.19. Principle of apparatus for measuring the forces of adhesion by the method of plane-parallel discs. 1,2) Discs; 3) test liquid; 4) compressing load; 5) counterweight; 6) thermostat.

For two pairs of bodies in contact, made of the same material, but having different dimensions [E (H) = const], we may write the relation

$$\frac{F'}{Q'} = \frac{F''}{Q''} = E\,(H) \qquad \text{(II.13)}$$

i.e.,

$$\frac{F'}{F''} = \frac{Q'}{Q''}$$

Hence, if we know the force of interaction F' and the dimensions of the bodies in contact for the two systems, we may calculate the force F" if the bodies are made of the same material.

In order to study the adhesion of particles in liquid media, particularly in solutions of electrolytes and surface-active materials, methods, involving crossed filaments and plane-parallel discs, are often used.

The method of crossed filaments was first used by Tomlinson [85] for determining the forces of adhesion between two intersecting glass or quartz filaments (some 60 μ in diameter) by reference to the maximum sag in one filament on slowly moving it away from the other, the length and rigidity of the filament under test being known.

Malkina and Deryagin slightly modified the Tomlinson method, eliminating the effects of vibration and achieving a smooth pull on the filament at the instant of detachment. These authors showed that for contact between two filaments of radius R_1 and R_2 meeting at an angle of $\pi/2$ the geometric factor equalled

$$Q = 2\pi\sqrt{R_1 \cdot R_2} \tag{II.14}$$

If $R_1 = R_2 = R$, then $Q = 2\pi R$.

If a film is deposited on the quartz surface, the adhesion of the filaments is due to molecular surface forces in the film material, and the quartz filaments themselves constitute a peculiar kind of dynamometer [87]. The principle of the improved apparatus is shown in Fig. II.18. The force of interaction between the filaments is measured by reference to the torsional angle of filament 1.

A similar method for estimating the adhesion of mutually perpendicular filaments (glass, quartz, or metal), 0.2-1.2 mm in diameter, was used by Fuks [88]. Fuks also used an apparatus (Fig. II.19) in which it was possible to determine the forces of interaction between colloidal particles by the method of plane-parallel discs (disc diameters 5-20 mm).

The principal part of the apparatus comprises steel or quartz discs 1 and 2 immersed in a test liquid 3. The distance H between the discs is determined by reference to the capacity of the plane condenser connected to disc 2. The lower limit of sensitivity of this method is 0.022-0.025 μ (in this case, the relative error is 12-18%).

Fuks proposed several forms of the apparatus; in one of these it was possible to measure the resistance of the boundary layer of liquid to thinning and the shear resistance in the liquid [88].

The determination of interaction between bodies in a liquid medium by means of measuring systems based on the principle of crossed filaments and plane-parallel discs is extremely tedious and requires the overcoming of a number of experimental difficulties (vibration, friction in contiguous parts, etc.). Still greater difficulties are encountered when studying the interaction of condensed bodies in a gaseous medium. These difficulties are associated with the greater sensitivity of the forces of interaction between the bodies to the gap separating the latter in gaseous media as compared with liquids. Hence, the direct methods of measuring interactions used in the case of liquid media become inapplicable in the case of gases.

Deryagin and Abrikosova [56] proposed using a feedback balance for this purpose (Fig. II.20). Since these authors were interested in determining not the total, but just the molecular interaction between the condensed bodies, the experimental difficulties were exacerbated by the electrification of the zone of interaction, and also the presence of adsorbed contaminations and dust particles on the surface.

When interaction takes place between the spherical quartz lens 1 and the chromium-coated plate 2, the balance arm 7 rotates relative to the fulcrum (prism 8). The deviation of the balance arm from the equilibrium position is proportional to the force of interaction; it gives rise to a current, which in turn produces an electrodynamic force opposing the molecular forces of attraction and compensating their effects. In the absence of the compensating arrangement (feedback) it is impossible to measure the interaction between the lens and the plate.

Deryagin and Abrikosova [56] determined the forces of interaction in air and in vacuum, varying the gap between the contiguous bodies from 0.1-1 μ, and from the resultant measurements estimated the adhesion in a liquid medium, since adhesion in a liquid medium usually involves gap widths of this order. The forces of interaction were not measured with the bodies in direct contact, which would have been very interesting for studying adhesion problems in air.

The modeling methods of [85-89] make it possible to study the adhesion of two bodies and obtain reproducible results expressed in terms of the width and properties of the gap separating the contiguous bodies.

However, modeling cannot replace the methods of determining forces of adhesion by the detachment of actual particles.

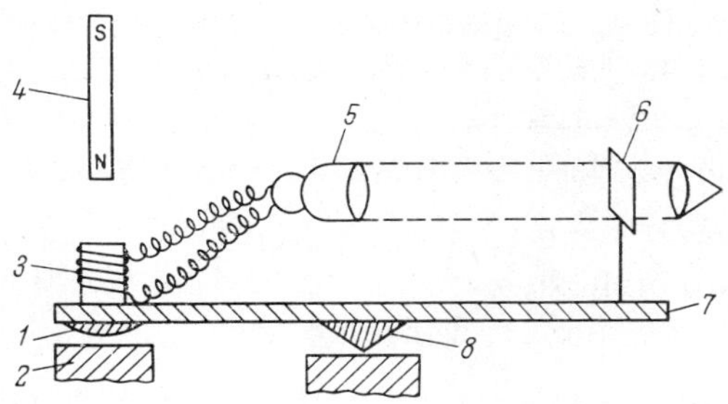

Fig. II.20. Arrangement of feedback balance. 1) Spherical lens; R = 10 cm and 25 cm; 2) plate (4 × 7 mm); 3) coil; 4) magnet; 5) photocell; 6) screen; 7) balance arm; 8) prism.

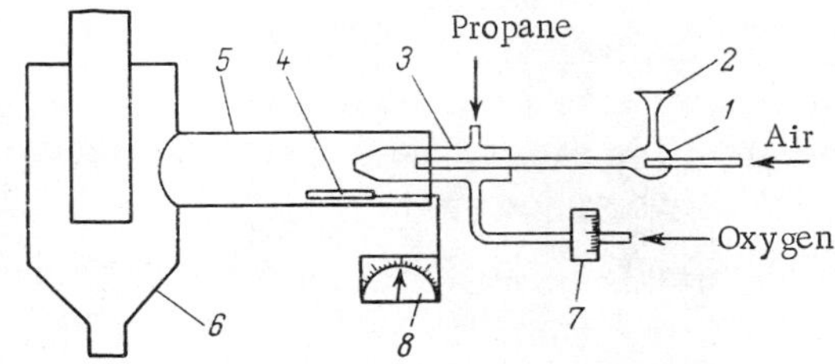

Fig. II.21. Arrangement of apparatus for fusing the dust. 1) Dust detector; 2) funnel; 3) large blowpipe; 4) thermocouple; 5) tube; 6) dust extractor; 7) rheometer (flow meter); 8) millivoltmeter.

A method of dimensional modeling [90] based on the equidistance of integral adhesion curves, has also been used. The essence of this method is that, by studying the known forces of adhesion between large particles, one may predict how small particles will behave, according to the particular properties of the bodies in contact. This method is only suitable in special cases.

§ 9. Methods of Preparing and Depositing Particles on a Surface

Production of Spherical Particles. In carrying out experiments on adhesive forces, it is important to eliminate the effect of particle shape on adhesion. For this purpose it is most convenient to use spherical particles. Usually, quartz and glass powders [91] with particles of spherical shape are employed. The process of preparing the particles comprises the following stages: the fragmentation of the original material, the fusing of the particles in special electric furnaces or in the flame of a gas burner, and the separation of the particles by fractions [92].

The particles are heated and melted in the flame of a gas burner or an electric furnace, becoming spherical under the influence of surface tension, and then cooled.

For melting the particles in the flame of a gas burner, the original dust is passed through a funnel into the air line and sprayed into the burner flame. In order to increase the flame temperature so as to melt the large particles (more than 50 μ in diameter), a mixture of oxygen and propane is used. However, even under these conditions not all the particles over 120 μ in size are melted. In order to determine the best melting conditions (air, oxygen, and propane flow, burner flame temperature) as well as the yield of the various resultant fractions, the apparatus illustrated schematically in Fig. II.21 was used [92].

Particles with diameters between 0.5 and 25 μ may be obtained in an electric furnace; larger ones cannot be heated right through.

In order to separate out a fraction of melted particles with constant particle size, one may use simple meshes, water or air separation (for example, in a column such as that shown in Fig. II.22), or other devices. The suggestion of certain authors [77] that the adhesive properties of the powder should be used in order to effect fractionization, can hardly be regarded as applicable, owing to the indeterminacy of the adhesive forces for even a single size of particles under the same conditions.

Methods have been described in the literature for obtaining spherical particles 0.2-100 μ in size from steel [93-95], polystyrene [96-98], rubber and plastic [99], and mercury, wax, and paraffin [100, 101].

Methods of Counting Numbers of Particles. In order to calculate adhesion numbers and estimate the effect of various factors on adhesion, we must have a method for counting the number of particles situated on various substrates. All existing counting methods may be arbitrarily divided into visual and automatic methods.

The visual method lies in counting the number of particles situated on a plane surface within a particular band or at several points on the substrate and falling within the field of view of the microscope. The visual method of counting particles, for all its simplicity, suffers from a serious disadvantage: the counting of particles depends on the individual characteristics of the observer [102]. This disadvantage may be partly eliminated by photographing the field of view of the microscope and counting the particles by reference to their projection on the negative or photographic plate [103]. It is also possible to project the field of view of the microscope on a screen. For this purpose one uses, for example, a photographic camera and doubly-refracting prism [104], or a projection camera and a system of mirrors [105].

The size of the particles may also be determined visually. Here the size of an irregular particle should correspond to the size of a spherical particle of equal mass. It is convenient to use this analogy when, for example, one is using the centrifugal or vibration method for detaching the particles. In this case the decisive factor is the mass of the particle: all particles undergo the same acceleration, and the detaching force acting depends on the mass of the particles. The sizes of irregular particles are usually determined by reference to at least three measurements [104]. Electron microscopes may be used for determining dust-particle sizes.

The number of particles may also be determined by means of automatic electronic counters in conjunction with a microscope; the action of these is based on the conversion of a light signal from the particle into an electric current. This method may be used for determining the number and sizes of particles between 1 and 200 μ in diameter (coal, diamond, and metal dust, etc.) [106].

One of the most convenient methods of determining numbers of dust particles on a substrate before and after an experiment is the method of tracer atoms. It is only possible to maintain the proportionality between the number of particles on the substrate and the number of pulses measured in counting systems for particles of more or less uniform size; it is quite impossible to draw any conclusions regarding the size distribution of the particles in a mixture of many sizes. For determining the number of active particles, radiographic as well as counter methods may be used [83].

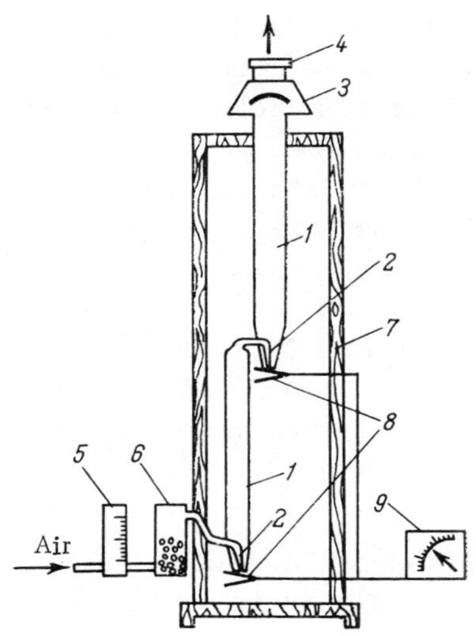

Fig. II.22. Arrangement of the apparatus for producing dust fractions of a single particle size. 1) Air columns; 2) test tubes; 3) trap (catcher); 4) filter; 5) rheometer; 6) drier; 7) frame; 8) vibrators; 9) LATR control system.

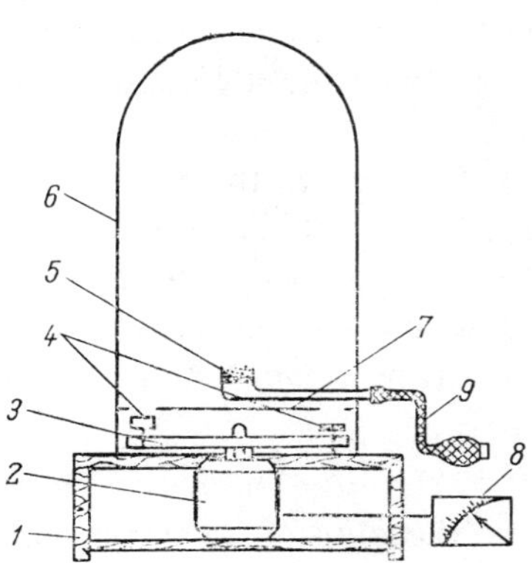

Fig. II.23. Apparatus for depositing dust on plates. 1) Stand; 2) electric motor; 3) disc for plates; 4) plates; 5) rubber membrane; 6) bell-jar; 7) screen; 8) LATR control system for varying the revolutions of the disc; 9) rubber tube with bulb.

Active particles may be obtained either by depositing a radioactive isotope on the surface of the particle from a solution [107-110] or by irradiating the dust in an atomic reactor [107, 111, 112]. In order to carry out activation by the first method, one uses [113] a solution of NaCl containing the radioactive isotope ^{24}Na, a solution of ^{35}S in kerosene [108], ^{204}Tl and ^{60}Co in nitric acid [83], etc.

By irradiating dust in a neutron flux [109, 112, 113], dust tagged with, for example, ^{24}Na, ^{32}P, and other isotopes has been obtained. For modeling radioactive dust formed by the processing of radioactive ores in isotope laboratories, nuclear reactors, etc., Semashko and Plievskii [114] used phosphoric acid (H_3PO_4) with a marker atom of ^{32}P. The acid particles had the following fractional composition: less than 60 μ, 40%; 60-80 μ, 45%; 80-100 μ, 15%.

A review of various methods of obtaining radioactive aerosols has been presented by Smirnov [115].

Particles activated by depositing a radioactive isotope on the particle surface from a solution may only be used for measuring adhesive forces in air, since, in a liquid the radioactive isotopes pass into solution, and this makes it impossible to determine the number of particles radiometrically.

When using fluorescent dust, the ratio of the final number of particles to the original number N/N_0 is proportional to the ratio of the intensity of the radiation from the different luminophores, and hence instead of the number of particles one may substitute the current recorded on the amplifier milliammeter [116].

For counting the initial and final number of particles one may use a colorimetric method based on the fusion of the powder with a substance subsequently monitored in a colorimeter. For example, the sample (particularly fabric) is treated with hydrochloric acid after first fusing with copper oxide (~30%); then a solution of dithyzone in carbon tetrachloride is added to the solution. Depending on the concentration of the copper dithyzonate thus formed, the solution varies in color from green to rose, and this is suitable for colorimetric measurement.

The number of potassium dichromate ($K_2Cr_2O_7$) particles, for example, may easily be determined iodometrically [117].

Preparation of the Surface and Deposition of Particles. The adhesive force depends on the degree of removal of contaminations from the surface. Research workers have used a variety of methods for cleaning surfaces and particles (while some have made no mention of such cleaning), and this is one reason why the results of adhesive-force determinations vary so much, even for contacts between the same type of bodies.

There are various methods of cleaning a surface. For example, glass surfaces may be cleaned with a chromic mixture and then water [81]; other recommendations include distilled water and acetone [80].

Steel surfaces are cleaned with silica gel, carbon, and B-70 gasoline [38], or activated charcoal and ethyl alcohol [13]; chromium—vanadium alloy—steel surfaces are cleaned with acetone [68]. The last stage in cleaning surfaces is treatment with an organic solvent. However, the preparation of pure solvents involves certain difficulties. In addition to this, contamination may easily be introduced when rubbing the surface. Hence, in order to secure a very good surface finish, it is better to use a flame [118] or glow discharge [119] in argon. In this way organic impurities undergo combustion.

The colored surfaces must not be treated with solvents, heating, or activated charcoal, as this would dissolve, soften, or damage the coatings. Contamination must be removed from colored surfaces with distilled water.

The particles may also be cleaned before depositing on the substrate. Glass, mica, aluminum, plexiglas, brass, and bronze are cleaned by washing with alcohol and distilled water and then drying at a temperature of 100°C [16].

The dust must be kept in jars with closed stoppers or desiccators at a specified humidity in order to avoid aggregation. Sometimes the dust is dried before an experiment.

The dust may be deposited on the surface by free settling. The special apparatus shown in Fig. II.23 may be used for depositing dust on a plate. A rubber membrane 5 is fitted to the end of the tube and a weighed quantity of dust is placed in the hollow of this. On pressing the bulb, the membrane is blown out, so that the particles are sprayed off and settle on the horizontal plate 4. Vertical and horizontal plates may be sprayed in this apparatus, either by free settling or with the dust meeting the plate at various specified velocities; the latter is achieved by rotating the disc 3 with an electric motor 2.

Chapter III

Adhesion in a Gaseous (Air) Medium

§10. Adhesion and Modification of Surfaces. Change in the Forces of Molecular Interaction

It is well known that adhesive forces are largely determined by surface properties [120, 121]. Hence, such forces may be varied by modifying the surfaces in question. Modification by means of alkylchlorsilanes is widely used for varying the adsorption of gases and vapors [122-124] in gas chromatography, for reducing the adhesion of films [125] and the adhesive power of solid surfaces [126, 127], for reducing the adhesion of ice [121, 128] and snow [129] particles, etc.

In our own investigations, and also those of other authors [130, 131], modification was achieved by treating previously moistened glass plates with a 1% benzene solution of the corresponding alkylchlorsilane. Then the samples were dried at a temperature of 120-130°C. As a result of such treatment, the surface was covered with a strong polysiloxane film, the hydrocarbon radicals being oriented in the direction of the surrounding medium, thus giving the surface a hydrophobic quality [130].

Figure III.1 shows our experimental values of adhesion numbers for spherical glass particles sticking to a glass plate as a function of the relative humidity of the air surrounding the dust-laden surface. We see from Fig. III.1 that modification (methylation by dimethyldichlorsilane) of one of the contiguous surfaces leads to a fall in adhesion (curves 2 and 3). Methylation of both surfaces (curve 4) still further reduces the adhesive forces.*

*The reduction in the adhesive forces after "hydrophobization" of the surfaces (i.e.,

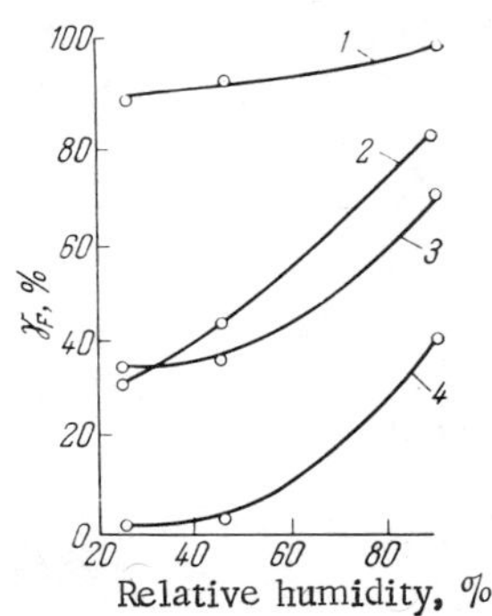

Fig. III.1. Adhesion number of spherical glass particles 70 ± 2 μ in diameter relative to a glass surface as a function of air humidity. 1) Ordinary glass; 2) hydrophobic substrates; 3) hydrophobic particles; 4) hydrophobic particles and substrates.

Hence, hydrophobization of the surface tends to reduce the adhesion of particles [121] and films [126, 127].

According to De Bruyne's rule, the greater the difference in the wetting capacity of the contiguous surfaces, the smaller is the adhesion. A difference in wetting power may be achieved not only by hydrophobization but also by hydrophilization of the surface, i.e., adhesion may also be reduced by changing the hydrophilic properties.

We made a special study of the effect of surface hydrophily on adhesive forces, using the pulse method for determining the adhesion of spherical glass particles to a glass surface. The results are set out below:

d_p, μ	30 ± 2	50 ± 2	70 ± 2
γ_F, %:			
to glass No. 23	93	83	76
to hydrophilic glass treated with liquid PK-10	81	77	70
to hydrophobic glass treated with silane	77	64	50

The wetting angle of a drop of distilled water of diameter 1700 μ equals 30° for glass No. 23, 18° for the hydrophilic glass, and 65° for the hydrophobic.

We see from these data that hydrophilization of the surface reduces the adhesion of the glass particles, although not so much as hydrophobization. There is a specific proportionality between the reduction in the adhesion and the change in the wetting angle.

In order to discover the reasons for the reduction in adhesive forces on a modified surface, let us consider the results of [131, 132] presented in Figs. III.2 and III.3. We see that the curves

making them hydrophobic) may take place as a result of a reduction in the capillary forces (see §13).

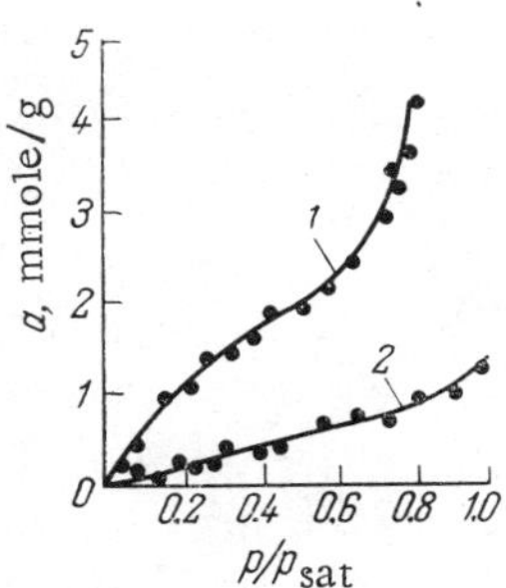

Fig. III.2. Isotherms for the adsorption of water vapor on silica gel. 1) Original; 2) modified with $(CH_3)_2SiCl_2$.

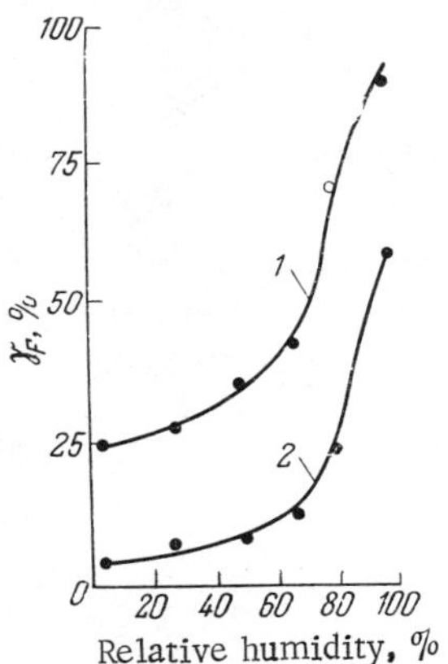

Fig. III.3. Adhesion number as a function of the relative air humidity for spherical glass particles 40-60 μ in diameter (detachment occurred at u = 5000 rpm). 1) Ordinary glass; 2) glass modified with $(CH_3)_2SiCl_2$.

representing the adhesion numbers as a function of relative air humidity (Fig. III.3) have the same form as the isotherms for the adsorption of water vapor [132] (Fig. III.2), while for the modified surface the curves are lower than for the ordinary substrate.

The data from [131, 132] presented in Figs. III.2 and III.3 agree with the results obtained by Luzhnov [133] when studying the force of static friction for Ukhtinsk channel-black powder and the water-vapor adsorption isotherms for the same materials (Fig. III.4).

The rise in adsorption in this case also is analogous to the rise in the coefficient of static friction and hence the static adhesive force. In considering the absolute value of the coefficient of friction determined by Luzhnov, it must be remembered that the author used powder with a wide spread of particle size, without indicating the dimensions of the particles.

Thus, on treating the surface with silanes the hydroxyl groups of the silica gel are replaced by methyl groups [131], which tends to reduce the dispersion interaction and adsorption of H_2O molecules on the silica gel [132] and also to reduce the adhesion of the particles to the modified glass in comparison with the ordinary glass surface.

The modification of a surface may change the adhesion of liquids as well as particles. The forces of intermolecular interaction tend to dominate the adhesion of liquids [134].

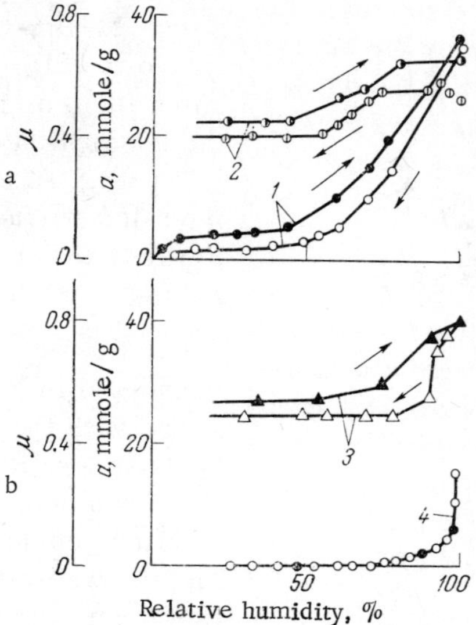

Fig. III.4. Coefficient of static friction of channel black powders (curves 2 and 3) and adsorption of water vapor on the same powders (curves 1 and 4) as functions of the relative humidity of the air. a) Hydrophilic; b) hydrophobic powders.

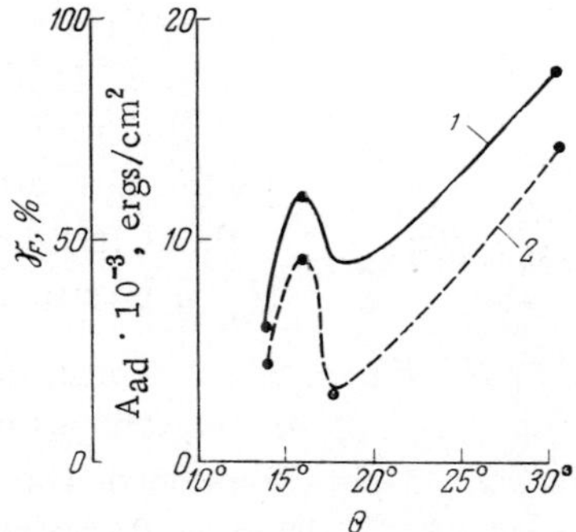

Fig. III.5. Adhesion number of particles (1) and work of adhesion of MT-16P oil (2) on ordinary glass surfaces and surfaces modified with various alkylchlorsilanes (dust $40 \pm 5\ \mu$, force of detachment $1.1 \cdot 10^{-2}$ dyn) as functions of wetting angle.

Figure III.5 illustrates the adhesion of particles to glass modified with silanes having hydrocarbon radicals of different lengths (curve 1) and the work of adhesion [135] of nonpolar MT-16P oil to the same substrates.* The similar behavior of the two curves suggests that the processes determining the adhesion of particles and the sticking of oil after modification of the surface are very similar.

Hence, the adhesion of particles changes after modifying the surface in analogy with adsorption and the adhesion of oil. This strongly suggests that one of the reasons for the adhesion of particles is the molecular interaction between the bodies in contact. The molecular component of the adhesive forces depends on the number of modifying reagent molecules situated on the surface and on the capacity of these molecules to suppress the effect of the properties of the original surface on adhesion.

Molecular interaction is not the only cause of particle adhesion. In order to elucidate the part played by molecular forces in the adhesion of practical systems (particles/surface) we must also consider the other factors underlying adhesion.

§ 11. Electric Forces Depending on the Properties of the Bodies in Contact

Contact Potential Difference. When particles come into contact with a substrate, the electrical charges situated on the surface of the particles attract equal and opposite charges on the substrate. This leads to the appearance of excess charges on the surface; the excess charges may be measured. When the dust particles are detached, the charge remaining on the substrate, equal and opposite to the charge on the particles, may be measured also. So far we have assumed that the charges associated with the contact and detachment of the particles are identical in sign and magnitude. However, experimental results show that the charge carried away by the particle on detachment is by no means equal in magnitude, nor on occasion even in sign, to the charge observed on contact [70, 136].

*This work was carried out by the author in company with Fuks and Timofeeva.

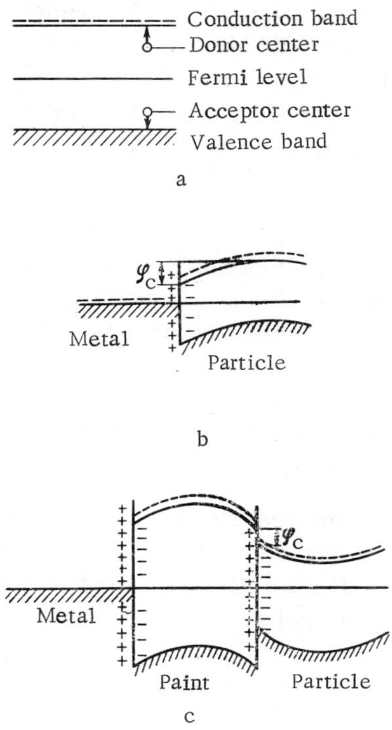

Fig. III.6. Energy levels of bodies in contact. a) Scheme of levels; b) contact with metal; c) contact with painted metal.

In considering contact phenomena it is usually assumed that up to the instant of contact the particles being deposited have not been in direct contact with each other or with the surface; this is equivalent to excluding the tribo-effect (the role of the latter is discussed in §35). In addition to this, no allowance is made for the development of a discharge between the particle and the surface on detachment of the particle. The detachment of a film may also produce a discharge of this kind; the extent of the discharge depends on the velocity of detachment, and has a considerable effect on the work of adhesion [1]. The effect of a discharge arising as a result of the detachment of a particle has not yet been studied, and we can only assume that the probability of a discharge developing between the particle and the surface diminishes with decreasing particle size [137].

Figure III.6a presents a schematic picture of the energy levels of bodies in contact [138, 139]. In addition to the principal energy bands, there are also impurity levels giving the surface donor or acceptor properties.

In order to study the effect of electrical forces on adhesion, it is convenient to consider two types of contact processes: those occurring between a semiconductor (dust particles) and a metal (substrate) and those occurring between a semiconductor (dust particles) and another semiconductor (glass, painted metal substrate) [70].

In the contact zone there is a leveling out of the Fermi levels and a curving of the conduction and valence bands (Fig. III.6b) with the simultaneous appearance of a contact potential difference φ_c at the boundary of the bodies in contact.

Figure III.6b shows the change in the energy levels of the semiconductor on contact with a metal when the Fermi level of the metal lies below the Fermi level of the particle material (glass). Here, hole-type conductivity develops in the metal, which becomes an electron donor, while the semiconductor (particles) becomes an acceptor.

Contact between a particle and a painted surface may be regarded as contact between two semiconductors, one of which is also in contact with the metal substrate (Fig. III.6c). At the boundary between the particle and the painted surface a contact potential difference φ_c develops. It may be supposed that the value of φ_c depends on the way in which the conduction band curves, and this, in turn, is related to the thickness of the layer of paint.

If the particle had no charge before contact, then, in order to eliminate the electric component of the force of adhesion it would be necessary to have $\varphi_c = 0$, i.e., no curvature of the conduction band. However, experiment shows that the particles are always charged, so that, in order to eliminate the electrical component of the adhesive force, we must have

$$U = \varphi_c \qquad \text{(III.1)}$$

where U is the contact potential difference observed when the particles touch the surface and φ_c is the potential difference developed as a result of the different work functions of the electrons in the contiguous bodies (associated with the difference in Fermi levels).

If Eq. (III.1) is satisfied, the charge observed on the particles after detachment should be zero (isoelectric point), and the electrical component will not affect the value of the adhesive force. Experimental results show that in individual cases this condition is in fact satisfied. For example, the charge determined on removing dust from a metal substrate (copper) equals zero [70].

In general, the value of the charge depends on the electrical conductivity of the bodies coming into contact.

On separating two metallic surfaces [140] the charge (q) is proportional to the contact potential difference φ_c, i.e.,

$$q = c\varphi_c \tag{III.2}$$

In turn, φ_c may be regarded as the potential jump associated with the plates of a parallel condenser:

$$\varphi_c = \varphi_2 - \varphi_1 = 4\pi\sigma_S \cdot H \tag{III.3}$$

where φ_2 and φ_1 are the potentials of the bodies in contact, σ_S is the double-layer charge density, and H is the distance between the bodies in contact.

If one or both of the bodies in contact are semiconductors or insulators (painted metal surface and particles), the electric field may penetrate inside the semiconductor. Then Eq. (III.3) takes the form

$$\varphi_2 - \varphi_1 - U_0 = 4\pi\sigma_S \cdot H \tag{III.4}$$

where U_0 is the potential drop inside the semiconductor [140]:

$$U_0 = \left(\varkappa_2 - \varkappa_1 + \frac{x^2}{2kT}\right) - \sqrt{\left(\varkappa_2 - \varkappa_1 + \frac{x^2}{2kT}\right)^2 - (\varkappa_2 - \varkappa_1)^2 - x^2} \tag{III.5}$$

Here, $\varkappa_2$ and $\varkappa_1$ are the electron work functions for the metal and semiconductor, and k is Boltzmann's constant.

The quantity x is determined by the expression

$$x = 2eH\sqrt{2\pi N_d \cdot kT\varepsilon} \tag{III.6}$$

where e is the elementary charge, N_d is the number of donor centers per unit surface, and ε is the dielectric constant.

Thus, we may expect that the potential difference (φ_c) arising when particles come into contact with a semiconductor is greater than on coming into contact with a metal for the same charge density. An increase in φ_c inevitably involves an increase in the adhesive forces.

Change in Adhesion Resulting from the Effects of Electrical Forces. The electrical forces arising on formation of a charge at the instant of contact between the

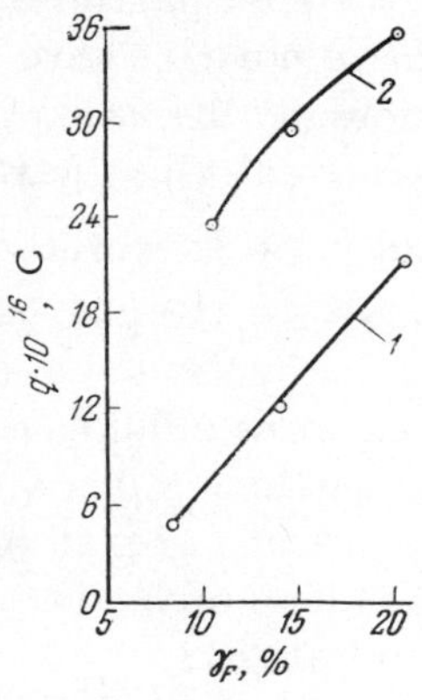

Fig. III.7. Adhesive force as a function of the charge determined on detaching glass spheres of various diameters (by the vibration method) from a painted metal surface. 1) d_p = 50 ± 5 μ, F_{det} = 3.6 · 10^{-3} dyn; 2) d_p = 70 ± 5 μ, F_{det} = 1.9 · 10^{-2} dyn.

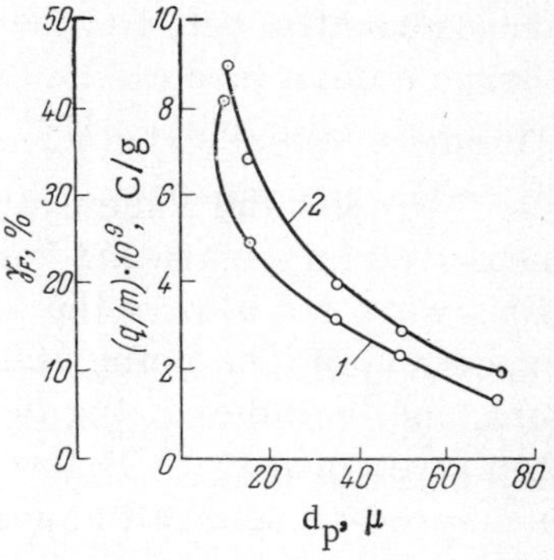

Fig. III.8. Adhesion number (1) and relative charge q/m (2) as functions of dust-particle dimensions (vibration method of detachment, detaching force 2.4 · 10^{-4} - 1.9 · 10^{-2} dyn.

particles and the surface may be measured in two ways. First, from the magnitude of the charges observed on detaching the dust particles from the substrate, and secondly by superimposing potentials on the substrate in order to neutralize the double-layer charge.

Let us first consider the first of these methods in more detail and pursue the effect of electrical forces on adhesion by reference to the charges on the particles. Below we present the results of a determination [11] of the charges on glass particles 40-60 μ in diameter on successive detachment from a surface painted with perchlorvinyl enamel:

$q \cdot 10^{15}$, C	1.9	2.3	4.6	4.1	6.8	7.1	8.5	11.0	12.6
No. of detached particles,* % . . .	7.3/7.3	19.5/26.8	14.6/41.4	7.3/48.7	12.4/61.1	7.3/68.4	17.0/85.4	7.3/92.7	2.4/95.1

*Numerator gives the number for a given force of detachment; denominator gives the running total.

We see from the data presented that the dust particles carrying the largest charges were the hardest to remove. There is a proportionality between the force of adhesion and the value of the charge determined on the detachment of spherical glass particles from painted surfaces [70] (see Fig. III.7).

Bredov and Kshemyanskaya [140] noted that the charges measured on separating the same surfaces under the same conditions were not always the same. Two plates were employed in the experiments, one being stationary, with dimensions 1.0 × 5.0 × 1.5 mm, and the other being in the form of a vibrator covered with a thin layer of metal. Below we present the values of the charges measured on separating germanium and gold plates:

Volume concentration of charge carriers $\cdot 10^{-14}$, cm^{-3} . . .	0.8	0.8	2	2	0.8	0.8
Contact potential difference, V. .	+0.51	+0.525	+0.47	+0.52	+0.64	+0.61
$q \cdot 10^3$, C	0.24	0.50	0.64	0.89	0.60	12.0

The indeterminacy in the charges arising as a result of the contact of the same particles with the same substrate may be explained by a local change in carrier concentrations, these being proportional to U_0 [see Eq. (III.4)]. In addition to this, it was shown in [70] that the relative charge q/m . . . (q is the unit charge of a dust particle on detachment, m is the mass of the particle) varied with dust-particle size in the same general way as the adhesion number γ_F varied with particle size under analogous conditions (Fig. III.8).

Patat and Schmid [20] came to the same conclusion after determining the charge associated with the detachment of a powder layer by inclining the dust-laden surface.

Below we present the charge and adhesive force as functions of particle size on removing a layer of silicon carbide (SiC) from steel surfaces in a nitrogen atmosphere:

d_p, μ	111	95	73	63	56	35	25
F_l (referred to 1 cm^2), dyn .	3	11	13	10	17	90	120
$q \cdot 10^{11}$, C/cm^2	1.1	3.9	5.7	6.8	12.1	135	274

We see from the data presented that the electrical charges increase as the particle size diminishes, so that the electrical component of the adhesive forces also becomes greater. The same tendency, though with a change in the sign of the charge, occurs on detaching aluminum oxide particles under the same conditions [20].

The value of the electrical component of the adhesive force may be calculated from the formula

$$F_e = 4\pi\sigma_S^2 \cdot S \quad \text{(III.7)}$$

Considering that $\sigma_S = q/S$, formula (III.7) may be written in the form

$$F_e = \frac{4\pi q^2}{S} \quad \text{(III.8)}$$

where σ_S is the double-electric layer charge density, q is the charge on a dust particle when the latter is detached from the substrate [11, 70], and S is the area of contact between the dust particle and the surface.

For glass particles 40-60 μ in diameter, the charge found on detachment from painted surfaces equals $1.9 \cdot 10^{-15}$ C and the calculated contact area is $2 \cdot 10^{-10}$ cm^2. The electrical component of the forces of adhesion equals 1 dyn, i.e., it is quite substantial [11]. On detaching particles from a metal surface the charge is much smaller than on detaching them from a painted substrate [70].

In this case,the electrical component of the adhesive force will have a value smaller than that corresponding to theory.

Below we present the adhesive force between a powder layer and a steel surface as a function of the voltage on the latter:

Substance	Aluminum oxide	Silicon carbide
d_p, μ	81	95
F_l (referred to 1 cm^2), dyn:		
grounded substrate	160	135
substrate at 420 V	120	126

We see from the data presented that the total adhesive force is greater for the grounded substrate.

The force of adhesion may be varied by eliminating or reducing the electrostatic component. For this purpose, in accordance with Eq. (III.7), we must reduce the charge density, which may be expressed in the form

$$\sigma_S = e\,(N_d - N_a) \qquad \text{(III.9)}$$

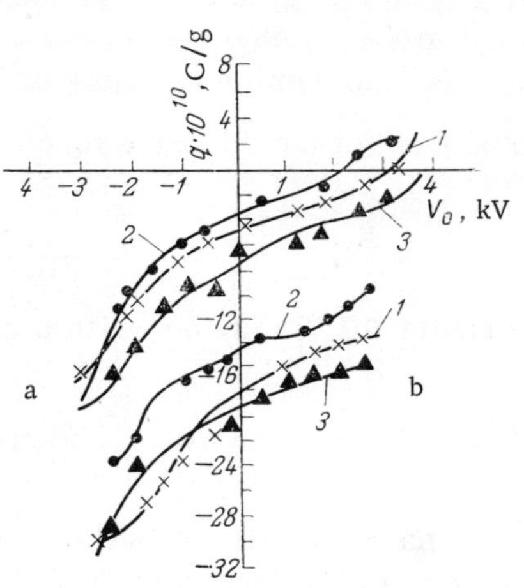

Fig. III.9. Charge of lead sulfide particles on a copper (a) or cadmium (b) substrate as a function of substrate potential. 1) Natural powder; 2) with metallic properties after vacuum treatment; 3) with metalloidal properties after treatment in sulfur vapor.

where N_d is the concentration of donor centers, and N_a is the concentration of acceptor centers on the surface.

Hence, in order to eliminate the electrical component, we must give the surface donor or acceptor properties [141]; this may be done by modification of the surface, namely, replacing certain surface molecular groups by others. Possible molecular groups may be placed [142] in a donor—acceptor series:

donor $-NH_4 > -OH > -OR > -COOR > -CH_3 > -C_6H_5 > =C= > -CN$ acceptor

in which each successive term is an acceptor with respect to the previous (donor) term. For example, in order to intensify the acceptor properties of a coating, the existing surface molecular groups must be replaced by one of the groups further to the right in the donor—acceptor series. The choice of any particular functional group of the donor—acceptor series for modification of a surface depends on the magnitude and sign of the charge on the powder, as measured when the latter is detached from the original substrate.

It should be mentioned that, in principle, it should be possible to make a combined modification of the surface in order to reduce (or increase) the dispersion interaction and the electrical component of the adhesive forces simultaneously. Thus, methylation gives a hydrophobic quality to the surface and promotes a reduction in the adhesion of particles. At the same time methylation intensifies the acceptor properties of the surface and may reduce the electrical component of adhesion, if on adhesion the surface reveals itself as a donor.

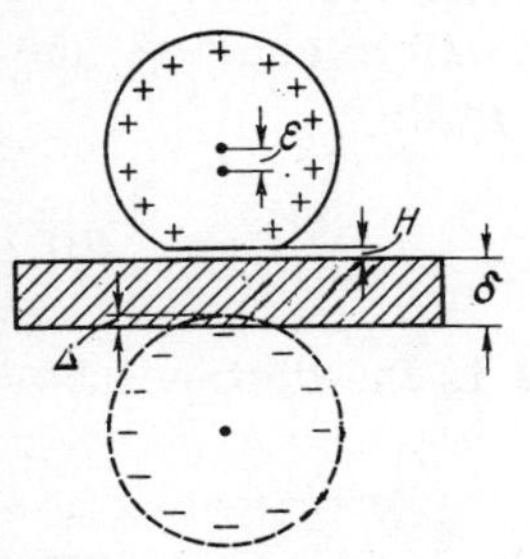

Fig. III.10. Mirror-image forces and displacement of equipotential charge.

In addition to this, we may change the contact potential difference by intensifying the metallic or metalloidal properties (donor or acceptor). For example, after heating lead sulfide powder in vacuo this substance acquires metallic properties; on heating it in sulfur vapor it becomes metalloidal [71].

Figure III.9 shows the change in the charge of lead sulfide particles (the normal type and those with metallic or metalloidal properties) with substrate potential.

In view of the fact that the electron work function for lead sulfide exceeds that for copper and cadmium, the particles obtain a negative charge in the absence of an external potential (intersections of curves 1 with the y axis); the work function for cadmium is smaller than for copper and, therefore, the difference in work functions of the lead sulfide—cadmium pair is greater than that of the lead sulfide—copper pair (curves a and b).

The communication of metalloidal properties to the powder (curves 3) leads to a rise in contact potential difference,and hence a rise in the adhesive forces and the charges observed on detachment of the particles. The communication of metallic properties to the powder (curves 2) leads to the opposite result.

§ 12. Electric Forces Arising Under the Influence of the Charge on the Particles

Coulomb Forces. The problem of determining the adhesive forces of charged particles to an uncharged surface may be reduced to a consideration of the Coulomb interaction between particles situated on both sides of the surface at equal distances from the latter. The charged dust particles induce equal and opposite charges on the surface (Fig. III.10). Image forces are thus set up. On interaction between the oppositely charged particles there is a displacement of the point of application of the equipotential charge in the direction of the zone of contact. The adhesive force resulting from the image interaction between the particle charges and the induced charge may be expressed thus [143, 144]:

$$F_{\text{im}} = \frac{Q^2}{l^2} \tag{III.10}$$

where Q is the charge on the particles, and l is the distance between the centers of the charges

$$l = 2\left(r + H + \frac{\delta}{2} - \Delta - \varepsilon\right) \tag{III.11}$$

r is the radius of a particle, H is the gap between the particles and the substrate (probably of the order of 10^{-6}–10^{-7} cm), δ is the thickness of the insulating layer (paint), Δ is the reduction in particle radius resulting from deformation at the contact zone, ε is the displacement of the center of the charge on the particle under the influence of F_{im}

$$\Delta = 1.55\sqrt[3]{\frac{F_{\text{ad}}^2}{E^2 \cdot d_{\text{h}}}} \tag{III.12}$$

and E is the reduced modulus of elasticity.

If the particle touches the metallic substrate ($\delta = 0$) and $r \gg H$, we may consider that $\varepsilon = 0$, $\Delta = 0$, and

$$F_{\text{im}} = \frac{Q^2}{(2r)^2} \tag{III.13}$$

For a painted surface under the same assumptions we obtain

$$F_{\text{im}} = \frac{Q^2}{4\left(r + \frac{\delta}{2}\right)^2} \tag{III.14}$$

It is quite a complicated matter to determine the displacement ε precisely. By using the approximate Dan solution, Balabanov [143, 144] showed that for

$$H = \alpha^2 r \tag{III.15}$$

where $\alpha \ll 1$ (α being a coefficient)

$$\varepsilon = \alpha r\sqrt{3} \tag{III.16}$$

Neglecting terms containing α^3, and considering that $\Delta = 0$ and $\delta = 0$, we find from formula (III.10) that

$$F_{\text{im}} = \frac{Q^2}{12 r^2 \alpha^2} \tag{III.17}$$

Since

$$\alpha^2 = \frac{H}{r} \tag{III.18}$$

we have

$$F_{\text{im}} = \frac{Q^2}{12 rH} \tag{III.19}$$

It is as difficult to use expression (III.19) as the original (III.10) since the gap between the contiguous bodies has not been determined. For a painted surface we may consider that $\delta \approx H$ and calculate the mirror-image forces from (III.19).

For calculating the image forces we must know the value of the charges on the dust particles suspended in the air and artificially charged in the high-voltage field. These data are presented in Table III.1.

Kunkel [145] found no explicit relationship between the charge and size of a particle. Subsequently, Bodenstedt [147] and Bogdanov [148] observed a proportionality between the sizes and charges of air-suspended particles for a specific range of particle size. There was in fact no great difference between the charges on particles of the same size but different materials. Thus, for particles of sugar, chalk, soot, wood, and grain, 10 μ in diameter, the charge may range from 600 to 1100 unit charges ($9.6 \cdot 10^{-17}$–$1.8 \cdot 10^{-16}$ C).

If we consider the actual values of the charges on the particles (Table III.1), we find that for quartz particles 10 μ in diameter on a metal substrate the image forces calculated from (III.13) are about (2-3) $\cdot$ 10^{-7} dyn, i.e., three or four orders smaller than the adhesive forces of glass particles on steel surfaces (see Table III.3).

In electrical filters, incorporating a corona discharge, the charge on one quartz particle, 115 μ in diameter, increases to $0.2 \cdot 10^7$ unit charges, or $0.32 \cdot 10^{-12}$ C. In this case, the image force is about 4.6 dyn, i.e., it considerably exceeds the ordinary values of the adhesive forces [13].

Table III.1. Charges on Particles of Different Diameters and Different Materials

Material	d_p, μ	q	
		Unit charges	C
I. Suspended in air			
Quartz [145]	10	600*	$9.6 \cdot 10^{-17}$
Quartz [146]	0.5-22	4-100	$(6.4\text{-}160) \cdot 10^{-19}$
II. In the field of a corona discharge [146] (15 kV for a current of $7.6 \cdot 10^{-5}$ A)			
Aluminum	235	1.2/1.7†	1.9/2.7
Zinc	182	1.7/1.1	2.7/1.8
Zinc	493	5.3/3.3	8.5/5.3
Zinc	175	6.7/0.9	1.1/1.4
Zinc	109	0.3/0.3	0.5/0.5
Zirconium	220	0.9/1.1	1.4/1.8
Quartz	115	0.2/0.2	0.32/0.32

*The sign of the charge depends on the method of dusting. Only 95% of the particles suspended in the air are charged; the particles may be charged either positively or negatively.

† In the numerator we record the calculated results and in the denominator the experimental data, multiplied by 10^{-7} for the unit charges and by 10^{12} for charges meatured in Coulombs

Thus, Coulomb forces raise the adhesion when the particles are preliminarily charged, for example, in electrical filters, in the course of electrical separation (see §41), in the deposition of powders in a high-voltage field (see §53), etc. In these cases, the electrical forces arising as a result of the charge on the particles have a decisive influence on the adhesive interaction.

Change in Coulomb Forces on Charge Leakage. The charge on the particles covering a grounded surface is not constant.* This charge falls as the time spent by the particle on the surface increases. The reduction in the charge, in turn, leads to a fall in Coulomb interaction. After contact with the surface, the particles are discharged through a resistance constituting the sum of the intrinsic and contact resistances of the particle. The change in the charge (Q) of an adhering particle situated

*We are considering the excess charge on the particles, not the charge associated with contact phenomena.

in an electric field as a result of charge leakage may be expressed by the equation

$$dQ = (I_H - I_p)\,dt + dQ_t \tag{III.20}$$

where I_H is the current in the gap between the contiguous bodies, I_p is the current on the surface of the particles, and dQ_t is the increment in the charge on the particles in time dt.

In the absence of external sources of charge on the particles, we clearly have $dQ_t = 0$. Considering that

$$I = \frac{Q}{RC} \text{ and } I = I_H - I_p \tag{III.21}$$

(where R is the resistance and C is the capacity of the contact zone), Eq. (III.20) may be written in the form

$$\frac{dQ}{dt} = \frac{Q}{RC} + \frac{dQ_t}{dt} \tag{III.22}$$

The increment in the particle charges in time t may be expressed by Pauthenet's formula [144]:

$$Q_t = (1 + 2B)\,E\,r^2 \cdot \frac{\beta t}{1 + \beta t} \tag{III.23}$$

$$\frac{dQ_t}{dt} = (1 + 2B)\,Er^2 \cdot \frac{\beta}{(1 + \beta t)^2} \tag{III.24}$$

where E is the field strength, $B = (\varepsilon - 1)/(\varepsilon + 2)$ (here, ε is the dielectric constant of the particle material), $\beta = \pi enk$ (e is the elementary charge, n is the number of ions in 1 cm^3, i.e., the density of ions at the given point in the field, and k is the mobility of the ions).

Eliminating the time from (III.23) and (III.24), we obtain

$$\frac{dQ_t}{dt} = \frac{(A - Q_t)^2}{A} \cdot \beta \tag{III.25}$$

where $A = Q_{max} = (1 + 2B)r^2E$.

We may rewrite Eq. (III.22) in the form

$$\frac{dQ}{dt} = \frac{Q}{RC} + \frac{(A - Q_t)^2}{A} \cdot \beta = 0 \tag{III.26}$$

After solving Eq. (III.26) we obtain [143, 144] an expression for calculating the limiting value of the charge remaining on the particle situated on the grounded surface:

$$Q = \frac{(1+2B)\,Er^2}{2RC\beta}\left[1 + 2RC\beta - \sqrt{1+4RC\beta}\,\right] \tag{III.27}$$

It follows from (III.27) that conducting particles* will be discharged when they fall on a grounded surface. The discharge actually takes place in a fraction of a second. Then the image forces and hence the adhesion will tend to zero. Only insulating or semiconducting particles will retain their charge. Experiments show that all materials having a transitional resistance of less than $10^3\ \Omega$ fail to retain their charge.†

Below we present some experimental data [143, 144] relating to the transitional resistance (R) of spheres 1 cm in diameter:

	R, Ω		R, Ω
Steel	0.2	Calcium	10^{12}
Graphite	1.0	Zirconium	10^{12}
Tungsten	10^{11}	Quartz	10^{13}
Gypsum	10^{11}	Paraffin	10^{15}
Ashes	10^{12}		

Considering that the surfaces of dielectric particles carry adsorbed layers communicating surface conductivity, i.e., $R \to \infty$, we may expect that charge leakage will always take place to a certain extent. Then the value of the image force will fall as the time spent by the adhering particle on the grounded surface increases.

Analying the data presented, we may conclude that, in the absence of particle charging, the image forces need only be considered for strongly charged conducting particles at the first instant of their contact with a grounded surface. For particles possessing insulating or semiconducting properties, the time of action of the image forces depends on the state of the medium (humidity, temperature) and the presence of adsorbed layers on the surface

*As regards conductivity, all substances may be divided into three groups: conductors, $R \to 0$, $Q \approx 0$; semiconductors, $0 < R < \infty$, $Q < A$; and insulators, $R \to \infty$, $Q = A$.

†The transitional resistance is determined not only by the internal resistance of the substance, but also, to a large extent, by the surface capacitive impedance.

of the bodies in contact. If the surface is nonconducting and nongrounded, and if other means of charge leakage such as ionization of the air are improbable, then Coulomb forces may produce adhesion of the particles for a considerable time.

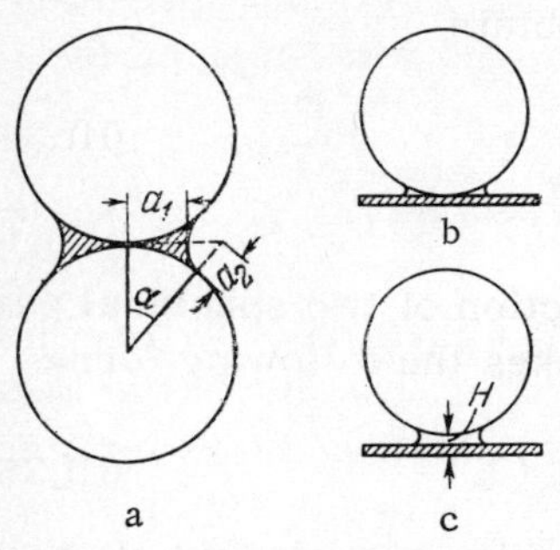

Fig. III.11. Capillary condensation associated with the contact of two particles (a) and of a particle with a surface (b, c). a,b) Without any intermediate layer; c) with an intermediate layer in the contact zone.

§ 13. Capillary Forces

Capillary Condensation. Condensation of water vapor takes place in the gap between bodies in contact (Fig. III.11a). The meniscus so formed first draws on the particle by way of surface tension ($F_{K,1}$) and, secondly, reduces the pressure of the liquid ($F_{K,2}$) by virtue of its concave shape, i.e.,

$$F_K = F_{K,1} - F_{K,2} \tag{III.28}$$

where

$$F_{K,1} = 2\pi a_1 \sigma \tag{III.29}$$

$$F_{K,2} = S p_K \tag{III.30}$$

S is the area of contact ($S = \pi a_1^2$); p_K is the pressure deficit (negative sign).

If a_1 and a_2 (Fig. III.11a) are the radii of the curvature of the intermediate water layer, then by Laplace's formula

$$p_K = -\sigma\left(\frac{1}{a_2} - \frac{1}{a_1}\right) = -\frac{\sigma(a_1 - a_2)}{a_1 \cdot a_2} \tag{III.31}$$

Substituting (III.29) - (III.31) into (III.28), we obtain

$$F_K = 2\pi a_1 \sigma + \pi a_1^2 \frac{\sigma(a_1 - a_2)}{a_1 \cdot a_2} = \frac{\pi a_1 \sigma (a_1 + a_2)}{a_2} \tag{III.32}$$

Expressing a_1 and a_2 in terms of the radius of the bodies in contact

$$a_1 = r(1 + \tan\alpha - \sec\alpha); \; a_2 = r(\sec\alpha - 1) \tag{III.33}$$

and substituting (III.33) into (III.32), we obtain

$$F_K = \frac{2\pi\sigma r}{\left(1 + \tan\frac{\alpha}{2}\right)} \tag{III.34}$$

As $\alpha \to 0$, formula (III.34) for the interaction of two spherical particles associated with capillary forces takes the following form:

$$F_K = 2\pi\sigma r \tag{III.35}$$

If one of the contiguous surfaces is plane, the height of the intermediate layer will be twice as small. Then,

$$F_K = 4\pi\sigma r \tag{III.36}$$

Expressions (III.35) and (III.36) are valid under conditions ensuring the total wetting of smooth surfaces, with an angle $\alpha \to 0$, i.e., for particles of relatively large size. For unwetted surfaces the right-hand sides of formulas (III.35) and (III.36) must be multiplied by $\cos\theta$ (θ is the wetting angle); then,

$$F_K = 2\pi\sigma r \cos\theta \tag{III.37}$$

$$F_K = 4\pi\sigma r \cos\theta \tag{III.38}$$

According to Eq. (III.38), capillary forces have their greatest effect on a hydrophilic surface when $\theta \to 0$, and their least effect on a hydrophobic surface when $\theta \to 90°$. This is in good agreement with the earlier relationship between the adhesion numbers and relative air humidity (see Fig. III.1).

Allowing for the effect of the roughness of the substrate, the expression for determining the adhesion due to capillary forces takes the following form:

$$F_K = 2\pi r\sigma \frac{\cos\theta_{app}}{\alpha} \tag{III.39}$$

and

$$\cos\theta_{app} = \alpha\cos\theta$$

where θ_{app} and θ are the apparent and true wetting angles, allowing for the roughness of the surface, α is the roughness coefficient

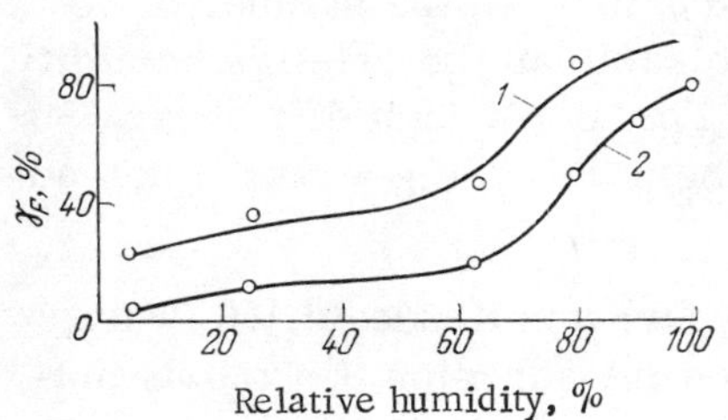

Fig. III.12. Adhesive force of glass particles (spherical) to a quartz substrate as a function of the air humidity. 1) d_p = 40-60 μ; F_{det} = 2.25 · 10^{-1} dyn; 2) d_p = 20-30 μ, F_{det} = 2.81 · 10^{-2} dyn.

of the surface, defined by the expression $\alpha = 1/\cos\beta$ ($\cos\beta$ is the average value of the cosine of the angle characterizing surface microrelief).

Equation (III.36) agrees with experimental data [77] for spheres of radius between 0.01 and 0.1 cm under conditions in which the particles are preliminarily pressed against the surface:

r, cm	0.02	0.04	0.055	0.88	0.10
F_{ad}, dyn:					
experimental	22	30	42	63	70
calculated from (III.36)	19.2	38.4	52.5	76	95.5

McFarlane and Tabor [81] recommend using (III.37) and (III.38) to determine the surface tension of water (or any other liquid) provided that the values of F_K and r are known. Experimental results obtained by these authors on determining surface tension by reference to the adhesion of glass spheres in the presence of a liquid held by capillary forces on glass surfaces are given below:

Liquid	Water	Glycerin	Decane	Octane
σ (dyn/cm) calculated from experimentally determined F_K and r	67.3	59	22.4	19.9
Actual [81]	72.7	63.5	25	21.8

We see from these data that the value of surface tension calculated from adhesion data is lower than the true value. The point is that the authors of [81] failed to allow for the disjoining pressure of the thin layer of liquid (Fig. III.11), which weakens the capillary interaction. This is why this method of determining surface tension by measuring the adhesive forces gave inaccurate results.

Relation Between the Adhesive Forces and Air Humidity. Experiments [149] show that capillary condensation starts appearing at a relatively low air humidity (above 65%).

Thus, for particles 20-30 and 40-60 μ in size, the number of residual particles remains almost the same as the relative humidity varies from 5 to 65% (Fig. III.12). For an air humidity above 65% there is a rise in the adhesion number, indicating a rise in the adhesive forces. *

The same effect was noted by Orr and Kozdecki [16] when studying the effect of air humidity on the adhesion of various materials, including the adhesion of glass particles to a glass surface. According to Luzhnov [133], a rise in the coefficient of friction (and hence the adhesive forces) occurs for a relative air humidity exceeding 60-70%, which agrees with our own results [149]. Hence one must reckon with the effect of capillary condensation on adhesive forces for a relative air humidity of 65-100%.

Capillary condensation takes place over a certain period of time, so that the effects of capillary forces are not felt immediately after the dust particle comes into contact with the solid surface. According to the experimental data obtained in [149], the rise in the adhesive forces ends about 30 min after glass particles 80-100 μ in diameter come into contact with the substrate in air at a relative humidity of about 100%.

Let us compare the forces calculated for adhesion under the influence of the capillary effect only [see (III.36)] with experimental results for the adhesion of spherical glass particles to steel surfaces as a function of particle size for a relative air humidity of 50-65%:

d_p, μ	40-60	20-30	10-20	5-10
F_{ad}, dyn:				
determined experimentally (for γ_F = 50%)	$2.1 \cdot 10^{-4}$	$2.1 \cdot 10^{-3}$	$6.1 \cdot 10^{-3}$	$1.3 \cdot 10^{-2}$
calculated from (III.36)	2.3	1.2	0.7	0.3

When the relative humidity of the air is no greater than 65%, the capillary forces play no part in adhesion.

Under conditions of capillary condensation, the adhesive forces are determined entirely by the capillary forces, which exceed all other adhesive components. The difference between the

*According to Corn [80], the adhesion of macroscopic particles increases uniformly (in direct proportion) as the relative humidity of the air rises.

capillary forces and the other components of adhesive interaction is particularly appreciable for particles more than 10 μ in diameter.

Theoretically we should expect [see Eqs. (III.36) and (III.38)] that the value of the capillary forces would depend on the particle size, the surface tension of the liquid formed as a meniscus by vapor condensation (see Fig. III.11), and also the capacity of the contiguous bodies to become wetted. Since the capillary forces are proportional to the particle dimensions, in cases in which the capillary component of the adhesive forces is dominant, the adhesion should be the same for all particles of the same size, while the difference between the adhesive forces of particles belonging to a particular fraction with a spread of particle sizes should not exceed the ratio of the dimensions of the extreme members of this fraction. For example, the adhesive forces calculated from Eq. (III.36) should be 4.52 dyn for particles 100 μ in diameter and 5.43 dyn for those 120 μ in diameter. Experimental results disagree with the calculated values. Actually the adhesive forces for particles 100-120 μ in diameter (for γ_F = 97-25%) fluctuate between 0.4 and 4.7 dyn, i.e., they vary by a factor of 12 over the fraction in question. Thus the scatter of the experimental data is much greater than would be expected, and hence the capillary effect fails to eliminate the indeterminacy of adhesive properties.

Even for a relative air humidity close to 100%, when adhesion is due solely to the capillary effect, the adhesive forces determined experimentally are still smaller than those calculated from Eqs. (III.36) and (III.38):

d_p, μ	40-60	80-100	100-120
F_K, dyn:			
calculated from (III.36)	2.3	4.1	5.0
calculated from (III.38)	2.0	3.7	4.3
F_{ad} (for γ_F = 97%), dyn	0.23	0.47	0.39
Adhesion determined experimentally:			
F_{det},* dyn	0.9	4.26	4.70
γ_F, %	41	22	25

*The force of detachment of the particles was measured after they had been in contact with the surface for 2 h at a relative humidity of 100%. The wetting angle of the glass surfaces was 30°.

It follows from the data presented that the forces holding the majority of the particles (more than 97%) at a relative humidity of 100% are an order smaller than the forces calculated for capillary condensation. Hence the majority of the particles are held on the surface by forces smaller than the capillary forces calculated from (III.36) and (III.38).

On the other hand, if we apply a detaching force close to the calculated value of the capillary forces, not all the particles become detached (γ_F = 22-41%).

As a result of our experimental data, we may conclude that the schematic picture presented earlier for capillary forces represents a particular case (see Fig. III.11b). It is clear that on raising the air humidity the space between the contiguous bodies will be filled with water (Fig. III.11c), and in the equilibrium state, when the disjoining pressure of the water film is balanced by the forces of interaction between the contiguous bodies, the gap (or film thickness) will take on an equilibrium value H. Clearly, the adhesion will depend on the value of H. If H is small, then the forces of molecular interaction between the contiguous bodies will be summed with the capillary forces; if H is large, the disjoining pressure of the thin layer between the contiguous surfaces considerably reduces the adhesion.* In accordance with this,†

$$F_{ad} = F_K - F_{disj} \tag{III.40}$$

Knowing the experimental value of the adhesive force F_{ad} and calculating the capillary forces F_K from Eqs. (III.36) and (III.38), we may also determine the disjoining pressure of the thin layer in the contact zone (referred to one particle).

However, as experience shows, if we apply a detaching force equal to the capillary forces or even greater than these, then even for a relative humidity close to 100% not all the particles are removed. Thus, for an applied force of 4.26 dyn, about 78% of the total number of spherical glass particles 80-100 μ in diameter are detached. (The surfaces are hydrophilic and incomplete wetting

*In colloidal systems [150] the disjoining pressure even exceeds the molecular forces.

†In accordance with Eqs. (IV.16) and (IV.10), F_{disj} is proportional to r^x ($x > 1$). Then

$$F_{ad} \simeq r(1 - r^{x-1}) \tag{III.40a}$$

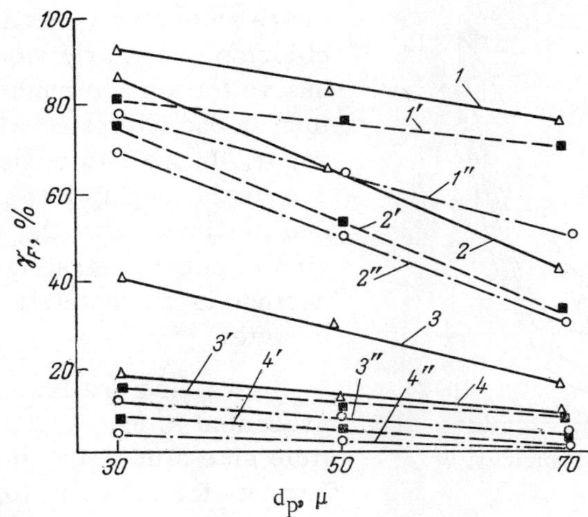

Fig. III.13. Adhesive force determined on detachment by the impulse method as a function of spherical glass particle size for substrates of ordinary glass (1,2,3,4), hydrophilized glass (1',2',3',4'), and hydrophobized glass (1", 2",3",4"). 1,1',1") In an atmosphere saturated with water vapor; 2,2',2") with acetone (or carbon tetrachloride vapor); 3,3',3") in air; 4,4',4") in vacuum.

is excluded.) Clearly, the value of H influencing the adhesion is determined not only by the wettability of the surfaces and the humidity of the surrounding medium but also by certain other factors such as, for example, the particle size, the surface roughness, etc.

Figure III.13 shows the adhesion number as a function of the dimensions of spherical glass particles (30, 50, and 70 μ). The adhesion of the particles formed the subject of one of our experiments carried out (by the impulse method) in an atmosphere saturated with water, acetone, or carbon tetrachloride vapor. The substrates included a simple No. 23 glass plate (curves 1-4), hydrophilized glass (curves 1'-4'), and hydrophobized glass (curves 1"-4").

It follows from the data presented that the adhesion of particles in vacuum is smaller than in air; water vapor increases the adhesive strength; in vacuum capillary forces are absent, while in air they appear on formation of a meniscus in the space between

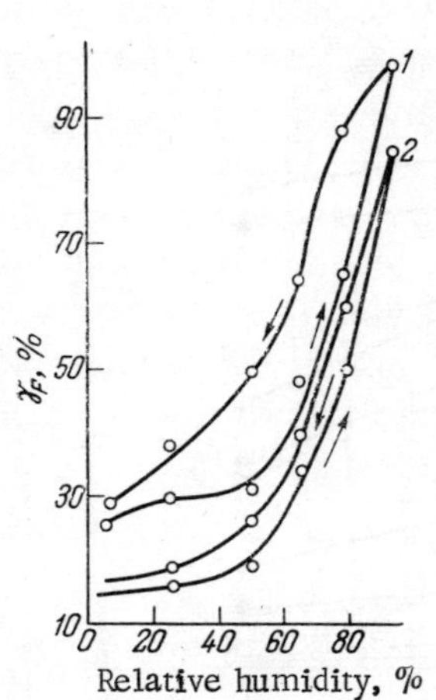

Fig. III.14. Hysteresis curves for the adhesive force of spherical glass particles 50 ± 2 μ in diameter sticking to a quartz substrate. 1) F_{det} = 0.1-4 dyn, t = 2 h; 2) F_{det} = 0.2-2 dyn, t = 24 h.

the contiguous bodies; in an atmosphere saturated with carbon tetrachloride or acetone vapor, the adhesive forces are much smaller than in one saturated with water vapor. In the latter case, the reduction in the capillary forces is due to the fact that the surface tension of carbon tetrachloride and acetone is much smaller than that of water. *

We also see from the results presented in Fig. III.13 that, as particle size diminishes the difference between the adhesive forces in atmospheres of water and carbon tetrachloride (or acetone) vapor are less noticeable, since the absolute value of the adhesive force increases with diminishing particle size, while the relative value of the capillary forces becomes smaller.

In all cases the adhesion (see Fig. III.13) of spherical glass particles to a glass substrate is greatest, while hydrophilization of the substrate (reducing the wetting angle from 30 to 18° with respect to a drop of distilled water 1700 μ in diameter) and to a greater extent hydrophobization (increasing the wetting angle from 30 to 65°) reduce the adhesive forces acting on the particles, provided that the surfaces of these have not been modified. Here it must be noted that, the more the wetting angle changes on modification of the surface as compared with that of the unmodified surface, the more does adhesion to this surface diminish. In view of the fact that the hydrophilic properties of the surface depend on the presence of functional molecular groups determining the dispersion interaction on the surface, we may suppose that the dispersion interaction does not depend on the humidity of the air.

* Water, σ = 72.3 dyn/cm; carbon tetrachloride, σ = 23.7 dyn/cm; acetone, σ = 25.7 dyn/cm.

After holding a dust-laden surface in an atmosphere saturated with carbon tetrachloride vapor for 24 h, 95% of all the particles 100-120 μ in size are retained for a detaching force of 1.53 dyn. The capillary forces calculated from Eqs. (III.35) and (III.36) for the condensation of carbon tetrachloride vapor in the contact zone equal 1.81 dyn. Apparently thin layers of nonpolar liquids do not have any disjoining effect. Hence, no reduction in the adhesive forces arising from the effects of a liquid interlayer in the contact zone are observed. This indirectly confirms the validity of Eq. (III.40) for estimating the value of the disjoining effect.

When studying the effect of moisture in the air on adhesion, certain hysteresis phenomena were also observed [149].

Figure III.1 shows the adhesive force of spherical glass particles 50 ± 2 μ * in diameter as a function of the relative humidity of the air. The lower branch of the hysteresis loop shows a rise in adhesion with increasing relative air humidity and the upper branch shows a reduction in adhesion with falling humidity (Fig. III.14). The failure of the adhesive force/humidity curves to coincide indicates that the processes of the capillary condensation and evaporation of moisture in the gap between the contiguous surfaces possess certain peculiarities characteristic of thin layers of liquid.

The hysteresis phenomenon was also observed by Luzhnov [133] when studying the static friction of soot powder (see Fig. III.4). In these experiments the upper branch of the hysteresis loop characterizes the change in the coefficient of static friction with increasing relative humidity and the lower with falling humidity. In the present case the failure of the curves to coincide is apparently associated with the replacement of dry by semi-liquid or liquid friction, which is always accompanied by a reduction in the friction coefficient.

Deryagin [151] developed a theory for determining the force of interaction between particles separated by a film of liquid in equilibrium with its vapor. The resultant equations allow for the effect on adhesion of not only capillary forces, but also the adsorption of vapor on the particle surface.

---

* The actual dimensions of the particles in the fraction differed rather more than ±2 μ, but only particles with diameters of 50 ± 2 μ were taken into account in the calculation.

Thus the capillary forces producing the adhesion of particles are the larger, the greater the surface tension of the liquid, the vapor of which surrounds the dust-laden surface, the greater the particle dimensions, and the better the wettability of the surface in contact. A liquid interlayer between the particles and the surface eliminates or greatly reduces the effect of electrical forces. The simultaneous action of capillary and electrical forces is practically excluded.

In an air medium, if the air humidity exceeds 65%, the capillary forces predominate over other components contributing to the adhesion.

§ 14. Dependence of the Forces of Adhesion on the Shape and State of the Surface

Possibility of Calculating the Area of Contact Between Particle and Surface. The force of adhesion depends on the area of contact of the particle with a plane surface, since the force of molecular interaction and the electrical component of the forces of adhesion are proportional to the area of the actual contact zone.

For elastic contact, in which a sphere of radius r is pressed to an ideally smooth surface by a force F_p, the radius of the area of contact may be calculated from the Hertz formula [1, 17]:

$$a = \sqrt[3]{0.75 r F_p \cdot \left[\frac{1-\mu_1^2}{E_1} + \frac{1-\mu_2^2}{E_2}\right]} \tag{III.41}$$

where a is the radius of the area of contact of the particle, F_p is the force of the applied pressure, μ_1 and μ_2 are the Poisson's ratios, and E_1 and E_2 are the elastic moduli for the materials in question.

If, when a microscopic particle came into contact with a plane surface, the force F_p equalled the adhesive force, then formula (III.41) could be used to determine the actual contact area of a smooth particle with a smooth surface.

So far, the actual contact area of a particle with a surface has not been measured, but we may provisionally assume the applicability of the Hertz formula to the calculation of this area. The effect of the area of contact on the adhesive force may as yet only be estimated indirectly. Thus, the adhesion of gold spheres 6–7 μ in diameter to a polyamide plate is greater than on a gold or a smooth quartz surface (0.25, 0.09, and 0.05 dyn are, respectively, required to detach 50% of the particles); this is because of the deformation of the contact area of the polyamide plate, since the deposition of dust on the test surfaces was accompanied by additional compression of the particles resulting from vibration [73].

Below we present some experimental values [13] for the adhesive force of spherical glass particles on various surfaces and also the contact areas calculated from formula (III.41), allowing for the elastic properties of the contiguous bodies:

d_p, μ	20–30	45–55	65–75	85–95
F_{ad}, dyn	$8.5 \cdot 10^{-4}$	$7.0 \cdot 10^{-3}$	$1.9 \cdot 10^{-2}$	$4.05 \cdot 10^{-2}$
S, μ^2:				
with glass substrate	$5.6 \cdot 10^{-4}$	$3.6 \cdot 10^{-3}$	$8.8 \cdot 10^{-3}$	$1.7 \cdot 10^{-2}$
with steel	$4.5 \cdot 10^{-4}$	$2.8 \cdot 10^{-3}$	$7.0 \cdot 10^{-3}$	$1.3 \cdot 10^{-2}$
with painted metal	$3.0 \cdot 10^{-3}$	$2.0 \cdot 10^{-3}$	$4.9 \cdot 10^{-3}$	$9.6 \cdot 10^{-2}$

It follows from the data presented that the area of contact has a considerable value (radius of contact of a particle tenth parts of a micron); on reducing the elasticity of the substrate (for example, in the case of painted surfaces), the area increases. However, formula (III.41) and the data presented are valid for the particular case in which smooth spherical particles stick to a smooth surface.

It should be mentioned that, according to investigations of Böhme et al. [64], the adhesive forces (and hence the area of contact) remain altered if the material of the particle and substrate are interchanged; for example, the adhesion of carbonyl iron powder to chromium–vanadium steel is the same as the adhesion of the steel powder to a carbonyl iron surface.

Effect of Particle Shape on Adhesion. The shape of the particles differs in practice from spherical, while the surfaces of contiguous bodies contain asperities causing variations in the area of contact, and hence the adhesion of the particles. The

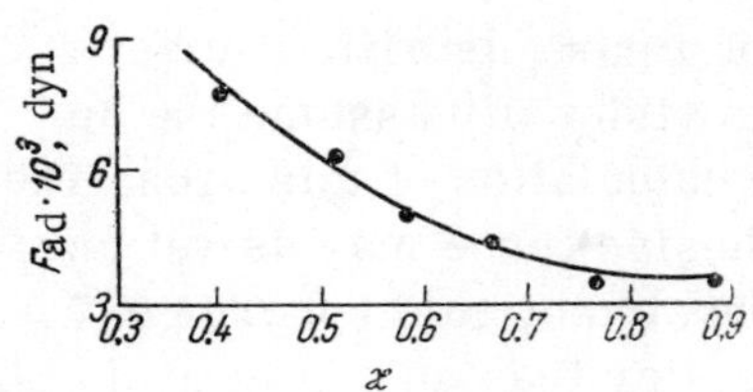

Fig. III.15. Adhesive force of particles with a double mean radius of 100-160 μ as a function of the sphericity factor.

effect of particle shape on adhesion may be taken into account by means of a "sphericity factor" $\varkappa$ (or the sedimentation radius of the particles) defined by reference to the change in the settling rate of particles of the form in question in a stationary medium as compared with spherical particles. Tekenov [83] determined values of $\varkappa$ for loess particles as a function of their shape: spherical, $\varkappa = 1.0$; isometric, $\varkappa = 0.9$; rounded, $\varkappa = 0.78$; soil, $\varkappa = 0.67$; elongated prismatic, $\varkappa = 0.59$; plane in the form of sheets and scales, $\varkappa = 0.42$.

These data are only valid for loess particles for which the double mean radius* exceeds 60 μ.

Figure III.15 shows the adhesive forces measured by the method of direct detachment for loess particles as a function of the sphericity factor. As the sphericity factor rises from 0.4 to 0.9, the adhesive force diminishes as a result of the reduction in the actual contact area of regularly shaped particles. For particles of irregular shape there is a greater spread of adhesive-force values than for spherical particles. Thus, for particles with a double mean radius of 180 μ, the adhesive force varies between $2.8 \cdot 10^{-3}$ and $1.4 \cdot 10^{-2}$ dyn.

Corn [80] noted that the sticking of filaments with fused ends of spherical shape (Fig. II.12b) was stronger than that of filaments with fused ends of other shapes (Fig. II.12a, c). This suggested that, in the first case, the contact area was greater. However, as these observations were made with large particles, 88 μ in diameter, we feel that the phenomenon in question may simply have been due to the greater mass and greater pressure of these particles as compared with particles of smaller dimensions.

Thus, the minimum adhesive force occurs for particles of isometric shape, approaching that of a sphere or regular polygon. In practice one is frequently concerned with spherical particles of

* As "double mean radius" we take the arithmetic mean value of two measurements in mutually perpendicular directions.

this type [153]. Thus, spherical particles of magnesium oxide, tin, and lead are obtained in an electric arc; emulsions are prepared from silver, latex, edestin [153], and other spherical particles; in nuclear explosions in podzol and sandy soils, soil particles are drawn into the zone of the explosion and melt under the influence of the high temperature, becoming spherical in shape [154-156]; ash particles often acquire spherical form as a result of the cooling of a melt or on melting [157].

The adhesive force of plane particles (i.e., particles with lengths and widths much greater than their thickness) is greater than that of isometric particles. Plane particles include kaolin, bentonite, mica, graphite, gypsum, etc. [153].

In addition to the shapes already mentioned, there are also particles of fibrous or acicular form (prisms, needles, fibers, etc.) having one dimension greatly exceeding the others. These include particles of zinc oxide (0.4-1 μ), asbestos [(0.3-3) · 100 μ or (0.5-1.5) · 1 μ], tobacco virus [(1.0-2.0) · (10-30) μ], etc.

We should expect that the adhesive force of acicular particles would be greater than that of plane and isometric ones, owing to the greater area of contact of the particles with the surface.

All that has been said so far refers to the adhesion of individual particles of an aerosol. Aerosols frequently occur in the form of aggregates formed of a large number (sometimes several millions) [158] of primary particles. Such aggregates may have linear or isometric shape. The adhesion characteristics associated with the primary particles are in general valid for aggregates of the same shape. However, not all the primary particles will be in contact with the surface. On detaching such an aggregate from a surface, the break will occur as a result of the rupture of autohesive as well as adhesive couplings, as in the case of a layer of powder.

Effect of Surface Roughness on Adhesion. The effect of the roughness of cast iron and steel surfaces on the adhesion of spherical glass particle was studied in more detail in [159] (spheres being used so as to eliminate the effects of particle shape). According to the experimental results (Fig. III.16), the adhesion number of glass particles with respect to a cast iron surface increases as the class of surface finish becomes numerically higher

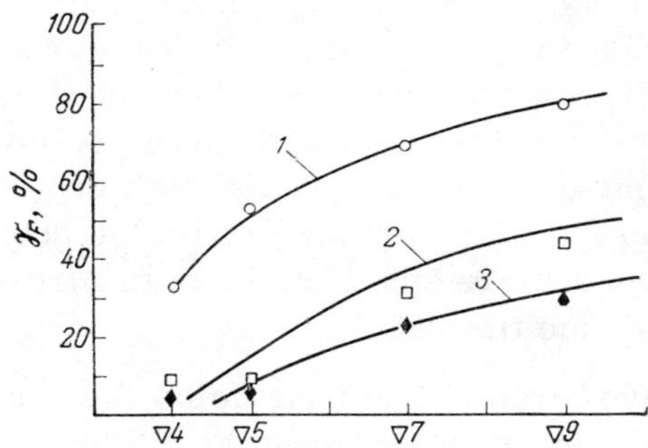

Fig. III.16. Adhesion number of spherical glass particles $40 \pm 5\ \mu$ in diameter as a function of the class of finish on cast iron surfaces for various detaching forces. 1) $F_{det} = 2.2 \cdot 10^{-2}$; 2) $9.3 \cdot 10^{-2}$; 3) $22.4 \cdot 10^{-2}$ dyn.

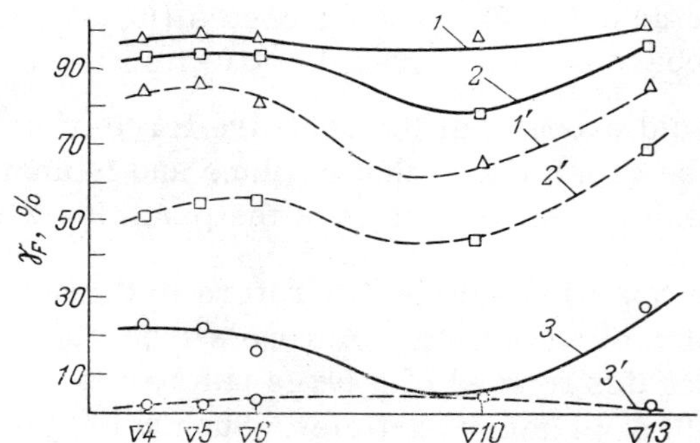

Fig. III.17. Adhesion numbers of spherical glass particles of various diameters on steel surfaces of various degrees of surface finish with a detaching force of 70 g (1,2,3) and 1150 g (1',2',3'). 1,1') $d_p = 20 \pm 5$; 2,2') 40 ± 5; 3,3') $70 \pm 5\ \mu$.

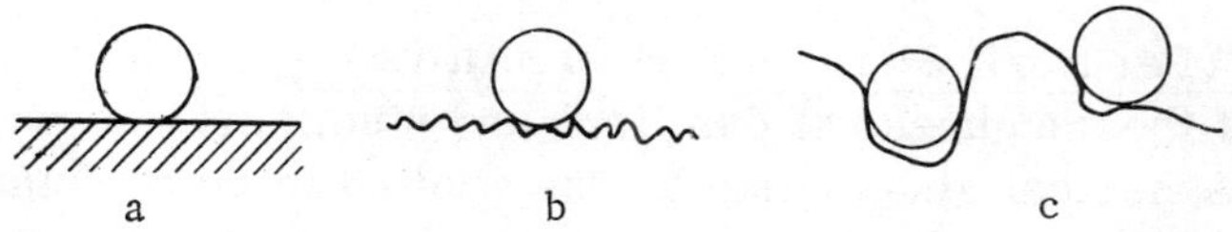

Fig. III.18. Types of substrate roughness associated with the adhesion of particles.

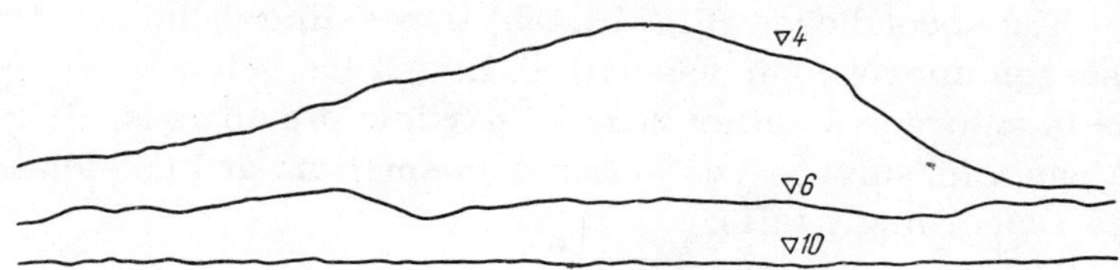

Fig. III.19. Profilograms of steel surfaces (magnification: horizontal, 1050×; vertical, 2000×).

(for the same force of detachment). However, it must be noted in this connection that the material of the substrate plays quite an important part. Thus, the integral adhesion curves for glass particles on steel surfaces with different degrees of surface finish are arranged in a much more compact manner.

Figure III.17 shows the variation in adhesion number as a function of the surface finish of steel surfaces for spherical glass particles with different detaching forces. We see from the data presented that the adhesion numbers (and hence the adhesion) reach maximum values for steel plates with a Class 13 surface finish.

As the surface quality worsens to class 10, the adhesion number falls; subsequently, it rises again. The roughness of the substrate has practically no effect on the adhesion numbers of small particles for a small detaching force (curve 1) or of large particles (70 μ in diameter) for a considerable detaching force (curve 3'). In the first case, almost all the particles are held on the surface, and in the second almost all are removed. Thus, the roughness of the surface has no effect on adhesion for the two extreme points on the integral adhesive-force curves.

From these results we may distinguish three cases characterizing the effect of surface roughness on the adhesion of particles (Fig. III.18).

The first case is possible (Fig. III.18a) when the contiguous surfaces are ideally smooth, for example, in the adhesion of spherical glass particles to a fused glass surface or to metal surfaces with a Class 13 finish. Only in this case may the contact area be calculated from the Hertz formula (III.41).

The second case (Fig. III.18b) is possible if the substrate possesses microscopic asperities, i.e., if the height of the projections is an order smaller than the particle dimensions. Here the true particle/surface contact area is smaller, and the adhesive force accordingly falls.

In the third case (Fig. III.18c), an increase in adhesive force takes place as a result of the macroscopic roughness of the surface, when the heights of the projections are comparable with the dimensions of the dust particles, for example, when the steel surface has a finish lower than Class 10. The true contact area again rises, and this leads to an increased adhesion number of the particles.

Of course, in practical conditions of the adhesion of particles to a plane surface, a variety of forms of surface roughness are possible.

Surface profilograms may be obtained by means of the "Kalibr VEI" profilometer-profilograph. On steel surfaces with Class-10 finishes, microscopic projections appear; these cause a fall in adhesion (see Fig. III.19). On plates with a Class-6 finish, macroscopic projections appear; the height of these becomes greater on Class-4 surface finishes, and this leads to a rise in adhesive force. The profilograms of Class-13 plates constitute practically straight lines. Hence, the profile of steel surfaces changes in two ways: first on the appearance of microscopic roughnesses and then on the appearance of macroscopic roughnesses (Class/10 and Class-6 surface finishes, respectively). This makes it clear why the adhesion number of spherical particles varies with substrate roughness (see Fig. III.17).

Cast iron surfaces possess greater microscopic roughness than steel surfaces with the same degree of surface finish. Thus, on working to a Class-4 finish the distance between the microscopic asperities on cast iron surfaces is a factor of 2-3 smaller than on steel. Hence, the adhesion of particles to cast iron surfaces is smaller than to steel surfaces.

We concluded from all this experimental material that adhesion to a microscopically rough surface is smaller than to smooth or macroscopically rough surfaces [159].

These conclusions agree with the data of Böhme et al. [73], who showed that the adhesive force of spherical gold particles 6-7 μ in diameter on a smooth quartz surface was smaller than on a rough one. Unfortunately, these authors do not indicate the size of the projections and the distance between them on the surface, so that the effect of the degree of roughness on the adhesive force of the particles cannot be estimated.

Corn [80] observed a fall in the adhesive force of fused glass particles to "Pyrex" glass (mean height of microscopic projections 2159 ± 127 A, 2921 ± 127 A, 4826 ± 127 A) with increasing microscopic roughness, and introduced a correction for roughness when determining the adhesive forces

$$\Delta F_{ad} = \frac{h_{av}}{8.3 \cdot 10^2} \qquad \text{(III.42)}$$

where ΔF_{ad} is the change in adhesive force due to the microscopic roughness in %, and h_{av} in the mean height of the microscopic projections, in A.

Equation (III.42) is only applicable for the conditions under which Corn's experiments were carried out (§ 7).

The adhesive forces may also vary with the roughness of the particle. Thus, the adhesive force between the particles of coke shape [152] possessing microscopic surface roughness and a plane surface is smaller than that of smooth spherical particles of the same material. This is because contact between the coke-shaped particles and the plane is effected at individual points, which reduces the contact area and hence the adhesive force. Coke-shaped powders include coal and silica gel particles [153] as well as particles of dust and ashes formed in nuclear explosions in carbonate soils (coral reefs), when the soil particles are decarburized and acquire a flocculent shape [157] similar to that of coke. In general, powders of coke shape may be obtained by combustion of the volatile components from the particle surface.

Variation of Adhesion with the State of the Surface and the Thickness of an Oil Layer. By the "state" of a surface we mean its cleanness, i.e., the presence or absence of foreign adsorbed matter such as surface-active substances, moisture, etc.

It is well known (see §§4 and 10) that layers of substances adsorbed on a surface may change the molecular interaction. Although adsorbed layers have no great influence on the electrical forces arising as a result of the charge of a double layer (see §11), nevertheless, if these layers engender a surface conductivity, we may expect a reduction in Coulomb interaction as the time of contact increases (see §12). Moisture on the surface promotes capillary condensation in the gap between the contiguous bodies (see §13).

The presence or absence of an adsorbed layer (layers) on the surfaces in contact (particle and plane) is determined by the degree of cleanliness of the surface. Unfortunately, many authors have not indicated the degree of cleanliness of the surface and the cleaning method employed; of those who have done so, many have used different cleaning methods, so that comparison of the results is difficult.

The literature contains very little information regarding the effects of various methods of cleaning contiguous surfaces on the adhesion of the latter. According to Akhmatov [38, 39], the adsorption cleaning of steel surfaces with silica gel leads to a sharper fall in the forces of friction (and hence the friction-dependent components of the adhesive forces) than cleaning with activated charcoal and B.70 gasoline.

Below we present the results of Fuks [118] on the relation between the adhesion of various oils to steel surfaces and the preparatory cleaning method:

Oil	S-220	MBP-12
Work of adhesion of the oil (referred to 1 cm^2) to steel surfaces cleaned in the following ways (ergs):		
by solvent*	1440	60,000
by adsorption†	2300	9,500
by glow discharge	690	8,700

We see from the data presented that the adhesion of the oil is weakest for surfaces cleaned by a glow discharge. (The method

* Soap solution, distilled water, and acetone.

† In accordance with GOST 7394-56 providing for cleaning with activated charcoal.

Table III.2. Adhesion Numbers (%) for Spherical Glass Particles 40–60 μ in Diameter for Different Detaching Forces (25–30 days after painting)

Coating \ F_{det}, dyn	$2.8 \cdot 10^{-4}$	$5.7 \cdot 10^{-4}$	$9.0 \cdot 10^{-4}$	$2.6 \cdot 10^{-3}$	$2.6 \cdot 10^{-2}$	$1.1 \cdot 10^{-1}$
Perchlorvinyl enamel VTU KU 518-58.	68.9	62.5	60.7	60.0	52.3	47.5
Perchlorvinyl enamel contaminated with Avtol (lubricating oil)* . .	100	100	100	100	98.7	98.7
Oil paint 4BO	100	100	100	100	98.7	98.7

* The Avtol was introduced on to the coating at a calculated dose of 1.4 g/cm^2.

of cleaning was developed by V. V. Karasev.) With this method of cleaning not only surface-active but also organic substances are removed from the surface.

We must particularly consider the effect of vacuum on adhesion. Depending on the conditions under which the particles are deposited on the surface, vacuum may either increase or reduce adhesion.

If the surface is sprayed in air, and then placed in vacuum, the adhesion in vacuum is in general smaller than the corresponding adhesion in air.

Thus it was found in [13, 79, 81] that the adhesion of spherical glass particles to a steel surface became smaller after placing the surface in a vacuum (particles deposited in air). According to Patat and Schmid [20], the adhesive force (referred to 1 cm^2) of a layer of aluminum oxide powder on a steel surface falls from 60.5 dyn in a nitrogen atmosphere at 760 mm Hg to 15.5 dyn in the same atmosphere at a pressure of 10^{-4} mm Hg.

If the particles are deposited on the surface in vacuum, then the adhesive force measured after holding the surfaces in vacuum is as a rule greater than the adhesion of the same surfaces in air [43, 48]. Thus the work required to split mica surfaces touching one another in vacuum increases by a factor of 25 as compared with that in air [160]. This may be explained by the vanishing of adsorbed moisture from the surface in vacuum, which increases the adhesion.

The presence of oil contamination on the surface increases the adhesion of particles as a result of the corresponding tackiness; this is illustrated clearly by the data of Table III.2 [11].

The effect of the degree of surface oiliness on the detaching force of the particles was also studied. (The surface of the plate was oiled by dipping in a solution of Avtol in carbon tetrachloride.) It was found that the manner in which the adhesion number varied with the thickness of the oil layer varied, the curve of γ_F against the degree of oiliness being divisible into three parts (Fig. III.20). First of all (part I), for a slight increase in oiliness (up to 0.5 mg/cm^2) there was a considerable rise in adhesion. Later (part II) the adhesion ceased to depend on the thickness of the oil layer (from 0.5-2.0 mg/cm^2); then (part III) the adhesion fell with increasing oiliness.

The mechanism of adhesion to an oily surface may be represented in the following way. The thicker the layer of oil, the more deeply may the particles penetrate into this layer. At a certain thickness of the oil layer, equilibrium is achieved between the forces tending to "sink" the particles (weight, inertia) and the resistance of the oil layer. Hence, over a certain range of oil-layer thicknesses the adhesive force is independent of the degree of oiliness. As the thickness of the layer of oil increases further, it acquires the property of fluidity, and adhering particles may be removed in company with the oil.

The adhesion of particles to an oily surface depends not only on the thickness of the oil layer but also on its viscosity [161], i.e., on the quantities of binding components in the oil (for example, for salt pastes, $MgCl_2$, $CaCl_2$, etc.).

In addition to this, the adhesion of a particle to a solid substrate covered with a tacky layer depends on the kinetic energy of the dust particle at the instant at which it comes into contact with the substrate (Fig. III.21).

We see from Fig. III.21 that the adhesive force on the particles depends on the depth of penetration of the particle into the oil layer, which in turn is determined by the settling velocity of the particles. The adhesive forces will naturally be different for plane and spherical surfaces, since, in the sticking of two plane solids separated by a tacky layer the contact area is constant,

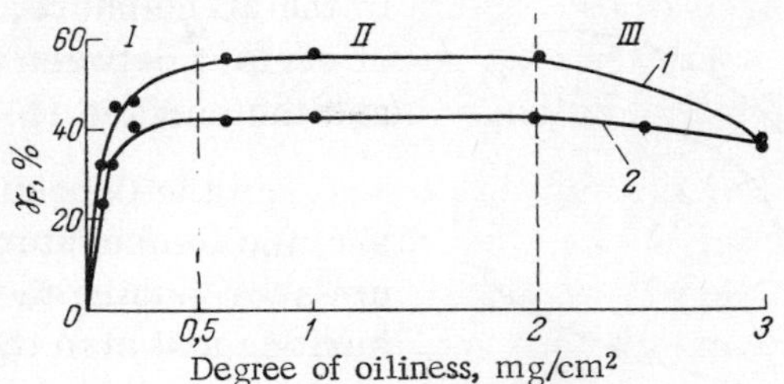

Fig. III.20. Adhesion number of particles 10-70 μ in diameter as a function of the amount of Avtol on a surface coated with perchlorvinyl enamel. 1) Ordinary surface; 2) surface modified with Chromolan.

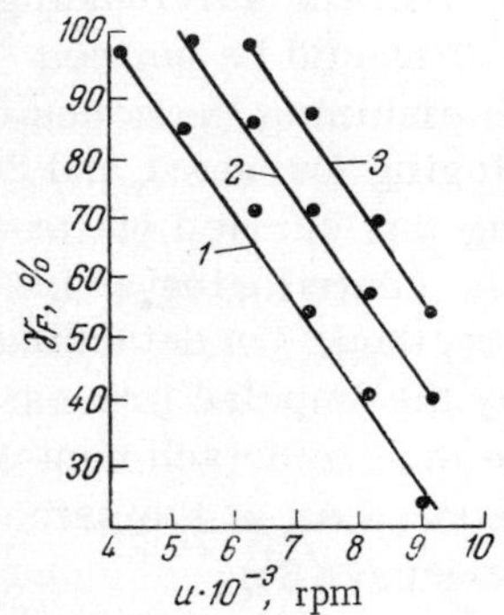

Fig. III.21. Variation of adhesion number with the number of revolutions of the centrifuge required for detaching particles with a wide spread of particle size settling on an oily surface (oil density 0.5 mg/cm^2) at different velocities. 1) v = 1.4; 2) 8.3; 3) 16.7 m/sec.

while in the sticking of two particles with a tacky interlayer the area of contact may increase as the particles penetrate into the layer.

In view of the absence of published data regarding the mechanism underlying the behavior of a tacky layer in the capture of dust, we must at present confine attention to qualitative considerations. It should be noted that the property of tackiness is used for the trapping of dust particles in self-cleaning oil filters, rough-cleaning automobile antidust filters, Rank filters made of metallic mesh, settling plates in konimeters, impactors, and similar apparatus [162]. In addition to this, adhesion to a tacky base may be used for studying dustiness and the composition of dust, particularly for trapping dust in the layers of atmosphere near the earth's surface [163].

§ 15. Influence of the External Medium on the Force of Adhesion

Factors on which adhesive forces depend include the temperature of the external medium, the presence of certain vapors

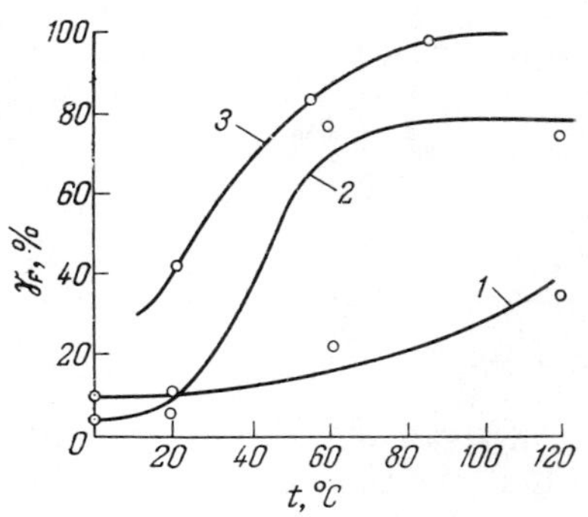

Fig. III.22. Variation in the adhesion of glass particles to glass (1) and perchlorvinyl enamel-painted surfaces (2,3) with air temperature. 1,2) Particles detached by centrifuging (d_p 70 ± 5 μ); 3) by the impulse method (d_p 30 μ).

in the atmosphere, and the period of contact between the particle and the surface [164].

In the deposition of dust in air, the temperature of the medium surrounding the dust-laden surface and also the temperature of the contiguous bodies themselves may vary over wide limits (for example, from −30 to +40°C).*

Figure III.22 shows the adhesive force as a function of the temperature of the surrounding medium. It should be noticed that, in determining the adhesion by centrifuging (curves 1 and 2) the dusting was carried out at a specified temperature in a thermostat and the centrifuging at 16-18°C; this constitutes a disadvantage of the method. On detaching a monolayer of particles 30 μ in diameter by the impulse process, when the whole process (from depositing the dust to detachment of the latter) is carried out in a single thermostat, i.e., at the same temperature, analogous results are obtained (curve 3).

In the impulse method the effect of air humidity is excluded, and hence the rise in the adhesive force with increasing temperature of the medium can only be explained by a change taking place in the properties of the contiguous bodies.

The rise in adhesive force on painted surfaces (Fig. III.22) is clearly explained by the development of surface tackiness, which increases with temperature.

It has been found experimentally that on placing a sample, previously coated with dust in air (at 20°C, relative humidity 50-60%), into a low-temperature medium (below 0°C) the adhesive forces increase sharply as a result of the "freezing" of the particles to the surface by virtue of moisture condensed between the bodies in contact. However, if the dust is actually deposited on the

*Under industrial conditions the temperature range may be much greater.

surface at a low temperature there is no rise in the adhesive force.

It was found in [13, 80] that the adhesive forces also increase with the time of contact. However, the results of different authors differ in respect of the period after which adhesion reaches a maximum (30 min according to [13] and 5 min according to [80]). This is evidently due to the method of measuring the adhesion in the two cases.

By analogy with the similar process in liquids, this phenomenon is called *aging* [13]. The reasons for aging have not yet been ascertained. According to Brandt [165], aging is due to the vanishing, over a period of time, of adsorbed gas layers situated in the space between the particle and the substrate. In our own opinion, this supposition is not very well founded. The work expended in disrupting the adsorbed layer in the contact zone is negligible in comparison with the adhesion energy, and the destruction of the adsorbed gas layers takes place almost instantaneously. In addition to this, the adhesive forces also increase with the period of particle/surface contact in a vacuum, i.e., in the absence of adsorbed gases and moisture. Hence we cannot consider the humidity of the medium [81] as the main cause of aging.

In view of the fact that in practice adhesion tends to be encountered not only in air, but also in various other gases and vapors, it is interesting to discover how the composition of the medium affects adhesion. Patat and Schmid [20] found that replacing air with nitrogen had no effect on the adhesion of aluminum oxide powder to a steel surface. However, it would hardly be correct to disregard the effect of the medium surrounding the dust-laden surface on adhesion entirely. In order to check these principles, the adhesion of glass particles (spherical) to glass of the same type as the particles in an atmosphere of ammonia and sulfur dioxide (SO_2) was studied (by the impulse method). The choice of SO_2 and ammonia as media was engendered by the fact that these substances form part of the atmosphere of chemical factories [166], and it was therefore particularly interesting to discover whether they affected the process of gas purification.

Below we present the adhesion numbers of glass particles 30 μ in diameter, determined by detaching these from a glass surface on the impulse principle:

Medium	Vacuum	Dry air	Moist air*
γ_F, %:			
without additives.	25	43	58
with added SO_2.	22	47	61
with added NH_3.	43	64	92

We see from these data that the presence of ammonia (the ammonia was dried on coming into the test bell-jar) leads to a considerable rise in adhesion.

The rise in adhesion in an ammonia atmosphere was also observed when studying the interaction of particles 70 μ in diameter with hydrophobized and hydrophilized glass surfaces (temperature, 18°C):

Medium	Vacuum	Ammonia atmosphere
γ_F, %:		
to glass No. 23	9	69
to hydrophobized glass . . .	1	57
to hydrophilized glass. . . .	3	61

Evidently, by virtue of a hydrogen bond, the ammonia is able to combine with the condensed and adsorbed moisture on the surface and thus influence the adhesive force [167].

The presence of sulfur dioxide in dry air has practically no effect on the adhesive force of dust. Even at temperatures of the order of 50-80°C, about 84-96% of the particles remain in an atmosphere of dry SO_2, i.e., approximately the same number as in air (particles detached by the impulse method).

Hence, if gases or vapors surrounding the dust-laden surface interact with the latter in the course of adsorption (NH_3 with moisture on the surface) the adhesion of the particles may change. Analogous processes take place in the agglomeration (autohesion) of particles [168].

*Temperature of medium, 18°C; relative air humidity, 65%.

§ 16. Dependence of the Forces of Adhesion on the Dimensions of the Particles

Scatter in the Values of Adhesive Forces. Experimental results indicate a spread in the values of the adhesive forces, i.e., a difference between the maximum and minimum adhesive force [13, 16, 64]. For macroscopic bodies this scatter or spread is no greater than 20% [80]; for microscopic particles it is greater, reaching one or two orders of magnitude, or even more. The distribution of microscopic particles with respect to adhesive forces is nonuniform. On the integral curve of adhesive forces [13] for particles 40-60 μ in diameter, two sections may readily be distinguished. The adhesive force in the first of these is commensurable with the weight of the particle. Here there is a sharp rise in the number of particles detached for an insignificant increment in the detaching force. The scatter characterizing the adhesive force in this section is no greater than 20%, i.e., it is the same as for macroscopic bodies.

In the second section the adhesive force exceeds the weight of the particles, and hence the number of particles remaining changes very little for a considerable change in the detaching force. With falling particle size the difference between the two sections of the integral adhesive-force curves becomes less marked.

The reasons for the spread in the adhesive forces of microscopic particles have not yet been finally explained. The spread might very probably be explained by differences in particle size. However, even on reducing the spread of particle sizes in the fraction, the scatter in adhesive forces is not eliminated. In addition to this, direct measurements of the diameters of the particles in the original and final fractions show that not all the particles of any specified diameter are removed by the same value of acceleration.

The adhesive forces are affected by the roughness of the contiguous bodies, since when the particles come into contact with the substrate the true contact area will be determined by the roughness of both substrate and particle. However, even the roughness of the contiguous bodies cannot be the main factor determining the spread in adhesive forces, since this spread remains very substantial even on surfaces of Class 13 finish for the adhesion of spherical glass particles obtained in the flame of a gas burner, i.e., ideally smooth particles.

The spread in the adhesive forces might be explained by the fact that when the dust particles come into contact with the surface the charges on similar particles are not identical. This is supported by the fact that the greatest adhesive forces correspond to strongly charged particles. However, the mean charge on particles 40-60 μ in diameter varies by about 5-6 times, whereas the spread in the adhesive forces is much greater in this case.

It has been suggested that one might explain the spread in adhesive force by reference to the energy inhomogeneity of the surface and the presence of active surface centers. Particles falling in such regions are held particularly firmly, so that it is most difficult to remove the last few particles.

Böhme [73] tried to eliminate the effect of the energy inhomogeneity of the surface on the adhesion by subjecting the dust-laden surface to preliminary shaking (frequency 10-3000 cps and amplitude 5-50 μ). However, shaking failed to remove the spread; in fact, it even tended to increase the adhesive forces. For example, the adhesion of gold particles 6-7 μ in diameter to gold and quartz surface increased by 20%.

One of the reasons for the spread in adhesive forces may well be the inadequate "purity" of the experimental conditions, i.e., poor cleaning of the surface and the presence of oil contamination on the latter, the existence of capillary moisture, etc.

Dependence of the Adhesive Forces on Particle Size. A key question in theory and practice is the dependence of the adhesive forces on the sizes of the particles. This dependence appears in various ways. Let us consider, for example, under what conditions the adhesive force is directly proportional to the particle size.

A relationship of this kind was observed by Bradley [43] in experiments on the adhesion of macroscopic bodies. Results for the adhesion of quartz spheres in vacuum are shown in Fig. III.23. Corn [80] also observed a direct proportionality between the adhesive force and the dimensions of macroscopic particles. In essence

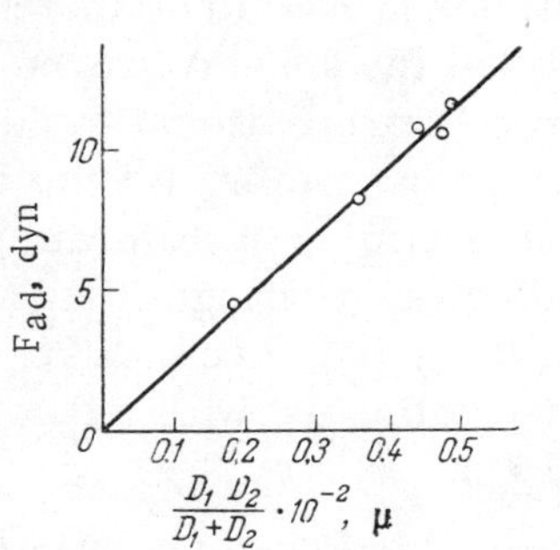

Fig. III.23. Adhesive force of quartz spheres as a function of the parameter $(D_1D_2)/(D_1 + D_2)$.

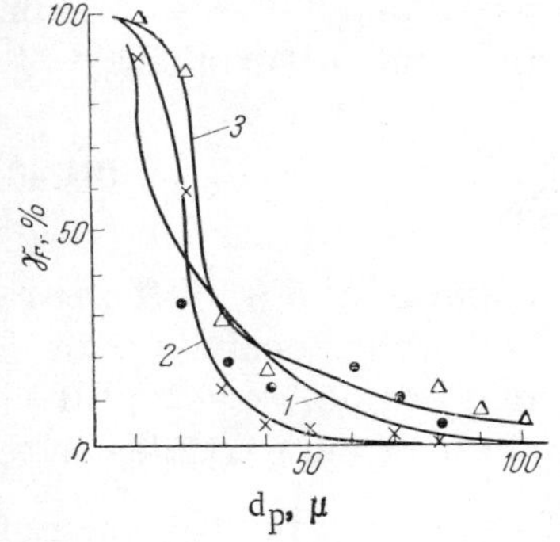

Fig. III.24. Adhesion number (detaching force 390 g) as a function of particle size. 1) Coal; 2) sand; 3) glass.

Bradley and Corn's experiments determined the interaction in vacuum between the fused ends of glass fibers and spherical and plane surfaces (autohesion and adhesion, respectively), the contiguous bodies having ideal, smooth, clean surfaces. The elimination of electrical charges and capillary forces in vacuum meant that only the molecular component of the adhesive forces was being measured in relation to the interaction between the fused ends of the filaments and the sphere or plane surface.

Thus, for macroscopic bodies, i.e., relatively large particles, the forces of interaction are proportional to the dimensions of the particles.

Experimental results indicate that the forces of adhesion depend in different ways on the dimensions of microscopic particles. In the majority of investigations it was found that the adhesive forces were inversely proportional to the size of the particles. In individual cases, however, particle size had no effect on the adhesive force.

In our own experiments we observed a rise in the adhesive force with falling particle size of microscopic glass spheres [13] in air.

A similar relationship was observed by Kordecki and Orr [16] when studying the adhesion of glass, sand, and coal particles to glass (Fig. III.24), and by Morgan [66] for quartz particles.

Böhme and his colleagues [64] found in one investigation that the adhesive forces of starch particles to a starch surface did not depend on their dimensions (for particles 7-21 μ in diameter). In another experiment [73], it was found that the adhesive force of gold particles 4-8 μ in diameter was directly proportional to the dimensions of the particles. In the same investigation, Böhme obtained integral adhesive-force curves for particles of different diameters; these curves intersected, indicating a change in the character of the relationship. However, these data are of a special character, since the laws were only established within the limits of the accuracy of the method (±10%).

The relationship between the adhesive force and the particle size may be expressed in terms of empirical formulas for certain specific substrate/particle systems.

By analyzing experimental data, a formula was obtained in [90] for the adhesion of spherical glass particles 20-100 μ in diameter to a surface painted with perchlorvinyl and polyurethane enamel:

$$\log F_{50} = k e^{-0.07 d_p} \tag{III.43}$$

where F_{50} is the force resulting in the retention of 50% of the particles in units of g, k is an experimental factor depending on the properties of the surface (for perchlorvinyl enamel, k = 20; for polyurethane, k = 11), and d_p is the diameter of the particles in μ.

Corn [80] proposes an empirical formula for the adhesion of the spherical fused ends of filaments:

$$F_{ad} = \beta d_p \tag{III.44}$$

where F_{ad} is the adhesive force in dynes, and d_p is the diameter of the spherical particles in microns.

The coefficient β is determined empirically as a function of the material of the surfaces in contact. For the adhesion of Pyrex glass particles to optical glass, $\beta = 0.017$; for quartz particles to optical glass, $\beta = 0.012$; and for quartz particles to Pyrex glass, $\beta = 0.0075$.

An empirical formula may also be obtained for the adhesion of a layer of powder. For the adhesion of spherical glass particles 10-60 μ in diameter to a steel surface with a Class 9 finish, the following formula was proposed in [21]:

$$F_l = \frac{10^5}{d_p^2} \qquad \text{(III.45)}$$

where F_l is the adhesive force of a layer of powder referred to 1 cm^2 (dyn), and d_p is the diameter of the dust particles (μ).

Patat and Schmid [20] propose an analogous formula in a different form:

$$\log F_l = B \log \frac{1}{d_p} + \log C \qquad \text{(III.46)}$$

Here B and C are constants depending on the properties of the contiguous bodies, and d_p is in cm. Thus, for the adhesion of silicon carbide to glass B = 2 and C = $7.9 \cdot 10^{-4}$, and formula (III.46) takes the form

$$F_l = \frac{7.9 \cdot 10^{-4}}{d_p^2} \qquad \text{(III.47)}$$

while for the adhesion of aluminum oxide to quartz, B = 0.7, C = 2.6, and

$$F_l = 2.6\, d_p^{-0.7} \qquad \text{(III.48)}$$

§ 17. Causes of Adhesion

After the foregoing detailed consideration of adhesive forces and the conditions affecting adhesive interaction, we must summarize all that has been said so far and on this basis discuss the causes of adhesion.

The molecular component of adhesive force manifests itself before direct contact takes place between the particles and the surface; it is due to the specific properties of the bodies coming into contact and depends on the particle dimensions [see Eq. (I.48)] and also on the true contact area. By changing one of these factors (by modifying the surface, changing the particle size, altering the surface roughness, etc.), we may change the molecular forces and thus also the forces of adhesion.

Electrical forces, which only arise when the particles come into contact with the surface, are the larger, the greater the contact potential difference. Hence modification of the surface with the aim of imparting the necessary donor—acceptor properties thereto enables us to change the magnitude of the electrical forces.

These forces are proportional to the area of contact [see (III.7)], which in turn is proportional to $r^{2/3}$ [see (III.41)]. The presence of moisture in the gap between the contiguous surfaces eliminates the possibility of electrical forces appearing.

Coulomb forces (image interaction) appear when the particles are charged in advance by a high-voltage field. Coulomb forces exceed molecular forces and the electrical forces associated with contact potential differences in value, and determine the adhesion of the particles. These forces cause interaction between charged particles and a surface when there is a definite gap between the contiguous bodies, and are inversely proportional to the square of the particle radius, i.e., $1/r^2$; they appear at the first instant of contact between the particle and the surface. The conductivity of the particle material and the contact zone, as well as moisture, cause charge leakage and lead to a reduction in the Coulomb forces, and hence the adhesion.

Capillary forces arise in the presence of a liquid meniscus in the space between the particle and the surface and for a relative air humidity exceeding 65%. Capillary forces (with or without making allowance for the disjoining effect of the liquid interlayer) are determined by Eqs. (III.34) and (III.40); they depend on the particle dimensions.

Capillary forces may be reduced by hydrophobization of the surface, thus reducing the adhesion of the particles.

The presence of an oil film on the surface causes adhesion as a result of surface tackiness. In this case the adhesive forces of the particles depend on the rheological properties and thickness of the oil film.

Under practical conditions there may be various combinations of and contributions from the various components producing adhesion. This to some extent explains the poor agreement between the results obtained by different authors under different conditions.

Table III.3. Adhesive Force of Particles Determined by Various Methods for Various Air Humidities

Substrate material	Particle material	d_p, μ	F_{ad}, dyn, for air humidity	
			50-60%	90%
Detachment of individual particles				
Pyrex [80]	Quartz (fused ends of filaments)	25	0.28	0.37
		36	0.3	0.55
		63	0.6	0.76
		88	0.88	1.38
Glass [81]	Glass	400	22	–
		800	30	–
Centrifuging for γ_F = 2%				
Glass [16]	Glass	50	0.37	1.83
	Sand	50	0.76	0.06
	Coal	50	0.55	0.94
Plexiglas [16]	Glass	50	1.44	1.97
Teflon [16]	Sand	50	0.65	1.28
Brass [16]	Coal	50	0.90	2.85
Centrifuging for γ_F = 50%				
Starch [64]	Starch	7-9	0.2	–
		13-15	0.2	–
		18-21	0.2	–
Gold [73]	Gold	4	0.07	–
		5-6	0.09	–
		7	0.1	–
		8	0.16	–
Vibration method for γ_F = 2%				
Steel [13]	Glass	40-60	1.64	–
Vibration method for γ_F = 50%				
Steel [13]	Glass	5-10	$1.33 \cdot 10^{-2}$	–
		10-20	$6.12 \cdot 10^{-3}$	–
		20-30	$2.15 \cdot 10^{-3}$	–
		40-60	$2.13 \cdot 10^{-4}$	–

Table III.4. Adhesive Force of a Powder Layer

Substrate material	Particle material	d_p, μ	F_l, dyn (referred to 1 cm²)
Steel [21]	Glass (spherical particles)	60-90	1.1
		40-60	21.7
		20-30	208.0
		10-20	370.0
Steel [20]	Aluminum oxide	324	37
		163	42
		97	67
		81	90
		68	85
		47	103
		35	143
		25	223
Steel [24]	Lime dust, ordinary	-	520
	nonwetting type		324
Glass [18]	Magnesite	200-300	39
		150-200	56
		88-150	83
		75-88	116
		60-75	169
Magnesite [18]	Magnesite	200-300	29
		150-200	40
		88-150	60
		75-88	89
		60-75	103

Each of the components of the adhesive forces depends on the particle size:

Component of adhesive force . . .	Coulomb	Electric	Molecular	Capillary
Dependence on dust-particle radius	$1/r^2$	$r^{2/3}$	r	$r(1-r^{x-1})$
Formula	(III.13)	(III.7); (III.41)	(I.48)	(III.40a)

The adhesion is only proportional to the particle dimensions when the molecular forces prevail. If electrical, capillary, and

Coulomb forces have to be overcome in addition to these in order to effect detachment, the total adhesive force may not be proportional to particle size.

The thermodynamic Deryagin theory of adhesion considers adhesion as an equilibrium and reversible process and the adhesive force as a function of the gap separating the contiguous surfaces. When this gap equals zero, the adhesive force is proportional to the dimensions of the bodies in contact [see (I.64)].

However, the adhesion process is not in fact reversible, since before the particle comes into contact with the surface some forces (Coulomb and partly molecular) are acting, while after contact others (molecular, electric, capillary, and Coulomb) have to be overcome in order to achieve detachment. The interaction of the particles with the surface resulting from forces other than molecular means that the process disobeys the conditions for which the Deryagin theory is valid. The adhesive force—particle size relationship may thus well differ from direct proportionality.

In conclusion, we shall give some generalized experimental data related to the adhesion of particles (Table III.3) and a layer of powder (Table III.4).

We see from the results presented that adhesion varies with particle size in different ways. It may be directly proportional, or even inversely proportional, to the particle diameter, or alternatively, it may be independent of particle size.

The particle material also has an effect on the adhesive force. Thus, for particles 50 μ in diameter made of different materials (glass, sand, coal) the adhesive force relative to a glass surface fluctuates from 0.06 to 1.83 dyn.

Results obtained by different methods of adhesion measurement (with the same materials and under the same conditions) are similar. Thus, the adhesive forces for glass particles sticking to a metal surface (brass and steel) are approximately the same, despite differences in the method of detachment (vibration and centrifuging); the adhesive forces measured with a relative air humidity of 50-60%, using the centrifuge method to remove the glass particles, are the same as those obtained by the method of detaching individual particles.

Chapter IV

Adhesion in Liquid Media

§ 18. Characteristics of Molecular Interaction in Liquid Media

Experimental data [10, 45] show that the forces of adhesion in air are several orders higher than those in a liquid medium. This difference is particularly noticeable for particles of small dimensions. Thus, in order to detach 50% of spherical glass particles 5-10 μ in diameter from a steel surface in air, it is necessary to apply a force of the order of 10^{-2} dyn, whereas, in water, less than 10^{-6} dyn suffices.

By way of example, Table IV.1 shows some results for the adhesion of graphite and quartz powders 5-11 μ in size to substrates of various materials in water [12, 15]. The adhesion number of the particles varies from 0-76%, depending on the properties of the bodies in contact and the applied detaching force. If oil is used as liquid medium (for example, highly purified mineral oil with η_{20} = 353.7 cS), then on using a detaching force of F_{det} = $1.6 \cdot 10^{-7}$ dyn for quartz particles 5 μ in diameter, γ_F = 9.5%, for a glass substrate and 18% for a steel substrate, i.e., the adhesion number is smaller than in the case of water. However, under the same condition the adhesion number for a paraffin substrate is larger (γ_F = 15%) in oil than in water. This once again underlines the fact that adhesion is affected by the wettability of the contiguous surfaces.

Capillary and electrical forces (§§11-13) which act in gaseous media hardly appear at all in liquids. Hence, the main contribution to the adhesive force arises from the molecular interaction of the contiguous bodies. In addition to this, repulsive forces due to the properties of the liquid media develop. Hence, in determining the causes of adhesion in liquid media we must give detailed

Table IV.1. Adhesion of Various Particles to Substrates of Various Materials in Water [59] (for $F_{det} = 7 \cdot 10^{-5}$ dyn)

Substrate material	Particle material	d_p, μ	γ_F, %
Glass	Quartz	11.0	0
	Graphite	7.5	20
Steel	Quartz	11.0	33
	Graphite	7.0	40
Bronze	Graphite	7.0	40
Glass [12]	Quartz	5.0	76*
Paraffin [40]	Quartz	5.0	0*

*For $F_{det} = 1.6 \cdot 10^{-7}$ dyn.

consideration to the attractive (molecular) and repulsive forces involved.

The exact value of the attraction between a glass sphere and a plane surface, i.e., the relationship between the molecular forces and the thickness of the intervening gap, was given by Deryagin and Abrikosov [56] (see §4). In addition to this, the force of molecular interaction may be estimated by using the van der Waals constant [see Eqs. (1.49) and (1.51)].

The van der Waals constant for the interaction of two solids (1 and 2) situated in a liquid (0) is given by the expression

$$\overline{A} = \overline{A}_{1,2} + \overline{A}_{0,0} - \overline{A}_{2,0} - \overline{A}_{1,0} \tag{IV.1}$$

For two similar bodies, for which $\overline{A}_{1,0} = \overline{A}_{2,0}$, Eq. (IV.1) takes the following form:

$$\overline{A} = \overline{A}_{1,2} - 2\overline{A}_{1,0} + \overline{A}_{0,0} \tag{IV.2}$$

where

$$\overline{A}_{1,2} = u_1 u_2 A_{1,2} + u_1 (1 - u_2) A_{1,0} + u_2 (1 - u_1) A_{2,0} + (1 - u_1)(1 - u_2) A_{0,0} \tag{IV.3}$$

$$\overline{A}_{1,0} = u_1 A_{1,0} + (1 - u_1) A_{0,0} \tag{IV.4}$$

$$\overline{A}_{2,0} = u_2 A_{2,0} + (1 - u_2) A_{0,0} \tag{IV.5}$$

Table IV.2. Effect of the Properties of Microscopic Particles and Substrate on Adhesion in Water

Substrate material	Particle material	d_p, μ	γ_F, %
Glass (No. 23)	Montmorillonite	1-2	93
	Kaolinite	1-2	80
Quartz	Quartz	2.5	75
	Graphite	2.5	0
	Mastic	2-3	0
Paraffin	Quartz	2.5	0
	Graphite	2.5	100

where u_1 and u_2 are the volume concentrations of the substances 1 and 2 in the liquid (0) and A and $\bar{A}$ are the van der Waals constants for one molecule and a group of molecules, respectively.

Putting expressions (IV.3)-(IV.5) into Eq. (IV.1), and making some simplifications, we obtain

$$\bar{A} = u_1 u_2 A \tag{IV.6}$$

The quantity $\bar{A}$ may be found either experimentally or by calculation.

Malkina and Deryagin [86] used their own method of crossed filaments for equilibrium adhesion in water to obtain a value of $\bar{A} = 1.3 \cdot 10^{-11}$ erg. However, this value of $\bar{A}$ cannot be regarded as sufficiently accurate, since the width of the gap separating the contiguous bodies was not measured, but rather assumed to be of the order of 10^{-7} cm. The value of $\bar{A}$ may be determined from Eqs. (1.28) and (1.34).

Below we present the calculated values of the constant $\bar{A}$ for the adhesion of $Al(OH)_3$ and $Fe(OH)_3$ particles to various surfaces in water [25, 26]:

Surface	$Al(OH)_3$	$Fe(OH)_3$	SiO_2	$CaCO_3$	Lead	Carbon
$\bar{A} \cdot 10^{12}$ for particles of:						
$Al(OH)_3$	1.26	1.42-1.45	1.27	1.40	1.27	1.70
$Fe(OH)_3$	1.42-1.45	1.77-2.00	1.41-1.45	1.68-1.76	1.71-1.85	1.70-1.77

The value of $\bar{A}$, and hence the adhesive forces associated with intermolecular interaction depend both on the properties of the liquid and those of the surfaces in contact.

Experimental results confirm the dependence of the adhesive forces on the properties of the contiguous surfaces in liquid media.

Table IV.2 presents some results [59, 169] relating to the adhesion of particles of various materials to various substrates in water. The value of γ_F was determined by turning the dust-laden plates into a vertical position after the particles had been in contact with the substrate for 30 min.

We see from Table IV.2 that, for the same degree of hydrophily (quartz particles to a quartz surface, kaolinite and montmorillonite to chemical glass No. 23) or hydrophoby (graphite particles to paraffin surfaces), the adhesion number in water reaches a maximum value. However, the adhesion of hydrophobic surfaces (graphite—paraffin) in water is greater than that of hydrophilic surfaces (quartz to quartz).

When the surfaces in contact differ sharply in wettability (hydrophilic quartz particles to a hydrophobic paraffin surface, mastic and graphite particles to a quartz surface), the adhesion reaches a minimum (the adhesion number is zero).

Figure IV.1 illustrates our own data for the adhesion of spherical glass particles 40-60 μ in diameter in distilled water to ordinary glass (curve 1) and glass modified with chlorsilanes and Chromolan (curve 2). Both in air (see §10) and liquids, the adhesion of glass particles to glass surfaces modified with chlorsilanes is less than the adhesion to ordinary glass.

§19. Hydrodynamic and Mechanical Factors

Variation in the Adhesion of Particles with the Conditions of Depositing Dust on Surfaces. Two cases of dust deposition on a surface are possible: deposition in air, with subsequent placing of the dust-laden surface in liquid ("air dusting"), and the direct settling of the particles on the surface in the liquid (liquid settling). Experience shows (Fig. IV.2) that after air dusting (curve 1) and holding the dust-laden surface in the liquid medium for not longer than 2 min all the particles remain on the surface, even for a detaching force of $2.8 \cdot 10^{-4}$ dyn,

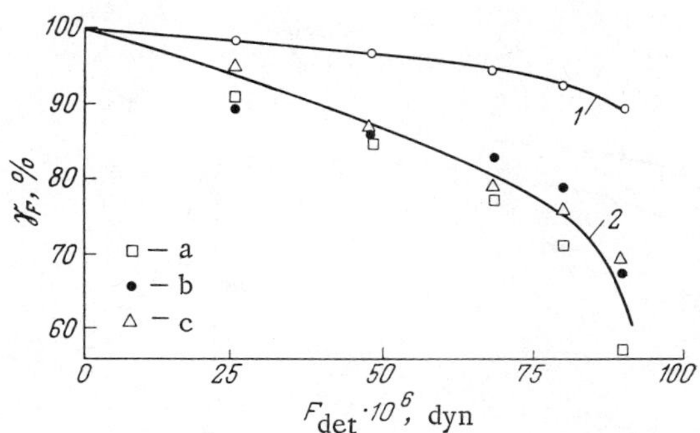

Fig. IV.1. Adhesion number as a function of detaching force. 1) Ordinary glass; 2) glass modified with Chromolan (a), dimethyldichlorsilane $(CH_3)_2SiCl_2$ (b), and methyltrichlorsilane CH_3SiCl_3 (c).

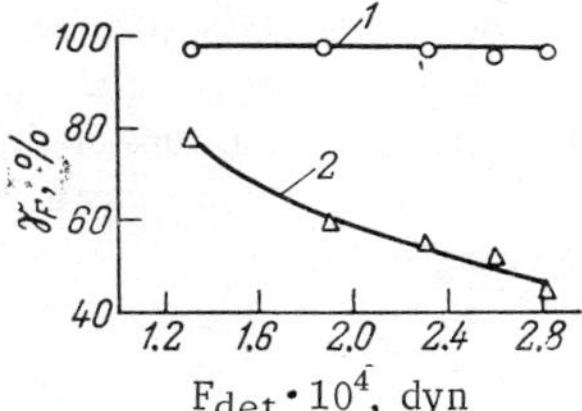

Fig. IV.2. Adhesion number of glass particles 70 ± 2 μ in diameter on a glass surface as a function of the detaching force (period spent by the "dusty" surface in the liquid no longer than 2 min). 1) Air dusting; 2) settling in distilled water.

while on setting the particles in distilled water and using the same detaching force after the same holding period only 45% of the particles remain (curve 2).

Fuks [59] directed attention to the dependence of the adhesion on the period of contact between the particles and the surface on settling in the liquid. At the initial instant the adhesion of the particles is a minimum (Fig. IV.3), but with increasing contact time adhesion becomes stronger, the rise ceasing some 60-90 min after the particle first came into contact with the surface. This phenomenon has become known as "aging." Fuks [59] studied the effect of contact time on adhesion specifically for the case of the liquid settling of particles.

Figure IV.4 shows the variation in the adhesion number with time for air dusting (curves 1) and liquid settling (curves 2) [170]. The adhesion reaches a maximum for a two-minute exposure in

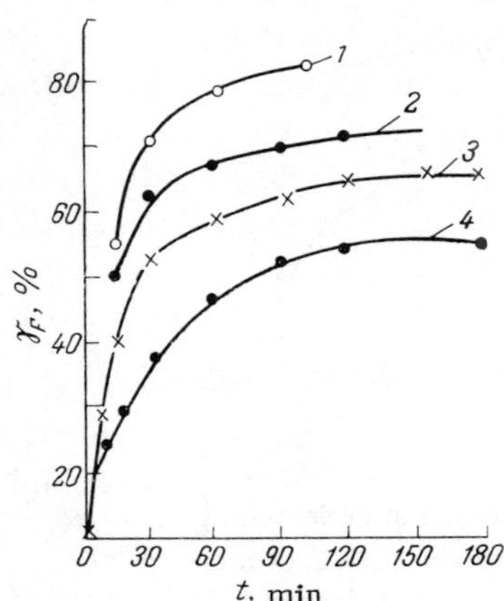

Fig. IV.3. Adhesion number as a function of the time of contact between the particles (detaching force $1.2 \cdot 10^{-5}$ dyn) and the surface for liquid settling. 1) Graphite in clean mineral oil on bronze; 2) graphite in aviation oil on glass; 3) graphite in Avtol 10 on glass; 4) carbon scale in Avtol 10 on glass.

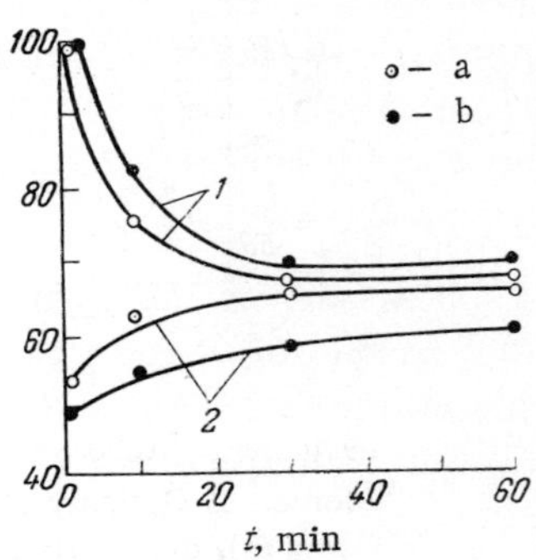

Fig. IV.4. Adhesion number on time of contact between glass particles 70 ± 2 μ in diameter and a glass surface in 0.1 M solutions of KCl (a) and $CaCl_2$ (b). 1) For air dusting; 2) for settling in liquid.

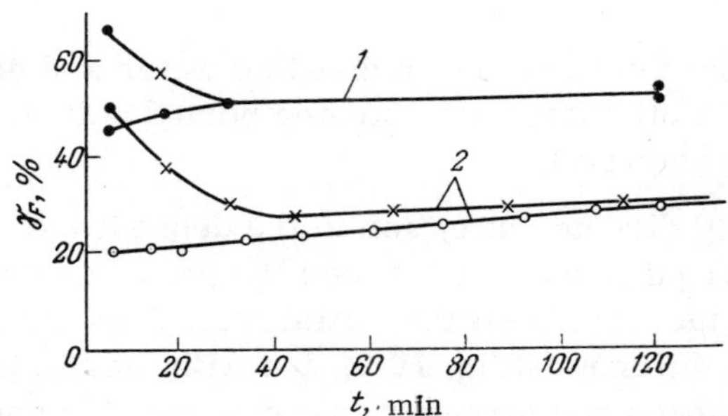

Fig. IV.5. Adhesion number as a function of the time of contact between glass spheres 70 ± 2 μ in diameter in distilled water and glass (1) and perchlorvinyl enamel-painted (2) surfaces. Upper branches of the curves, air dusting; lower branches, liquid settling.

the case of air dusting; as the time spent by the dust-laden surface in the liquid increases to 30 min, adhesion diminishes. On increasing the time to 60 min or more there is hardly any further change in adhesion. For particles settled in the liquid the position is the opposite. As the time spent by the substrate in the liquid medium increases adhesion does likewise; the maximum adhesion in this case roughly corresponds to the minimum adhesion in the case of air dusting.

Thus at the initial instant (in our own experiments after a 2-min period of contact between particles and the substrate in the liquid) the adhesion for air dusting is always greater than the adhesion for liquid settling.

An analogous law holds for molar and centimolar solutions of KCl and $CaCl_2$; however, the equilibrium value of the adhesion numbers for the same detaching force moves upward in the first case and downward in the second.

The increase in adhesion in distilled water (Fig. IV.5) for liquid settling (lower branches of curves 1 and 2) depends less on the time spent by the substrate in the liquid than in electrolyte lutions (see Fig. IV.4). The adhesion of glass particles to a glass surface in distilled water is greater (Fig. IV.5, curve 1) than to a surface painted with perchlorvinyl enamel (curve 2).

A rise in the forces of interaction between contiguous surfaces with increasing contact time is also observed on using methods simulating the adhesion interaction between particles. Thus Malkina and Deryagin [86], using the crossed-filament method (see § 8), showed that in aqueous media the adhesive forces between quartz filaments increased from zero at the initial instant to a certain equilibrium value close to the corresponding adhesive force in air. The authors explained this phenomenon by asserting that the layer of liquid separating the contiguous surfaces was gradually pressed out in the course of time under the influence of the compressive forces applied.

A similar $F_{ad} = f(t)$ relationship was obtained by Fuks [88] for crossed filaments.

The final value of the adhesive force of glass filaments in distilled water does not depend on the applied pressure (between 14 and 1400 dyn), although the time required to establish the equilibrium value of the adhesive force does depend on the load. For a load of 14 dyn, equilibrium is reached in 24 h and for 230 dyn, in 5-6 h.

Hydrodynamic Factor. In order to discover the reason for the dependence of the adhesive forces on contact time, let us consider the hydrodynamic phenomena taking place when the bodies approach or recede from each other. For this purpose it is customary to employ adhesion-simulating methods (see § 8), in particular the method of plane-parallel discs. The hydrodynamic factor due [88] to the motion of the liquid in the gap between the contiguous surfaces, determining the change in adhesion with contact time for the interaction of plane-parallel discs, may be represented by the Stefan–Reynolds equation

$$t = \frac{3\pi\eta R^4}{4F}\left(\frac{1}{H_1^2} - \frac{1}{H_0^2}\right) \quad \text{(IV.7)}$$

where t is the time during which the distance between the discs changes from H_0 to H_1 ($H_1 > H_0$), η is the dynamic viscosity of the liquid, R is the radius of the discs, F is the force causing the mutual approach or recession of the discs. As $H_1 \to \infty$, formula (IV.7) simplifies:

$$t = -\frac{3\pi\eta R^4}{4FH^2} \quad \text{(IV.8)}$$

When a sphere of radius r is detached from (or approaches) a plane surface, we may use Taylor's formula:

$$t = \frac{6\pi\eta r^2}{F}\log\frac{H_0}{H_1}\text{m} \quad \text{(IV.9)}$$

For the deposition of particles we shall understand the force F to mean the interaction of the particles with the surface, and for the detachment of particles (for example, in a centrifuge) we shall understand it to mean the value of the centrifugal force. In the case of detachment the force F is directed oppositely to the force of adhesion.

In accordance with Eq. (IV.9), the detaching force for the particles is inversely proportional to the time of its application [59], i.e., the longer the force acts, the less does its magnitude have to be in order to remove a specified number of particles. A relationship of this kind was observed [59] on detaching particles by centrifuging in a liquid medium (Fig. IV.6).

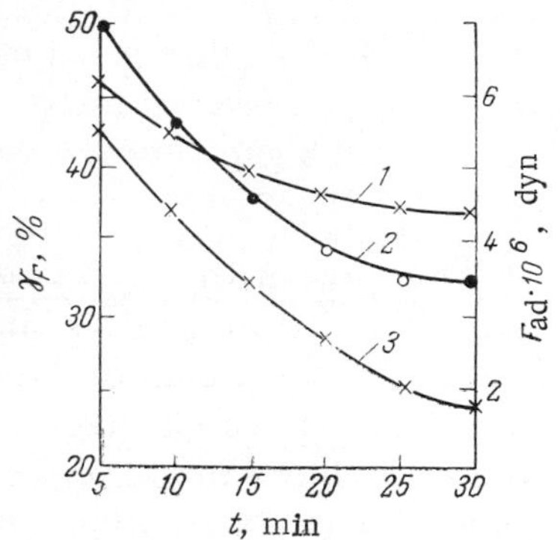

Fig. IV.6. Adhesion of particles as a function of centrifuging time. 1,2) Adhesion numbers for scale on steel in aviation oil and purified mineral oil (F_{det} = $2.2 \cdot 10^{-5}$ dyn; 3) minimum adhesive force of graphite in MK aviation oil on bronze.

This kind of relationship no longer occurs in air; the detaching force manifests itself instantaneously. We have frequently observed that, if particles are not removed at the very first instant in air, their adhesion undergoes no subsequent change.

Equations (IV.7)-(IV.9) are valid for comparatively large distances between the contiguous bodies (disc—disc, particle—plane), i.e., when the molecular interaction between the solids is weakened and the boundary properties of the liquid fail to appear. Thus for purified mineral oils and a gap width exceeding 0.3 μ, the interaction of plane-parallel discs is due to the hydrodynamic factor [88]. For a small gap between particles and surface, molecular forces [see Eq. (I.49)] and the disjoining pressure of the liquid layer will both exert an effect.

On the basis of the material so far considered, we may suppose that the hydrodynamic factor operates in a liquid medium during the initial period of particle adhesion, i.e., in the case of kinetic adhesion. As the particles approach the surface, the liquid is forced out and the layer between the contiguous bodies is reduced to an equilibrium thickness; this corresponds to a transition from kinetic to static adhesion. In the case of air dusting with subsequent placing of the dust-laden sample in the liquid medium, the contact zone is wetted and filled with liquid, forming a liquid interlayer of equilibrium thickness.

Hence, the hydrodynamic factor has an effect on the interaction of particles with a surface during contact and detachment.

Furthermore there is an inverse-proportional relationship between the forces of adhesion (or detachment) and the period spent by the dust-laden surface in the liquid medium (or the period of application of the detaching force). However, this relationship only holds for limited periods and is later broken. It thus follows that the time for the establishment of the equilibrium thickness of the liquid interlayer depends not only on the hydrodynamic factor associated with the flow of liquid into or out of the gap between the bodies, but also on other factors.

Mechanical Properties of the Boundary Layer of Liquid. It is well known that the mechanical properties, including viscosity, of a liquid are different in the main bulk of the liquid and in the thin boundary layer; each liquid has a characteristic limiting thickness of the latter, below which the liquid passes into a quasi-solid or quasi-crystalline state. Thus, according to Akhmatov [39], the thickness of such layers equals 0.08 μ for myristic acid, 0.058 μ for oleic, and 0.05-0.1 μ for high-molecular unsaturated fatty acids.

Adsorbed layers of liquid formed on the surface of contiguous bodies also possess properties differing from those of the liquid in the bulk state (for example, high elasticity, shear strength, and viscosity) [171]. So far it has not been established either theoretically or experimentally how a change in the structural-mechanical properties of a liquid affects adhesion; however, there is no doubt that these factors should have an effect not only in the case of mutual displacement of the bodies (friction) but also when the bodies approach or recede from each other.

It is known that the viscosity of the boundary layer (η_b) exceeds the volume viscosity η_{vol} by several times (not more than five) [172]. According to Eqs. (IV.7)-(IV.9), an increase in viscosity leads to an increase in the time required for the bodies to approach one another and hence affects kinetic adhesion.*

The properties of adsorbed layers of liquids may be estimated not only by the ratio η_b/η_{vol}, but also by a nondimensional coefficient of boundary condensation [88]

$$\psi = \frac{t_{det} \cdot F_{det}}{\eta} \qquad \text{(IV.10)}$$

*A dependence of t on η was observed when measuring the forces of interaction by particle-adhesion simulation methods.

where t_{det} is the time of detachment, F_{det} is the force of detachment per unit area, and η is the dynamic viscosity of the liquid.

The product $t_{det} \cdot F_{det}$ has the dimensions of dynamic viscosity. The ratio of this quantity to the dynamic viscosity in the main volume characterizes the change in the mechanical properties of the liquid in the process of separating the bodies. For steel discs the coefficient of condensation* (for t_{det} = 10 min and F_{det} = 3900 dyn/cm^2) has the following values:

Benzene	$1.2 \cdot 10^3$	Bone oil	$2.05 \cdot 10^7$
Turbine oil	$1.06 \cdot 10^7$	Stearic acid in turbine oil, 0.1% solution	$3.12 \cdot 10$
Machine oil MBP-12 . . .	$1.9 \cdot 10^7$		

The time to establish the equilibrium state of the forces of interaction between plane-parallel discs is directly proportional to both the ratio η_b/η_{vol} and the coefficient of boundary condensation, i.e., the viscosity has an effect on the kinetic adhesion. However, viscosity has little effect on the absolute value of the equilibrium adhesive force, i.e., on static adhesion. This may be seen by analyzing Eq. (IV.8). On the one hand, in fact, for R = const,

$$\frac{t}{\eta} \cdot H_0^2 F = \text{const}$$

On the other hand, if we measure the time for the plane-parallel discs to approach one another, we find that the ratio t_{exp}/t_{theor} [where t_{exp} is the time for the approach of the discs measured experimentally, and t_{theor} is that calculated from Eq. (IV.8) with $\eta = \eta_{vol}$] is proportional to the ratio η_b/η_{vol}, i.e., we may write

$$\frac{t_{exp}}{t_{theor}} = \frac{t_b}{t_{vol}} \sim \frac{\eta_b}{\eta_{vol}} \text{ i.e., } \frac{t_b \cdot \eta_{vol}}{t_{vol} \cdot \eta_b} = \text{const}$$

which was confirmed experimentally in [88].

This effect of viscosity on the time required to establish equilibrium has been observed not only by simulation methods based on a model of particle adhesion, but also by the direct measurement of the forces required to detach adhering particles. Thus, the adhesion of graphic particles 5-6 μ in diameter to a steel sur-

*The coefficient of boundary condensation depends on the size of the discs and the experimental conditions.

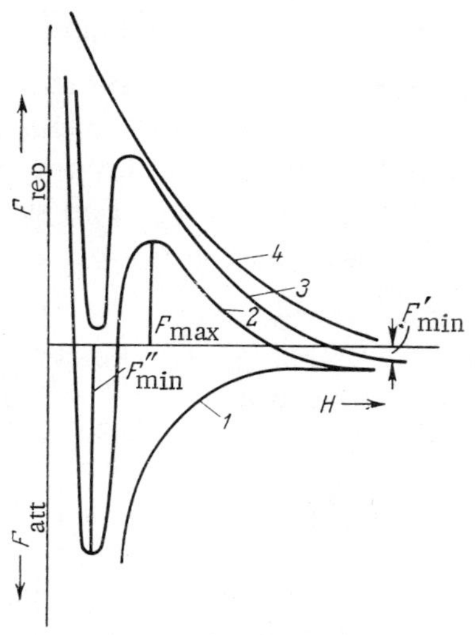

Fig. IV.7. Interaction between two particles as a function of the distance between their surfaces. 1) Attraction; 2,3) resultants of attraction and repulsion; 4) repulsion.

face varies very little on raising the viscosity of the oil from 5 to 200 centistokes under equilibrium conditions, although the viscosity of the oil varies by a factor of 40 [59]; the adhesion numbers for a detaching force of $5.1 \cdot 10^{-5}$ dyn are 52-55%. For static adhesion, when an equilibrium thickness of the liquid interlayer between the contiguous bodies has been established, the adhesion is determined by factors other than viscosity.

§ 20. Disjoining Pressure of a Thin Layer of Liquid

Resultant of the Forces of Attraction and Repulsion. The fact that the adhesive forces in liquids are lower than those in air indicates the presence of repulsive forces in addition to the forces of molecular attraction.

When contact is established between a particle and a plane surface in vacuum, the force of molecular interaction [see Eqs. (I.49), (I.61), and (I.62)] fall off with distance (curve 1, Fig. IV.7). In a liquid medium repulsive forces falling off more slowly with distance appear (curve 4). The variation in the resultant force of interaction with distance is expressed in the form of curves 2 or 3 if we consider that, for repulsion between the two particles, the ordinate is positive and for attraction negative. For relatively wide gaps between the contiguous bodies, the forces of molecular attraction, which fall off with increasing H (Fig. IV.7, curve 1) on a power law [see Eqs. (I.44)-(I.51)], slightly exceed the forces of repulsion. Under certain conditions (in electrolyte solutions) the forces of repulsion are greater. Curve 2 corresponds to the case in which the repulsive forces exceed the attractive forces at moderate distances, and curve 3 to the case in which this holds for any distances between the particles. The repulsive forces give rise to a potential (force) barrier (F_{max}) preventing the mutual approach of the particles. The force barrier appears at moderate distances of the same order as the effective thickness of the ionic atmospheres.

The forces of interaction (attraction and repulsion) depend not only on the properties of the bodies in contact and the layer separating these, but also on the external applied force. This force determines the thickness of the gap between the bodies. If the compressing force is no greater than F_{max} (height of the force barrier) the adhesive force will be relatively small and equal to F'_{min} ; if the compressive force exceeds F_{max} the adhesive force will be equal to F''_{min}.

Concept of Disjoining Pressure. In liquid media additional forces of repulsion appear, owing to the so-called disjoining pressure of the thin layer between the surfaces of the contiguous bodies. The disjoining pressure is due to a particular characteristic of thin layers of liquid, consisting of a difference between their thermodynamic (and chemical) potentials and those of the bulk phase. The disjoining pressure of thin layers of liquid, particularly of the single-component type, may be expressed in the form

$$P(H) = -\Delta\mu$$

where P(H) is the disjoining pressure, and $\Delta\mu$ is the excess of the chemical potential referred to unit volume.

The concept of the disjoining pressure, which occurs for a liquid interlayer between solid surfaces [173, 174] and also an air bubble pressed against a surface [174, 175], has subsequently been extended to a wide range of problems such as the stability of lyophobic colloids [176], lubricating effects, the equilibrium and motion of moisture in soils, and other processes [177, 178].

The disjoining pressure of a thin plane-parallel layer of liquid situated between two identical or different phases equals the pressure P(H) with which the layer of liquid acts in a state of equilibrium on the bodies bounding it, tending to separate them. The greater the pressure between the two bodies, the smaller is the equilibrium thickness of the layer of liquid and the greater is the disjoining pressure. For the adhesion of particles in a liquid

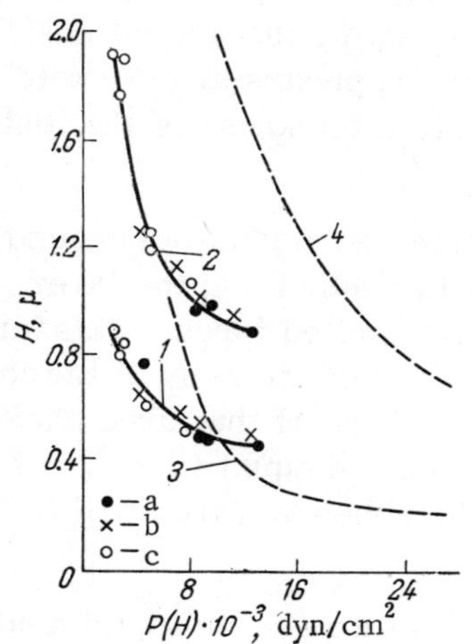

Fig. IV.8. Disjoining pressure P(H) as a function of the thickness of the liquid film. 1,2) For particles of related soils of different granulometric compositions (a, sandy loam; b, loam; c, clay) in water; 3) for plane mica surface in water; 4) for plane steel surfaces in paraffin oil.

medium the disjoining pressure equals the force (referred to unit area, usually 1 cm^2, or one particle) with which the thin layer of liquid acts in a state of equilibrium on the particles, tending to separate them.

The methods used to determine the disjoining pressure in [178, 179] yielded isotherms relating the value of the disjoining pressure to the thickness of the liquid layer separating the boundary phases. Isotherms of this kind are shown in Fig. IV.8 [180].

For different soils (sandy loam, loam, and clay) of approximately the same granulometric composition (curves 1 and 2), the dependence of P(H) on H is practically the same. Figure IV.8 also shows the results of Deryagin [173] (curves 3 and 4) giving the relationship between the value of the disjoining pressure and the thickness of the liquid layer separating plane mica and steel surfaces.

We see from the data presented that the laws giving the relationship between the disjoining pressure and the thickness of the liquid layer are analogous for particles and plane surfaces. The surfaces of the soil particles in contact are not parallel to each other. This results in a nonuniform thickness of the water interlayer. For this reason, the average thickness of the water layer cannot be measured directly; a certain "reduced" thickness equivalent in its disjoining effect to the nonuniformly thick layer of water between the irregularly shaped particles has to be determined instead.

The average thickness of the water interlayer between the soil particles responsible for autohesion may be determined from the semiempirical formula [180]

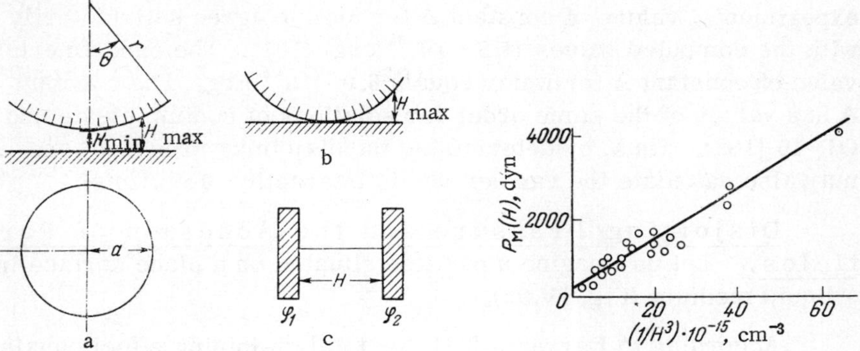

Fig. IV.9. Layer of liquid between a particle and a surface (a,b) and between two planes (c).

Fig. IV.10. Molecular component of the adhesive force as a function of the thickness of the water layer.

$$H = \frac{200 \cdot w\rho_1}{S_1(1-n)\,\rho_2} \tag{IV.11}$$

where w is the humidity of the soil, n is its porosity, S_1 is the total surface of unit volume of soil, and ρ_1 and ρ_2 are the apparent densities of the soil and the water.

The value of the disjoining pressure for a liquid layer situated between an air bubble and a solid substrate may be calculated from the Laplace formula:

$$P(H) = P_\sigma \text{ and } P_\sigma = \frac{2\sigma}{R}$$

The quantities P_σ (the pressure sustained by the thin layer of liquid and balanced by the disjoining pressure), σ (the surface tension of the water), and R (the radius of the bubble) are determined experimentally.

The disjoining pressure counteracts the molecular forces bringing the two bodies together (Figs. IV.7 and IV.9). This fact enabled Sheludko [179] to calculate the molecular forces from the known value of the disjoining pressure of free liquid films. The experimentally determined relationship $P_M(H) = A/H^3$ constitutes a straight line (Fig. IV.10) the slope of which equals the van der Waals constant [see Eq. (I.51)]. For aniline with the addition of a solution of dodecyl alcohol the value of A = $3.5 \cdot 10^{-13}$ erg. These experimental values of constant A for aniline agree satisfactorily with the computed values ($4.8 \cdot 10^{-13}$ erg) [181]. The experimental value of constant A for water equals $6.0 \cdot 10^{-4}$ erg. The constant A has values of the same order for solutions of sodium oleate and OP-10 [182]. Thus, by determining the disjoining pressure, one may also calculate the van der Waals interaction (for films).

Disjoining Pressure and the Adhesion of Particles. Let us imagine a particle situated on a plane surface in a liquid medium (Fig. IV.9a).

According to Deryagin [57], the total disjoining effect equals

$$F_{disj} = 2\pi r \int_{H_{min}}^{\infty} P(H)\, dH \tag{IV.12}$$

The radius (a) of action of the disjoining effect is given by the equation

$$r + H_{min} - H_{max} = a \cot\theta \tag{IV.13}$$

Hence,

$$a = \frac{r + H_{min} - H_{max}}{\cot\theta} \tag{IV.14}$$

$$H_{max} = r + H_{min} - a \cot\theta \tag{IV.15}$$

Denoting

$$r + H_{min} - H_{max} = b$$

we obtain

$$b = \sqrt{r^2 - a^2}; \quad a = \sqrt{r^2 - b^2}; \quad H_{max} = r + H_{min} - \sqrt{r^2 - a^2} \tag{IV.16}$$

In the particular case (Fig. IV.9b) in which there is direct contact between the particle and the solid surface, i.e., $H_{min} = 0$

$$H_{max} = r - \sqrt{r^2 - a^2}; \qquad a = \sqrt{2rH_{max} - H_{max}^2} \tag{IV.17}$$

The overall disjoining effect may be calculated from Eq. (IV.12) if the boundary conditions, i.e., the value of H_{min} (since the disjoining effect is not manifested when $H_{max} \to \infty$), and the

Table IV.3. Thickness of a Layer of Liquid and the Corresponding Disjoining Pressure in Various Media

Surfaces in contact	Medium	H, μ	P(H), dyn/cm²
Two air bubbles [176]	10^{-3} N solution of sodium	0.26	2,200
	oleate in 10^{-3} N NaCl	0.41	350
	In 10^{-4} N NaCl	0.48	2,300
		0.77	350
Two air bubbles [184]	Solution of undecyl acid:		
	0.00002%	0.050	1,500
		0.12	500
	0.006%	0.10	2,200
		0.15	800
Cleavage plane of mica (muscovite) [174]	Water:		
	once distilled	0.041	188,000
		1.02	4,300
	twice distilled and	0.55	13,700
	filtered*	2.1	4,300
	Paraffin oil	0.24	93,000
		2.0	4,300
Two mica flakes [185]	Water	1.4	3,900
		5.0	2,400*
	Solution of acetic acid:		
	5.05 M	1.5	5,100
		6.5	3,900
	0.125 M	2.2	7,400
		7.0	5,400
Two steel discs [88]	Mineral oil	0.022	39,000,000
		0.2	78,500

*Limiting value on reaching with a change in distance produces no fall in pressure.

form of the P(H) relationship, are known. According to Deryagin [183],

$$P(H) = AH^{-n} \tag{IV.18}$$

where n = 2–3.

Substituting Eq. (IV.18) into (IV.12) and integrating, we obtain

$$F_{disj} = 2\pi r A \left| \frac{1}{(-n+1) H^{n-1}} \right|_{H_{min}}^{H_{max}} \tag{IV.19}$$

Allowing for (IV.18),

$$F_{\text{disj}} = \frac{2\pi r}{(n-1)} [P(H)_{\min} \cdot H_{\min} \cdot H_{\max} - P(H)_{\max}'] \qquad \text{(IV.20)}$$

where $P(H)_{min}$ and $P(H)_{max}$ are the values of disjoining pressure corresponding to H_{min} and H_{max}.

By using Eq. (IV.20), we may calculate the forces acting on the particle due to the disjoining pressure. For this we must know H_{min} and H_{max} and the corresponding values of P(H). These quantities have not been determined experimentally for the adhesion of particles, but have been determined for contiguous surfaces (Table IV.3). Considering that the minimum and maximum values of the thickness of the liquid layer shown in Table IV.3 correspond to H_{min} and H_{max}, while the disjoining pressure for these layers corresponds to P(H) for the adhesion of particles, we may make an approximate calculation on the basis of Eq. (IV.20) and thus estimate the order of magnitude of the force due to the disjoining pressure (F_{disj}).

Using existing data for the interaction of mica cleavage planes in once-distilled water [174] and for the interaction of two air bubbles in a 10^{-3} N solution of sodium oleate in a 10^{-3} N solution of NaCl [176] (Table IV.3), we may calculate F_{disj} from Eq. (IV.20) for the adhesion of particles in water and the corresponding solution of sodium oleate. At the same time we may use Eq. (IV.16) to calculate the radius a of the circle (see Fig. IV.9b) inside which the disjoining pressure operates.

In the calculation we assume that n = 3, while H_{min} and H_{max} are independent of the size of the particles (10–40 μ in diameter).

As a result of the calculation we obtain

d_p, μ	10	20	30	40
$F_{disj} \cdot 10$, dyn:				
in sodium oleate solution .	0.9	1.8	2.7	3.6
in water	7	14	21	28
a, μ:				
in sodium oleate solution .	1.7	2.4	3.0	3.5
in water	4.3	6.2	7.7	8.9

As already noted, in the equilibrium state the adhesive force in a liquid medium equals the molecular interaction of the contiguous bodies after subtracting the disjoining pressure (referred to one particle). Hence, in order to estimate the contribution of the disjoining pressure in the case of the adhesion of particles in a liquid medium, we may compare the value of F_{disj} with the F_{ad} of the particles in air (see page 113 and Table III.3).

It follows from this comparison that the value of the disjoining pressure is comparable with the adhesive force and its molecular component in air, and hence F_{ad} is greatly reduced when the dust-laden surface is placed in liquid.

<u>Causes of the Disjoining Pressure.</u> The disjoining pressure arises from several causes: the molecular (van der Waals) action of the solid phase on the boundary layer of the liquid (molecular component), and the formation of a double electric layer at the interface between the two phases. The electrical component, in turn, includes ionic and diffusion components:

$$P_E(H) = P_I(H) + P_D(H) \tag{IV.21}$$

Thus the total disjoining pressure equals

$$P(H) = P_I(H) + P_D(H) + P_M(H) \tag{IV.22}$$

where the indices I, D, and M, respectively denote the ionic, diffusion, and molecular components of the disjoining pressure.

The electrical component of the disjoining pressure may be expressed as the resultant of ponderomotive forces and osmotic pressure in the layer and in the main volume of the electrolyte [186], i.e.,

$$P_E(H) = F_p + P_o \tag{IV.23}$$

The ponderomotive force (F_p) is associated with the inconstancy of the electric field in a direction normal to the surface of the bodies in contact (see Fig. IV.9c), and is given by the following (reduced to unit surface area):

$$F_p = \frac{\varepsilon E^2}{8\pi} \tag{IV.24}$$

where ε is the dielectric constant of the solution, and E is the field strength

$$E = \frac{\partial\psi}{\partial H}$$

ψ is the potential of the surface [187], and H is the distance between the surfaces.

The osmotic pressure is due to the nonuniform distribution of ions near the boundaries of the surfaces:

$$dP_o = kTd\,[\Sigma n_I] \qquad \text{(IV.25)}$$

where n_I is the number of ions in unit volume of solution.

Then, clearly,

$$P_E(H) = \frac{\varepsilon E^2}{8\pi} + kT\,\Sigma\,n_I \qquad \text{(IV.26)}$$

In order to simplify the calculations we may consider the following cases.

1. The layer of liquid is situated in the middle between the bodies in contact (the gap equals 0.5 H), where the concentration of ions is constant, i.e., $d[\Sigma n_I] = 0$; then,

$$P_I(H) = \frac{\varepsilon}{8\pi}\left(\frac{\partial\psi}{\partial H}\right)^2 \qquad \text{(IV.27)}$$

2. The layer of liquid is situated on the surface of the plates at constant surface potential, i.e., with $\partial\psi/\partial H = 0$; then

$$P_D(H) = kT\Sigma n_I \qquad \text{(IV.28)}$$

In this case, n_I is the number of ions per unit volume of the solution for $H \approx 0$.

In formulas (IV.26)-(IV.28) the relation between the ionic component and the gap H between the contiguous bodies, which determines the adhesion, is expressed in implicit form.

For particular cases we may establish a direct relation $P_I(H) = f(H)$ for relatively large values of a potential ($\varphi > 100$ mV),* and a value of H small compared with the thickness of the atmosphere h_I, which is given by the formula

*Potential of the solution.

$$h_{\rm I} = \sqrt{\frac{\varepsilon kT}{4\pi e^2 \Sigma z_{\rm I} n_{\rm I}}} \qquad \text{(IV.29)}$$

where k is Boltzmann's constant, e is the charge on the electron, and z_I is the electrovalence of the ion in unit volume of solution.

This relationship may be expressed by the formula

$$P_{\rm I}(H) = \frac{\pi}{2}\varepsilon\left(\frac{kT}{ze}\right)\frac{1}{H^2} \qquad \text{(IV.30)}$$

For relatively large values of H, and for a symmetrical electrolyte, when $z_1 = z_2 = z$ and $n_1 = n_2 = n$

$$P_{\rm I}(H) = 64\, c_{\rm I} n_{\rm I} kT \exp\left(-\frac{H}{h_{\rm I}}\right) \qquad \text{(IV.31)}$$

where c_I is the ion concentration in mole/cm^3.

For weakly charged surfaces [178]

$$P_{\rm I}(H) = \frac{\varepsilon}{8\pi}\cdot\frac{\psi^2}{h_{\rm I}^2\cdot ch\left(\frac{Hh_{\rm I}}{2}\right)} \qquad \text{(IV.32)}$$

In order to calculate $P_I(H)$ from (IV.31) and (IV.32), we must know the thickness of the ionic atmosphere, which for a binary, univalent electrolyte (NaCl, KCl) may be calculated from Eq. (IV.29) with z = 2. Hence, by expressing the electrolyte concentration as $c_I = n_I/N$ (N is the total number of ions in unit volume of electrolyte) we obtain

$$h_{\rm I} = \sqrt{\frac{\varepsilon kT}{8\pi e^2 N c_{\rm I}}} \qquad \text{(IV.33)}$$

A knowledge of the thickness of the ionic atmosphere is essential in order to estimate not only the ionic but also the diffusion component of the disjoining pressure.

In certain cases, in particular for solutions with concentration of over 0.5 N, the ionic component is usually smaller than the other components of the disjoining pressure, and may be neglected when estimating the adhesive interaction between two solid bodies. Then, in accordance with formula (IV.21), $P_E(H) \approx P_D(H)$, the electrical component of the disjoining pressure arises from the interaction of the diffuse layers formed at the surfaces of the contiguous bodies.

In the Deryagin—Landau theory [187], a limiting case is considered; in this theory the region in which the diffuse layers overlap is so small that the deformation of the layers may be neglected. Then the formula for determining the electrical component of the disjoining pressure has the form

$$P_E(h) = 2\pi h_I kT\left[ch\left(\frac{e}{kT}\varphi_{0.5\,H}\right) - 1\right] \quad \text{(IV.34)}$$

$$h_I = \sqrt{\frac{\varepsilon kT}{8\pi e^2 N c_I}} \cdot \ln\frac{B_0}{B_{0.5\,H}} \quad \text{(IV.35)}$$

$$B_0 = \frac{\exp\left(\frac{e}{2kT}\varphi_0\right) - 1}{\exp\left(\frac{e}{2kT}\varphi_0\right) + 1} \quad \text{(IV.36)}$$

$$B_{0.5\,H} = \frac{\exp\left(\frac{e}{2kT}\cdot\frac{\varphi_{0.5H}}{2}\right) - 1}{\exp\left(\frac{e}{2kT}\cdot\frac{\varphi_{0.5H}}{2}\right) + 1} \quad \text{(IV.37)}$$

where e is the charge on the electron, φ_0 is the potential of the diffuse electric layer, and $\varphi_{0.5\,H}$ is the potential in the middle of the gap between the contiguous bodies.

For $\varphi_0 > 50$ mV, formula (IV.34) may be replaced by a simpler one, namely,

$$P_E(H) = 64\,nk\,TB_0^2 \cdot \exp\left(-\frac{H}{h_I}\right) \quad \text{(IV.38)}$$

Let us return to Eq. (IV.22). Let us compare the electrical and molecular components of the disjoining pressure, considering, with a certain amount of assumption, that these produce the full disjoining pressure, i.e.,

$$P(H) \approx P_E(H) + P_M(H) = 64\,nkTB_0^2 \cdot \exp\left(-\frac{H}{h_I}\right) - \frac{A}{H^3} \quad \text{(IV.39)}$$

Here we must distinguish three cases [188] characterized by the concentration of the electrolytes.

1. For a low electrolyte concentration and great thicknesses of the ionic atmosphere, when $P_E(H) \gg P_M(H)$, the total disjoining

pressure is positive and may be calculated from the formula

$$P(H) = 64\,nkTB_0^2 \cdot \exp\left(-\frac{H}{h_{\mathrm{I}}}\right) \qquad (\mathrm{IV.40})$$

The Deryagin—Landau theory [187] in this case enables us to estimate the electrical component of the disjoining pressure.

2. For a fairly high electrolyte concentration, when $h_{\mathrm{I}} \ll H$ and $P_E(H) \ll P_M(H)$, the total disjoining pressure equals

$$P(H) = -\frac{A}{H^3} \qquad (\mathrm{IV.41})$$

3. For medium electrolyte concentrations, when $P_E(H)$ and $P_M(H)$ are of the same order, calculation of the disjoining pressure is more difficult.

We see from Eq. (IV.39) that the disjoining pressure depends on the extent of the gap separating the bodies in contact. Fuks confirmed this relationship experimentally by the method of plane-parallel discs for solutions of $CaCl_2$, KCl, and NaCl with a concentration of less than 5 mg-equiv/liter.

By varying the electrolyte concentration, we may achieve the vanishing of the force barrier (see Fig. IV.7, curve 2), as a result of which adhesion of the particles takes place. The concentration above which the appearance of a force barrier is possible (threshold concentration) may be determined from the equation [188]

$$c_{\mathrm{t}} = C \cdot \frac{\varepsilon^3 (kT)^5}{A^2 \cdot e^6 \cdot z^6}\, f(\beta) \qquad (\mathrm{IV.42})$$

where C is a constant, A is the van der Waals constant, $f(\beta)$ is a function depending on the asymmetry of the electrolyte, i.e., the ratio of the charges on the cations and anions. The remaining notation is the same as that used earlier.

The equilibrium thickness of the layer, and hence the disjoining pressure, depend considerably on concentration [189]. Below we present values of the equilibrium thickness of a layer of liquid (H) corresponding to specific values of concentration (c) of KCl in solution (for free films) [179]:

$c_{KCl} \cdot 10^4$, mole/liter. .	1.13	2.05	3.13	5.20	8.10	20.2
H, μ.	0.139	0.114	0.104	0.083	0.067	0.05

The equilibrium thickness of the layer for dilute solutions of electrolytes is inversely proportional to the valence of the electrolyte [179]:

$$\frac{H_1}{H_2} \sim \frac{z_2}{z_1}$$

In view of this, the adhesive forces in dilute electrolyte solutions may be varied by varying the valence of the electrolyte. However, for medium electrolyte concentrations and the simultaneous action of $P_E(H)$ and $P_M(H)$, the adhesive forces depend little on the valence of the ion [see Eq. (IV.39)]. Under these conditions, for 0.01 M KCl solutions, the equilibrium thickness of the layer is slightly larger than when only electrical forces act [$P_M(H) = 0$] [179]. The electrical component of the disjoining pressure and the methods of calculating this are valid for relatively low electrolyte concentrations. In cases in which the electrolyte concentration suppresses the diffuse layer of ions at the interface of the solid phases, the discussions presented here lose their meaning.

In conclusion, it should be noted that, in accordance with (IV.22), the components of disjoining pressure considered, particularly the molecular and ionic components, appear at comparatively short distances from the boundary of the solid phase (a few hundreds of angstroms).

The disjoining pressure may receive contributions from not only primary [Eq. (IV.22)] but also secondary causes, including: the presence of solvate layers [173, 174] and their properties, the hydration of the ions [46], and also the effect of an adsorbed monolayer (or layer) of liquid, oriented with respect to the surface, on subsequent polymolecular layers [164, 171].

§ 21. Adhesion in Solutions of Electrolytes

As already considered, the presence of electrolytes in the solution should have a considerable effect on the adhesion of powder particles.

The dependence of the adhesion on the concentration and valence of the electrolyte cation was determined experimentally by Buzach [8, 14, 190-195], Fuks [12, 15, 59, 196], and ourselves [45,

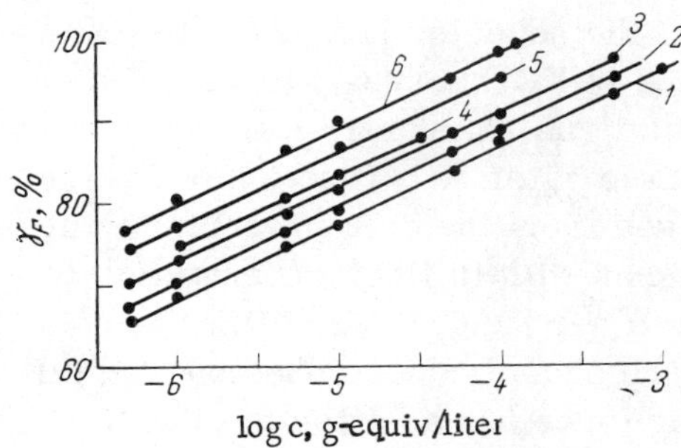

Fig. IV.11. Adhesion number of quartz particles to a glass substrate as a function of the concentration of various electrolytes. 1) LiCl; 2) NaCl; 3) KCl; 4) $PbCl_2$; 5) $CaCl_2$; 6) $BaCl_2$.

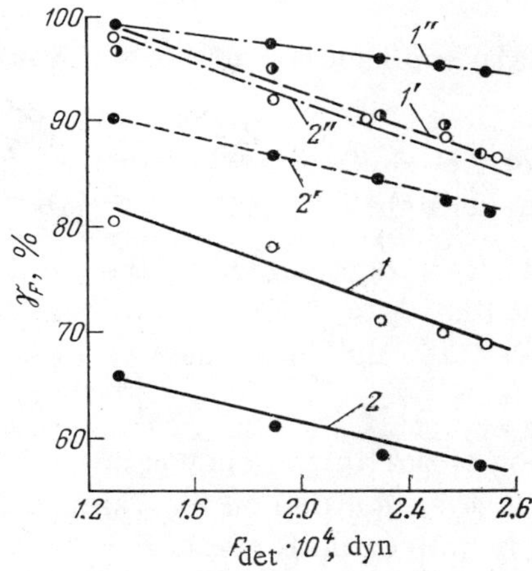

Fig. IV.12. Adhesion number of spherical glass particles 70 ± 2 μ in diameter as a function of the detaching force in 0.01 M (1,1',1") and 0.001 M (2,2',2") solutions of: 1,2) KCl; 1',2') $CaCl_2$; 1",2") $AlCl_3$.

76, 170]. Fuks' experiments, in contrast to Buzach's and our own, were made with weakly concentrated electrolytes and chiefly with uni- and divalent cations.

Figure IV.11 shows the dependence of the adhesion number of quartz particles with respect to a glass surface on the concentration of electrolytes with different cation valences [12]; Fig. IV.12 represents the adhesion number of glass particles adhering to a glass surface in KCl (curves 1 and 2), $CaCl_2$ (curves 1' and 2'), and $AlCl_3$ solutions (curves 1" and 2") with concentrations of 0.01 and 0.001 mole/liter as a function of the applied detaching force [45]. We see from the data presented that, on the one hand, adhesion diminishes with falling concentration for all the electrolytes and, on the other hand, the smaller the concentration of the solutions the more sharply does adhesion [45] fall with increasing applied detaching force. In addition to this, for solutions with concentrations from 0.01 M to 0.001 M the adhesion rises with increasing cation valence.

The adhesion number is related to the concentration of the electrolytes [12, 15, 169] by the following expression:

$$\gamma_F = K_1 + K_2 \log c \tag{IV.43}$$

where K_1 and K_2 are coefficients, and c is the concentration of the electrolyte.

The value of the coefficient K_2 is expressed by the tangent of the slope of the straight lines $\gamma_F = f(\log c)$.

The quantity K_1 is determined by the section of ordinate cut off by the straight line $\gamma_F = f(\log c)$. For the chlorides of $Li^{\cdot}$, $Na^{\cdot}$, $K^{\cdot}$, $Rb^{\cdot}$, $Mg^{\cdot\cdot}$, $Ca^{\cdot\cdot}$, $Ba^{\cdot\cdot}$ this quantity lies between 8.75 and 10.0.

The adhesion of particles depends not only on the concentration of the electrolytes, but also on the valence of the cations [45]; this is particularly noticeable for solutions with $c = 10^{-2}$–10^{-3} mole/liter. Figure IV.13 shows the dependence of the adhesion number on concentration within these limits for uni-, di-, and tervalent cations. As the valence of the cations diminishes, the detachment of the particles becomes easier. Thus, for a detaching force of $2.7 \cdot 10^{-4}$ dyn, 56% of the particles remain in a solution of 10^{-3} mole/liter of KCl, while 88% remain in an $AlCl_3$ solution of the same concentration. For a greater concentration of the solutions (0.1-1 mole/liter), with the same detaching force, almost all the particles remain on the dust-laden substrate. It is thus impossible to determine the dependence of the adhesive forces of these particles (for solutions with a concentration greater than 0.1

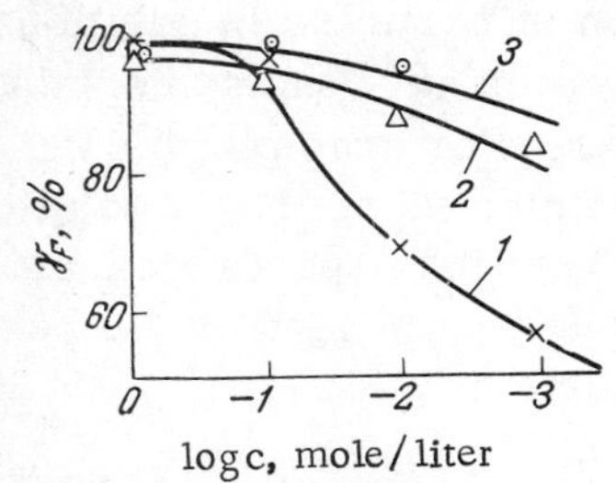

Fig. IV.13. Adhesion number of spherical glass particles 70 ± 2 μ in diameter with respect to a glass surface as a function of the concentration of the following solutions: 1) KCl; 2) $CaCl_2$; 3) $AlCl_3$.

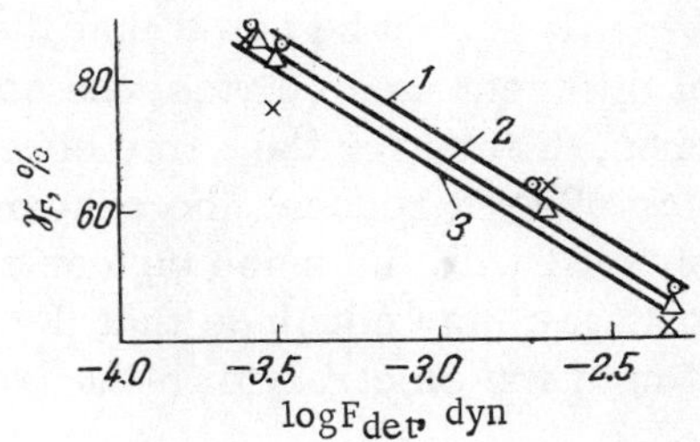

Fig. IV.14. Adhesion number as a function of detaching force for 0.1 M solutions of: 1) KCl; 2) $CaCl_2$; 3) $AlCl_3$.

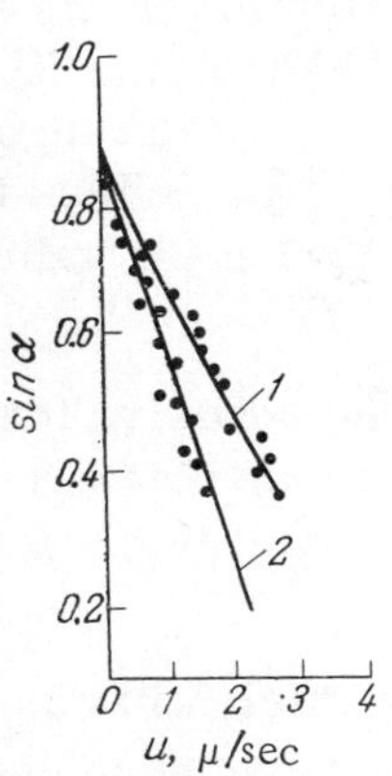

Fig. IV.15. Angle of detachment of quartz particles as a function of the velocity of electrophoresis in solutions of chlorides (Buzach). 1) Univalent (LiCl, NaCl, KCl); 2) divalent ($CaCl_2$, $SiCl_2$, $BaCl_2$).

mole/liter) on the nature of the electrolyte by the method of inclining the surface in question.

In order to determine the dependence of the adhesion number on the valence of the cation, we set up a series of experiments with solutions of KCl, $CaCl_2$, and $AlCl_3$ of concentration 0.1 mole per liter in which the adhesion number of particles 70 ± 2μ in diameter to a glass surface was determined by the centrifuge method. It was found as a result of the experiments (Fig. IV.14) that adhesion diminished with increasing cation valence. Thus in more concentrated solutions the dependence of the adhesion on the cation valence is the reverse of that found in dilute solutions [45, 170].

It should be noted that the adhesion of particles in solutions of different electrolytes, the cations of which have the same valence, differs for the same concentrations. For example, a solution of LiCl reduces the adhesion of particles more than a solution of NaCl with the same concentration. From the experimental results we may conclude that the adhesive forces depend on the position of the electrolyte in the lyotropic series:

$$\mathrm{LiCl} < \mathrm{NaCl} < \mathrm{KCl} < \mathrm{RbCl} < \mathrm{CsCl}$$

for univalent cations, and

$$\mathrm{CaCl_2} < \mathrm{SrCl_2} < \mathrm{BaCl_2}$$

for divalent cations.

In these series the adhesive force in a solution of the same concentration increases for each successive electrolyte. The lyotropic series coincides with a series of electrolytes arranged in order of diminishing thickness of the residual layer of liquid between plane-parallel discs [46]. This fact confirms the validity of the theoretical premises (see §20).

However, whereas for dilute solutions the adhesive force depends on the position of the electrolyte in the lyotropic series, for concentrated solutions, although this characteristic is preserved, it becomes less explicit.

Sometimes changes in the adhesion of particles in electrolyte solutions are associated with a change in the value of the ζ potential. However, the ζ potential can only indirectly characterize the change in the adhesive force, since it is itself a function of the thickness of the diffuse layer of adsorbed ions.

Figure IV.15 shows the adhesive force (or rather $\sin\alpha$, where α is the slope angle at which all the particles are removed) as a function of the velocity of electrophoresis u,* for univalent (curve 1) and divalent (curve 2) cations [8]. The first curve lies above the second, i.e., for the same ζ potential the adhesion in solutions of univalent cations is greater than in solutions of divalent cations; this may be explained by the different thicknesses of the

*It is well known that the value of the ζ potential is proportional to the velocity of electrophoresis, i.e., $\zeta = ku$.

liquid boundary layer. For a potential equal to zero, the adhesive force in solutions of univalent and divalent cations should be the same. In this case, for specific forces of molecular interaction, the disjoining pressure of the thin layer of liquid is also the same.

For ter- and quadrivalent cations, the maximum value of the ζ potential corresponds to the minimum adhesive force.

The ζ potential is associated not only with the state of the diffuse layer and the thickness of the ionic atmosphere h_I, but also with the hydrodynamics of the electrokinetic phenomena by which it is determined.

The adhesive forces also depend on the pH of the medium in which the dust-laden surface is situated. For example, the adhesion number for the interaction of glass particles 40-60 μ in diameter to a Class 9 surface with a detaching force of $9.0 \cdot 10^{-5}$ dyn varies as follows with the pH of the medium (alkalinity created by NaOH, acidity by HCl):

pH	2.5	7.0 (distilled water)	7.2 (tap water)	8.0	10.0
γ_F, %	27	45	41	4	6

Thus, in an alkaline medium, the particles are removed more easily.

§ 22. Adhesion in Solutions of Surface-Active Substances

Variations of adhesion in liquid media in the presence of organic acids and alcohols and surface-active substances (SAS) possessing detergent properties were studied in [8, 59, 76].

The results of some experiments [8] into the adhesion of quartz particles in solutions of various organic acids are presented in Fig. IV.16. For weakly concentrated solutions (concentration up to 1 mole/liter) the maximum adhesion occurs in solutions of formic acid. On increasing the velocity of electrophoresis (and hence the ζ potential) adhesion is reduced. Thus, the minimum adhesion of quartz particles occurs in a solution of propionic acid, the velocity of electrophoresis being greatest for this.

The adhesion of particles falls with increasing density and thickness of the diffusion and adsorbed layers of the acid solutions, and this in turn depends on the position of the saturated acids in the homologous series.

For a concentration higher than 7 mole/liter, the situation is the opposite to that shown in Fig. IV.16. Adhesion reaches its maximum value in the solution of propionic acid.

Adhesion in aqueous solutions of alcohols is also determined by the position of the alcohol in the homologous series.

Figure IV.17 illustrates the antagonism between the action of electrolytes (in the present case aluminum chloride solution) and alcohol on adhesion [8]. As the $AlCl_3$ increases to 0.1 mole per liter in the absence of alcohol, the value of $\sin\alpha$ rises, i.e., the sine of the limiting angle of slope at which the quartz particles become detached from the glass surface increases. For $c_{AlCl_3} >$ 0.1 mole/liter, the adhesion (i.e., $\sin\alpha$) falls slightly.* For an $AlCl_3$ concentration close to 0.1 M the character of the relationship between $\sin\alpha$ and c_{AlCl_3} changes and a bend appears in the curves.

The addition of ethyl alcohol in quantities not exceeding 50% to the $AlCl_3$ solutions (curves 2 and 3) has hardly any effect on the relationship between $\sin\alpha$ and c_{AlCl_3}. For alcohol concentrations of 80 and 96% (curves 4 and 5) the character of the relationship is reversed.

In concentrated solutions of $AlCl_3$ (~0.4 mole/liter) the presence of ethyl alcohol has only a slight effect on adhesion.

The addition of such electrolytes as KCl and $BaCl_2$ in quantities not exceeding 0.1 mole/liter to solutions of propyl, ethyl, and methyl alcohols leads to an increased adhesion. We noted something rather similar (Fig. IV.18) when studying adhesion in solutions of surface-active substances (commercial products containing small quantities of electrolytes). The fact that the adhesive forces in solutions of T Duomin, NT Armak, and sodium oleate exceed the expected values may clearly be attributed to the presence of such electrolytes.

*We remember that Buzach [8] deposited his particles in the liquid. In our own experiments the particles were deposited in air and no maximum of this kind was observed.

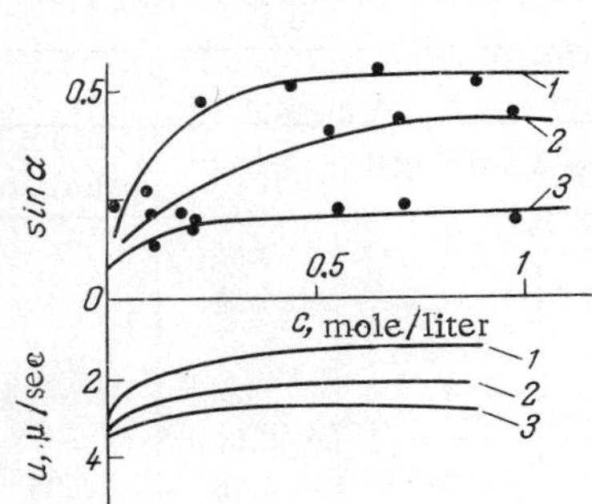

Fig. IV.16. Angle of detachment of quartz particles and rate of electrophoresis in acid solutions as functions of concentration. 1) Formic; 2) acetic; 3) propionic.

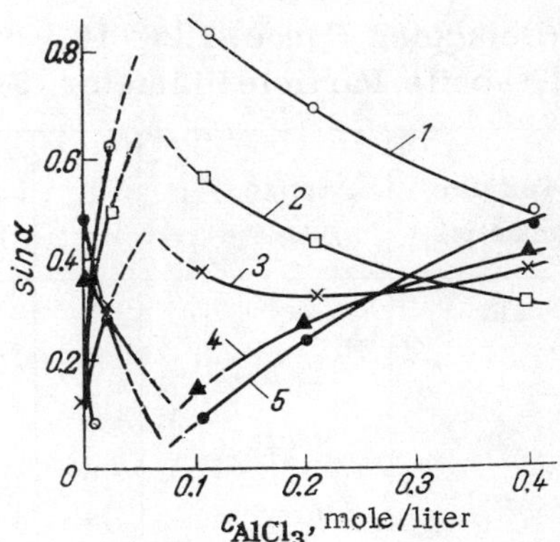

Fig. IV.17. Angle of detachment of quartz particles as a function of $AlCl_3$ concentration in aqueous solutions of ethyl alcohol. 1) In absence of alcohol; 2) $[C_2H_5OH]$ = 20%; 3) 50%; 4) 80%; 5) 96%.

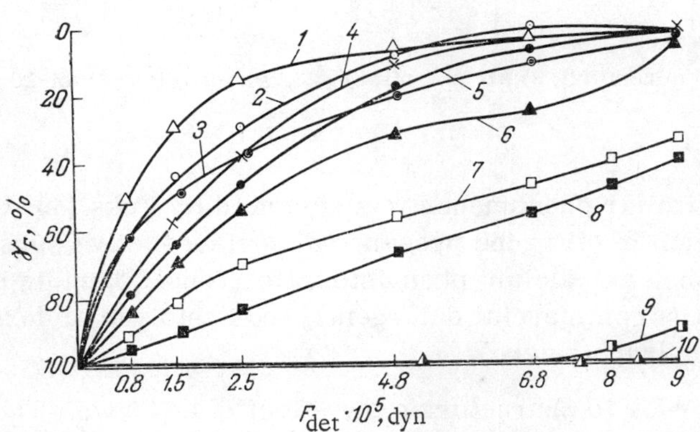

Fig. IV.18. Adhesion number as a function of detaching force for glass spheres 40-60 μ in diameter from steel plates in solutions of surface-active substances. 1) Étomid NT/25; 2) Merzolyat D; 3) Sulfonol; 4) Sulfonate; 5) Arkvad 18; 6) OP-10; 7) Gardinol; 8) distilled water; 9) sodium oleate; 10) Duomin T; Armak NT.

Table IV.4. Effect of Additives on the Adhesion of Scale and Graphite Particles to Metal Surfaces
(Detaching Force, $1.26 \cdot 10^{-4}$ dyn; Scale Partical Diameter, 10–12μ; Graphite Particle Diameter, 5–6μ; Additive Concentration, 1%)

Particle material	Surface material	Oil*	$a\gamma_F/\gamma_F$			
			Paranoks	Santolyub	Tenol	Barium palmitate
Scale	Steel	1	0.93	–	0.83	–
		2	0.93	–	0.92	–
		3	2.35	2.45	2.00	2.15
	Bronze	1	2.70	–	2.25	–
		2	0.80	–	0.90	–
		3	0.48	–	0.75	0.35
	Glass	1	0.05	0.31	0.05	–
		4	0.50	0.78	0.75	–
Graphite	Steel	1	1.05	1.40	1.90	–
		2	–	1.42	–	–
		3	0.53	0.70	0.67	0.30
	Bronze	1	0.45	0.85	0.48	–
		2	1.30	–	1.10	–
		3	–	0.98	–	–
	Glass	1	0.10	0.68	0.25	–
		4	0.28	0.28	0.35	–

*1) Purified mineral oil; 2) AU axle oil; 3) MK aviation oil; 4) Avtol-10.

A similar phenomenon was observed by Fuks [59] when studying adhesion in oil in the presence of surface-active substances (additives) such as calcium phenylstearate (Tenol), barium palmitate, Paranoks (a commercial detergent), and Santolyub (chloroderivative of alkylnaphthalene).

In order to characterize the effect of additives, Table IV.4 presents the ratios of the adhesion numbers in oil with and without additives (${}_a\gamma_F$ and γ_F, respectively) for a constant force of detachment.

We see from the data presented that the introduction of additives sometimes has no effect on adhesion and sometimes produces a sharp change in adhesive interaction. As a rule, adhesion to glass samples with hydrophilic properties falls sharply on introducing additives into the liquid. The sticking of particles to metal surfaces covered with oxide films and possessing hydrophobic properties increases in the presence of additives.

There is an optimum concentration of surface-active substances for which a maximum reduction in adhesive force is achieved. This is confirmed by our experimental results for aqueous solutions of OP-10 (Fig. IV.19). On raising the concentration of surface-active substances to 1% adhesion diminishes (curves 2 and 3). Further raising the concentration (above 1%) causes micelle formation and has little effect on adhesion. A similar effect is exerted on the adhesion of particles in an aqueous medium by the complex-forming substance sodium hexametaphosphate (curves 1 and 4). Our experiments based on sodium hexametaphosphate with a tracer phosphorus atom show that, like certain condensed polyphosphates [197] this substance may collect on the surface of the contiguous bodies and thus facilitate the detachment of particles.

Experiments carried out in [76] revealed some of the special characteristics of the variations in adhesion number observed in solutions of surface-active substances (SAS) for various methods of depositing the particles on the surface.

Figure IV.20 shows the variation in the adhesion of particles 40 μ in diameter in solutions of various SAS, particularly Sulfonol and OP-10, as a function of the time spent by the painted surface in an aqueous medium for various methods of depositing the particles on the surface, namely, air dusting followed by placing the dust-laden surface in the liquid (curves 1 and 1') and liquid settling (curves 2 and 2').

The equilibrium state in solutions of SAS is reached in 5-10 min, i.e., three to five times more quickly than in an aqueous medium (see Fig. IV.5). This is not only due to the wetting power of the solution. In solutions of SV-102 and DB wetting agents, the method of depositing the particles has no effect on the adhesion. Whereas in solutions of OP-10 and Sulfonol the equilibrium state is reached in 5-10 min, in solutions of SV-102 and DB it is reached in under 2 min.*

*In the centrifuging method employed, the adhesive force is determined 2 min after putting the surface in the aqueous medium.

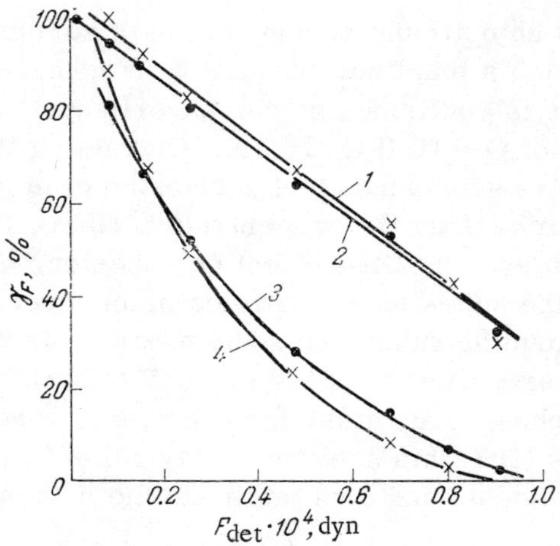

Fig. IV.19. Adhesion number of glass particles 40-60 μ in diameter with respect to steel surfaces in aqueous solutions of OP-10 and sodium hexametaphosphate (SHMP) as a function of the detaching force. 1) 0.0001% solution of SHMP; 2) 0.0001% solution of OP-10; 3) 1.0% solution of OP-10; 4) 1.0% solutions of SHMP.

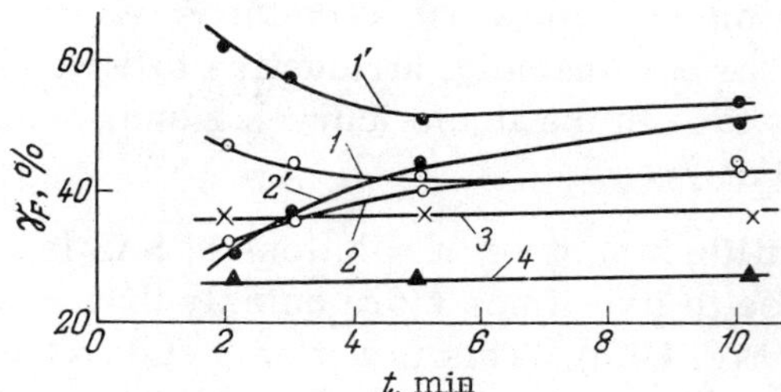

Fig. IV.20. Adhesion number of glass particles 40 ± 2 μ in diameter as a function of the time of contact with painted surfaces in SAS solutions. 1,1') 0.1% solutions of Sulfonol and 0.25% solution of OP-10 for air dusting; 2,2') the same for liquid settling; 3) 0.5% solution of SV-102; 4) 0.1% solution of DB.

Clearly, in this case, the time required to form the adsorbed layer is of decisive importance. It is known that in solutions of SV-102 and DB adsorption equilibrium is reached in tenth parts of a second [198]. This means that the dynamic and static coefficients of surface tension differ very little from one another. Hence, in solutions of SV-102 and DB the adhesion is independent of the means of depositing the particles on the surface (Fig. IV.20, curves 3 and 4).

In solutions of OP-10 and Sulfonol the values of the static and dynamic coefficients differ sharply from one another at the instant of forming the new phase, and only become equal after about 20 min. The adhesive forces in OP-10 and Sulfonol solutions are therefore different.

The effect of SAS solutions on adhesion has also been verified under other conditions. Thus, in [76], the adhesion number was determined in relation to the detachment of spherical glass particles 30 μ in diameter (by the immersion method) from an oily surface after spending various periods in the liquid:

t, sec	2	5	10	30	60	120
γ_F, %:						
in 0.1% Sulfonol solution	73	68	63	47	39	38
in 0.25% DB solution	57	35	31	30	28	28
in 0.25% SV-102 solution	61	40	38	37	39	38

It follows from these data that, under these conditions also, solutions of DB and SV-102 manifest their effects for a shorter period than a Sulfonol solution. Hence, the variation in adhesion with time of particle contact is determined by the properties of the SAS, particularly the values of the static and dynamic surface-tension coefficients. This fact is of particular practical significance, for example, in dust-catching and in the washing of vehicles.

§ 23. Dependence of the Forces of Adhesion on the Shape of the Surface and on the Temperature of the Aqueous Medium

The effect of the roughness of the substrate on the adhesive force is not so great in an aqueous medium as it is in the adhesion of dust to steel surfaces in air [159]. Thus the adhesion of glass

Table IV.5. Adhesive Forces of Needle-Shaped Particles *

Substrate material	Particle material	Length of particles, μ	Medium	F_{ad}, g units	
				faces	ends
Paraffin	Paraffin	4.8–5.2	Water	0	0.071
		2.4–2.6	Aqueous solution of $CaCl_2$ (1 mg-equiv/liter	0.23	1.03
Glass	Mercury iodide	2.5	Water	0.023	4.64
		1.0	Aqueous solution of $CaCl_2$ (5 mg-equiv/liter	0.017	6.24

*The particles were detached by means of a force directed tangentially to the surface.

particles to surfaces of Class 9 finish in water differs little from that of similar surfaces finished to Class 13 standard. All that has been said of the effect of particle shape on air adhesion (see §15) remains true to a certain extent for liquid adhesion.

The number of experiments in which the effect of particle shape on the adhesive forces has been investigated is extremely limited. Table IV.5 presents some results of Fuks et al. on the sliding of irregularly shaped particles, as determined by the inclined-surface method.

We see from the data presented that the detachment of needle-shaped particles depends on their position on the surface.

Sometimes, for practical purposes, it is important to know the change in adhesion resulting from a change in the temperature of the liquid medium.

As in air (see §15), the adhesive forces in distilled water increase with increasing temperature (Fig. IV.21). The rise in adhesion with increasing temperature in this case is clearly due to the fact that the forces of molecular interaction become stronger when the particles interact with the surface at distances greater than the diameter of the molecules (see §4).

In electrolyte solutions (LiCl, NaCl, KCl) adhesion also increases with rising temperature (from 20 to 60°C); this was shown in [46] by the plane-parallel—disc method. A rise in adhesion with increasing temperature was also observed by the crossed-filament

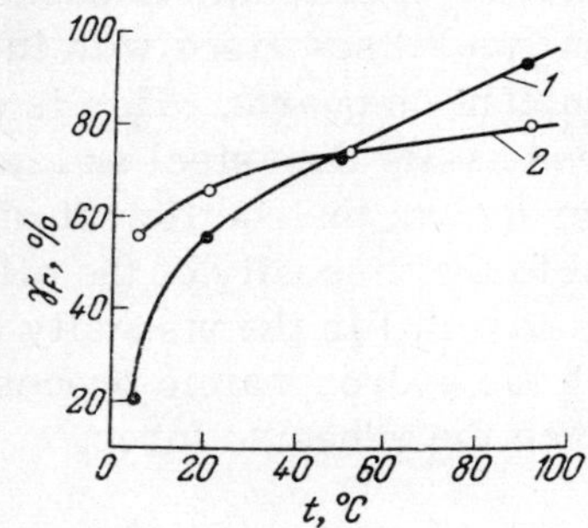

Fig. IV.21. Adhesion number as a function of temperature for the interaction of glass particles 40 ± 2 μ in diameter with glass (2) and perchlorvinyl–enamel-painted surfaces (1) in distilled water (detaching force 3.0 · 10^{-5} dyn).

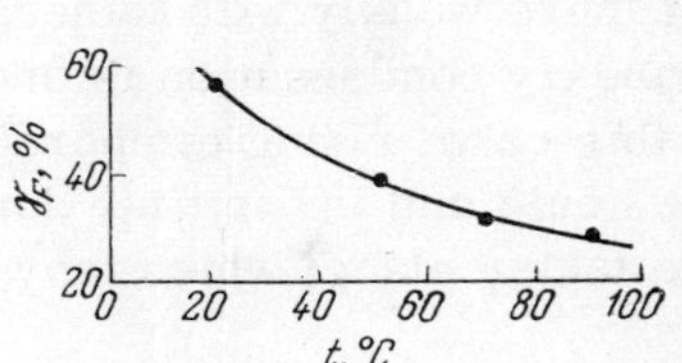

Fig. IV.22. Adhesion number as a function of temperature for glass particles 40 ± 2 μ in diameter adhering to a glass surface in an aqueous solution of OP-10 (detaching force 3.0 · 10^{-5} dyn).

method in [86]; for a 10^{-2} N solution of $MgSO_4$, there was a sharp jump in adhesive force between 33 and 43°C, after which the adhesive force remained almost constant. In addition to this, the time required for the equilibrium value of the adhesive force to be reached (aging time) fell with increasing temperature.

In the present case the rise in the adhesive forces may be explained by a reduction in the disjoining pressure of the liquid layer in the contact zone which takes place as a result of the reduction in the equilibrium thickness of the layer with increasing temperature of the medium.

On replacing water by aqueous solutions of SAS (in particular a 0.1% solution of Sulfonol or a 0.25% solution of OP-10) there is a reduction in the adhesion of particles to glass surfaces with increasing temperature (Fig. IV.22).

Hence, the removal of particles from glass and painted surfaces in air and distilled water will be more effective at low temperature (between 0 and 20°C), and in aqueous solutions of Sulfonol and OP-10 at high temperatures. Hence, in practice, for example, for washing motor vehicles one uses ordinary water at 15–20°C or aqueous solutions of detergents after heating to 60°C.

The variation in the adhesive forces with temperature depends not only on the properties of the liquid but also on those of the bodies in contact. Thus, the coefficient of boundary condensation [see Eq. (IV.10)] of mineral oil on steel discs rises with increasing temperature but remains constant on quartz. This is evidently because quartz surfaces wet less easily than steel and interact more weakly with mineral oil. For quartz the coefficient of boundary condensation is proportional to the viscosity of the oil. In this case, rising temperature leads to a fall in the viscosity of the liquid and in the time during which the hydrodynamic processes are taking place, which may even reduce the adhesive force.

§24. Effect of Particle Size on the Forces of Adhesion in Liquid Media

Let us consider the relationship between the adhesive force and the dimensions of the particles in liquid media.

According to Buzach [8], as the radius of quartz particles increases from 6 to 93 μ, the slope of the dust-laden surface at which detachment takes place in distilled water first falls slightly to pass through a minimum for particles 13 μ in radius and then rises again. This engendered the idea that the slope could be identified with the adhesive force, and led to the erroneous conclusion that the relationship between the adhesive forces and the particle size was analogous to the relationship between the slope and the radius (r). There is in fact an appreciable spread of particles with respect to adhesive force; at a specific angle of inclination of the surface a certain proportion of the particles are removed (whence the concept of "adhesion number"). The very small particles studied by Buzach, however, cannot be detached even on turning the plate through 90°. Buzach deposited his particles in air. Under these conditions he should have made allowance for the period of contact (see §19) but this he failed to do. Unfortunately, this error of Buzach in determining the relationship between the adhesion number and the particle size has passed into handbooks on colloid chemistry (Zhukov, Sheludko, and others).

The measurement of adhesion by the inclined-surface method (see §6) and the estimation of adhesion from the specific sticking force (see §2) give a relative estimate of the intensity of adhesion.

In other experiments [45, 164, 170] on the adhesion of particles in liquid media, in contrast to those of Buzach, the adhesion—

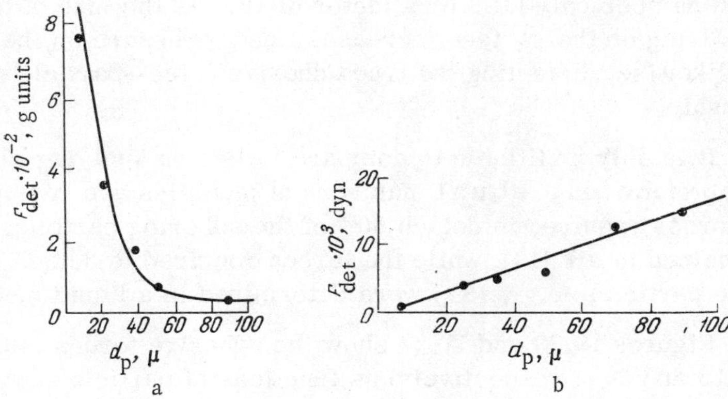

Fig. IV.23. The detaching force, expressed in units of g (a) and in dynes (b), of spherical glass particles adhering to a steel surface in distilled water as a function of particle size.

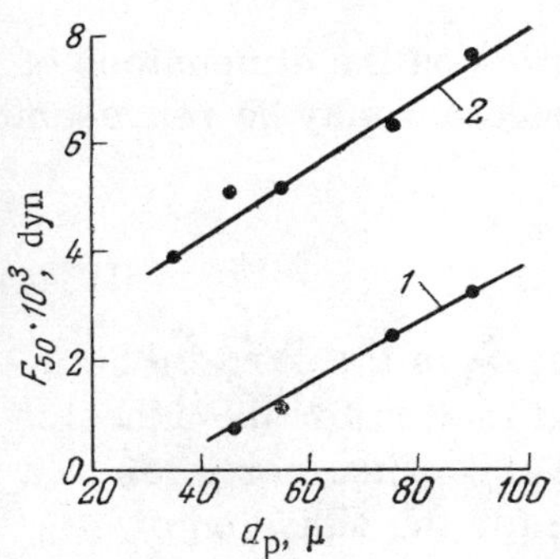

Fig. IV.24. Adhesive force relative to various surfaces as a function of the dimensions of the adhering spherical glass particles. 1) Glass; 2) painted steel surface.

diameter relationship was studied for plates (steel surfaces with a Class 9 finish) immersed in the liquid medium (distilled water) after first depositing the particles in air. The results are given at the bottom of this page.*

The relation between the adhesive forces and the dimensions of the particles cannot be fully revealed by the method based on inclining the dust-laden surface. As the detaching force rises by three orders of magnitude ($7.6\cdot 10^{-6}$ dyn for particles of 5-15 μ and $1.0\cdot 10^{-3}$ dyn for particles of 100-120 μ), the ad-

d_p, μ	100-120	80-100	60-80	40-60	30-40	20-30	5-15
Force numerically equal to the weight of particles, dyn.	$1.0\cdot 10^{-3}$	$5.5\cdot 10^{-4}$	$2.6\cdot 10^{-4}$	$9\cdot 10^{-5}$	$3\cdot 10^{-5}$	$1.2\cdot 10^{-5}$	$7.6\cdot 10^{-6}$
γ_F, %	8	12	15	26	60	89	99

*The particles were detached by rotating the surface through 90°.

hesion number only falls by a factor of 12. As the size of the particles lying on the surface increases, their pressure on the surface does likewise, distorting the true adhesive force—particle size relationship.

It is only justifiable to compare forces on applying which *approximately equal* numbers of particles are removed. The forces required to detach 50% of the adhering particles were determined in air [13], while the forces required to detach almost all the particles (e.g., 98%) were determined in a liquid medium.

Figures IV.23 and IV.24 show the adhesive forces [45] (for γ_F = 10 and 50%, respectively) as functions of particle size. We see from these data that the adhesive forces expressed in absolute units (in dynes) are directly proportional to particle size for d_p = 7.5-90 μ, independently of the substrate material. If the adhesive force is expressed in relative units (units of g) there is an inverse proportionality between F_{ad} and d_p.

The dependence of the adhesive forces on the dimensions of spherical glass particles in an aqueous medium may be represented by the empirical formula

$$F_{ad} = kd_p \tag{IV.44}$$

where F_{ad} is the adhesive force in dynes, d_p is the particle diameter in microns, and k is an experimental factor; for the adhesion number corresponding to the adhesion of 10% of the particles to a steel surface (Fig. IV.23), $k = 1.6 \cdot 10^{-4}$; for the adhesion of 50% of the particles to a glass surface, $k = 3 \cdot 10^{-5}$ and for a painted surface, $k = 10^{-4}$ (Fig. IV.24).

Fuks [164] came to analogous conclusions regarding the dependence of the minimum adhesive force F_{min} on particle size (on the basis of earlier work [12, 15, 59]; he found that the minimum particle size for which a direct proportionality between F_{ad} in dynes) and d_p persisted was 5 μ.

According to Fuks, the minimum adhesive force (for γ_F = 98%) was proportional to the diameter of the quartz particles (over the range 5-50 μ), and approximately equal to

$$F_{min} = fd_p$$

where f is a coefficient equal to 10^{-7} dyn/μ, and d_p is the diameter of the quartz particles in microns.

In a liquid medium, for which there is a liquid interlayer between the contiguous bodies, and the effects of capillary, electric, and Coulomb forces are excluded (see §§11-13), adhesion is due solely to the molecular forces (the disjoining pressure opposes adhesion). The value of the molecular forces is directly proportional to the dimensions of the particles [see Eqs. (I.47) and (I.49)]. For an aqueous medium the experimental results relating to the dependence of the adhesive forces on particle dimensions coincide with the theoretical data, and Deryagin's thermodynamic theory of adhesion is practically confirmed (see §5).

Chapter V

Adhesion of Particles to Paint and Varnish Coatings

§25. Characteristics of the Adhesion of Particles to Paint and Varnish Coatings

The dust always present in the air tends to settle on the surface of buildings, motor vehicles, railroad trucks, and other such objects [167]. Adhering dust worsens the external appearance of paint coatings, intensifies corrosion, accelerates the aging of paints and enamels [199], and may have an abrasive effect on being removed under dry conditions [200]. Dust contained in industrial refuse and capable of collecting oxides of sulfur and nitrogen may stick to painted surfaces and, in the presence of atmospheric moisture, damage not only the coating but also the surface subjected to painting or enameling. The cleaning of surfaces from adhering dust requires labor and material means.

Any change (usually a reduction, but in certain cases an increase) in the dust-holding capacity of paint coatings is of undoubted theoretical and practical significance.

In order to solve problems of dust holding we must discover what properties of paint and varnish coatings determine the adhesion of dust to these. Properties important in this connection include: roughness, moisture resistance, elasticity, physicochemical properties (hydrophoby and hydrophily as well as tackiness), and electrical properties.

Under conditions in which the coatings are situated in liquids, the adhesion of particles to the coatings depends on the properties of the medium. This question has been treated in detail in Chapter IV.

Table V.1. Brief Characteristics of the Main Classes of Paint Coatings

Class of coatings	Characteristics of the surface	Adhesion*
I	Even, smooth, without visible defects	+ −
II	Even, smooth, but with visible defects (channels, streaks, etc.)	− +
III	Smooth, with visible defects; substrate roughness visible	+
IV	Visible defects, substrate roughness visible	++

*Arbitrary notation: +, indicates intensification; ++ , considerable intensification; −, reduction of particle adhesion; − + or + −, indicate that either intensification or reduction of adhesion are possible.

In view of the variety of forces giving rise to adhesion and the differing properties of paint coatings and particle surfaces, it is impossible to give any single prescription or recommendation for changing the dust-holding capacities of coatings. The problem of increasing or reducing dust-holding powers must be treated separately in each specific case.

Changing the dust-holding capacity of coatings has only been considered by a very few authors. We shall therefore now and then have to confine attention to existing data or else consider adhesion characteristics by analogy with other phenomena of the same kind.

The effect of substrate roughness, which determines the true contact area of the dust particles, on the adhesive force was considered in some detail in Chapter III. In order to reduce the adhesion of particles, paint coatings should possess microscopic roughness without containing macroscopic asperities. Let us consider the main classes (with respect to external appearance) of paint coatings [201], their possible roughness, and the expected adhesion of dust particles to the coatings (Table V.1).

Class I includes coatings which admit polishing and burnishing (furniture, lacquer, automobile nitro-enamel, etc.) leading to the removal of microscopic roughness and to a rise in the adhesive force (see §14). Without polishing and burnishing, the adhesive force of the particles may be reduced.

If the dimensions of the projections (asperities) on coatings of Class II are commensurable with the dimensions of the particles the adhesion of the latter increases. Thus, coatings of Classes I and II may either increase or reduce adhesion, depending on the method of finishing and the dimensions of the particles.

The roughness of Class III coatings is affected by the unevenness of the original surface. For example, the roughness of the surface of a phosphate film treated with A-1-N lacquer remains comparatively large owing to the presence of asperities more than 30 μ high in this film [202]. Defects (cracks, cleavages, etc.) in coatings of Classes III and IV also tend to increase particle adhesion.

Lacquers copy surfaces better than enamels.

Hence adhesion to surfaces painted with lacquers depend on the roughness of the original substrate more than do those coated with enamels. The unevenness of surfaces may be increased as a result of pigments contained in the enamels. The roughness of such coatings increases with the steeping of the film-forming material [203].

It should be noted that the separation of coatings into classes by reference to their external form is quite arbitrary and depends not only on the properties of the coating material, but also on the presence of various impurities, the painting technology and quantity, and other factors.

Coatings tend to swell under the influence of water. Thus the weight of perchlorvinyl enamels (PKhV-14, 15, 26) increases by 25% after 100 h in water [204]. The swelling leads to the appearance of loose material, blisters, whitening, and breaks in continuity, which inevitably results in increased dust-holding capacity.

The adhesion of dust is influenced by adsorbed and condensed moisture; the first of these factors provides conditions for the formation of hydrogen bonds [131], and the second gives rise to capillary forces between contiguous bodies.

Counteraction to swelling depends on the rate of diffusion of water molecules through the film, i.e., the vapor permeability of the coatings. A nonpolar hydrophobic rubber or paraffin film has a low vapor permeability, while a hydrophilic gelatin or skin-glue film has a high permeability. There will be a minimum amount of condensed and adsorbed moisture on rubber or paraffin films; there will thus be little swelling and few breaks in the continuity of the coating, so that dust-holding will hardly increase at all.

Below we present some results [204] relating to the vapor permeability of various films (amount of water passing through the film in 120 h):

	mg		mg
Skin glue	3900	Tung oil and calcium resinate lacquer	200
Gelatin.	730	Tung oil and phenol resin lacquer	107
Linseed oil.	570	Paraffin.	10
Polymerized linseed oil .	286		
Rubber	232		

It should be noted that the smaller the vapor permeability, the greater the hydrophoby of the coating, which not only counteracts the swelling of the latter, but also affects the forces of adhesion (see §27).

Recently more and more attention has been paid to atmosphere-resistant and moisture-impermeable coating materials [205], for example, in apparatus construction (instrument making), enamels constituting copolymers of vinylchloride with vinyl acetate; in agriculture, alkyd-styrene enamels; and for the painting of light motor vehicles [206], polyacrylate enamels.

Epoxy enamels (particularly É-10M) and melaminoalkyd lacquer enamel [205] are distinguished by high moisture-resistant properties.

On the other hand, cellulose esters and cellulose lacquers are hygroscopic [206].

§26. Adhesion of Atmospheric Dust

The adhesion of atmospheric dust is largely determined by the conditions of contact between the dust particles and the surface and by the elastic properties of the paint coating.

In view of this, let us consider the adhesion of dust resulting from the free settling of particles as a function of the elastic prop-

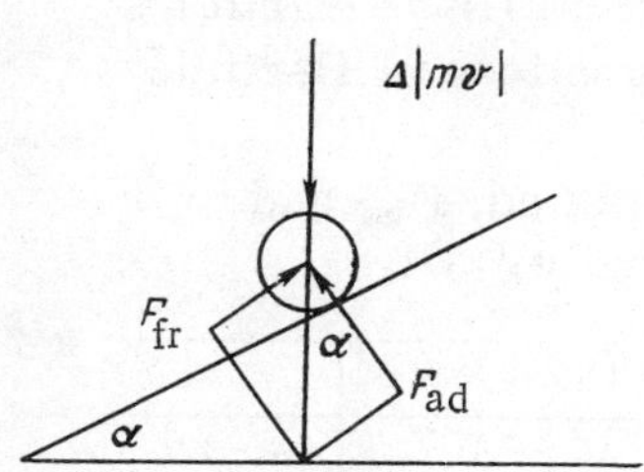

Fig. V.1. Components of adhesive forces for freely falling dust particles.

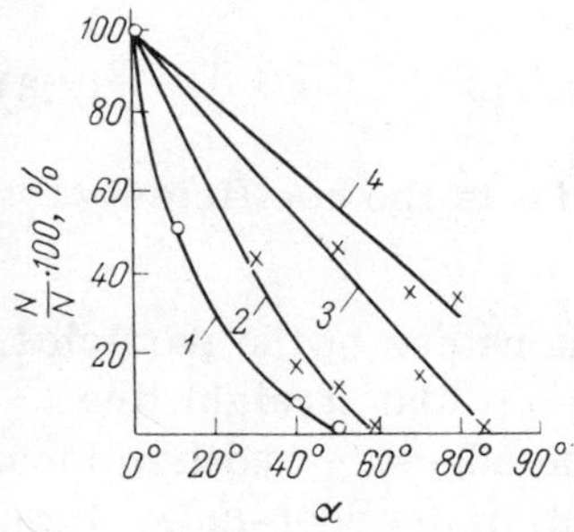

Fig. V.2. Number of spherical glass particles adhering to a perchlorvinyl-enamel-coated surface after falling freely, expressed as a function of the slope of the surface. Particle diameters: 1) d_p = 90; 2) 70; 3) 40; 4) 15 μ.

erties of the coating in the presence of surface oil contamination and also under conditions of precipitation (rain).

Adhesion for Free Particle Settling. Dust particles situated in air are acted upon by the force of gravity which compels them to settle. When dust particle settle they come into contact with surfaces and are held on the latter by adhesion and friction.

The rate of free settling for spherical particles may be calculated from the Stokes formula:

$$v_{\text{free}} = \frac{2}{9} \cdot \frac{r^2 g (\rho_1 - \rho_2)}{\eta} \quad \text{(V.1)}$$

where r is the radius of the dust particle; ρ_1 and ρ_2 are, respectively, the densities of the dust and the air; η is the absolute viscosity of the medium (for air, $\eta = 17.3 \cdot 10^{-5}$ g · cm^{-1} · sec^{-1} at t = 20°C).

Let us suppose that the dust-laden surface is placed at an angle to the horizontal plane, that the particles fall vertically, and that there is no deformation of the contact zone (Fig. V.1).

From the momentum equation, we have

$$\Delta |mv_{\text{free}}| = (F_{ad} + F_{fr})\, t \quad \text{(V.2)}$$

where m is the mass of the particles, F_{ad} is the adhesive force determined experimentally [13], F_{fr} is the force of friction, and t is the time of contact.

Projecting these forces on the vertical axis and taking the final velocity of the particles as zero, we obtain

Table V.2. Adhesive Force of Spherical Glass Particles and Slope of the Steel Surface as Functions of Particle Diameter
(for γ_F = 50%; v_{free} and α calculated; F_{ad} and μ determined experimentally [13])

d_p, μ	v_{free}, cm/sec	μ	F_{ad}, dyn	α, °
90	63.5	0.43	$7.2 \cdot 10^{-4}$	42
50	18.2	0.60	$2.13 \cdot 10^{-4}$	56
25	4.5	0.73	$2.15 \cdot 10^{-3}$	77
15	2.0	0.90	$6.12 \cdot 10^{-3}$	90

$$mv_{free} = \left(F_{ad} \cos\alpha + P\mu \sin\alpha \right) t \qquad (V.3)$$

where P is the weight of the particles, and μ is the coefficient of friction.

The coefficient μ depends on the dimensions of the particles. This coefficient, which is equal to the slope of the straight line characterizing the relationship between the adhesive and frictional forces and the mass of the powder on rotating the dust-laden surface, may be determined experimentally (Fig. I.6, p. 20).

According to Malyshev [207], the contact time may be estimated thus:

$$t = \frac{6.11 \cdot 10^{-4}}{v^{1/5}} \qquad (V.4)$$

Substituting Eq. (V.4) into (V.3) and taking v = v_{free}, we obtain

$$a / \sin\alpha = b \cdot \cot\alpha + c \qquad (V.5)$$

where

$$a = mv_{free}^{6/5};\ b = 6.11 \cdot 10^{-4} F_{ad};\ c = 6.11 \cdot 10^{-4} P\mu \qquad (V.6)$$

The data required to calculate the coefficients a, b, and c are presented in Table V.2. The same table gives the results of a graphical solution of Eq. (V.5), with an estimate of the limiting values of slope angle.

We see from the table that the limiting slope increases with diminishing particle size. The data presented here refer to steel surfaces. However, the situation is very much the same for the free settling of dust particles on a painted surface. It should be remembered that we have neglected the elasticity of the bodies coming into contact (this will be considered subsequently).

We carried out an experimental investigation in order to determine the limiting slope of a surface above which no adhesion of freely falling particles took place. Figure V.2 presents some experimental data relating to the adhesion* of freely falling spherical glass particles as a function of the slope of a surface coated with perchlorvinyl enamel. For particles of diameter 90, 70, 40, and 15 μ, only 50% of the particles striking the surface remained adhering to the latter at slopes exceeding 10, 22, 42, and 48°, respectively. In practice, freely falling particles cannot reach a surface placed vertically; hence, under experimental conditions, the slope α has always to be rather less than 90° in order to achieve contact with the particles.

The results here presented are valid for particles falling freely in air. The settling rate of particles smaller than 10 μ in diameter is insignificant (order of a few cm/sec). Such particles cannot fall vertically; they are only able to come into contact with the surface as a result of inertial, diffusion, and Brownian motion.† In this case, not only vertical surfaces, but also plates inclined at $\alpha > 90°$ may become dusty.

Adhesion of Particles as a Function of the Elastic Properties of the Coating. Dust particles suspended in the air move in the wind and on striking a surface may stick to the latter; the same may occur when a moving object strikes particles suspended in the air.

The adhesion of a dust particle on striking the surface of an object normally was considered by Gillespie [208, 209]. If the kinetic energy is greater than the energy of adhesion, i.e., $E_k > E_{ad}$, the particles will glance off. For $E_{ad} > E_k$, the particles will ad-

*The adhesion is expressed in percentages showing the proportion N_a of particles adhering to the surface referred to the number N coming into contact with the surface.

†See Fuks' monograph for a description of the motion of particles in air.

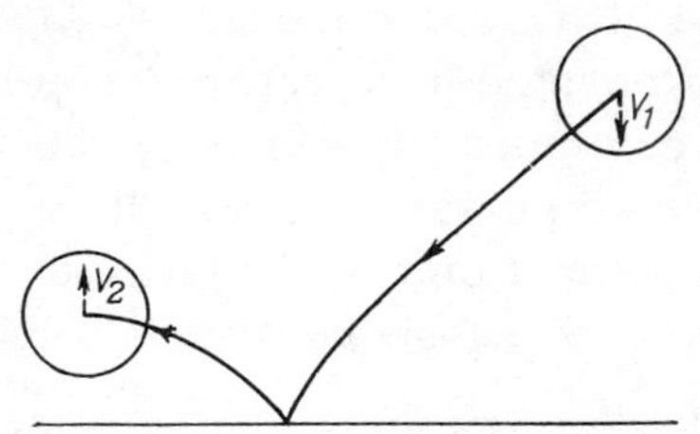

Fig. V.3. Rebound of a particle from a surface.

here to the surface. The adhesion energy may be calculated from the equation

$$E_{ad} = 6\pi\varepsilon_{ad} \cdot Hr + \pi\varepsilon_{ad} \cdot a^2 \qquad (V.7)$$

where ε_{ad} is the adhesion energy per unit contact area, H is the distance between the contiguous bodies, r is the radius of a dust particle, and a is the radius of the contact area.

Equation (V.7) allows for the molecular interaction of the bodies in contact, Since the value of H cannot be determined exactly, Gillespie's calculations must be regarded as approximate.

Gillespie's results for the adhesion of spherical quartz particles 2 μ in diameter to quartz plates are as follows:

v, cm/sec	0	1	10	100
$a \cdot 10^4$, μ	0	1.8	4.5	11.3
$E_k \cdot 10^9$, ergs	0	0.045	4.5	450
$E_{ad} \cdot 10^9$, ergs	6.8	6.8	6.8	6.8

Jordan [128] considered the conditions for the sticking and rebounding of vertically-incident particles (Fig. V.3). According to Bradley [43], the interaction between two macroscopic quartz particles of diameter d_p at a distance of H from each other equals

$$F_{ad} = -212\, d_p \qquad (V.8)$$

while the energy equals

$$F_{ad} = -212\, Hd_p. \qquad (V.9)$$

For sodium pyroborate

$$(Na_2B_4O_7)\ F_{ad} = -434\ d_p.$$

The condition for the rebounding of dust particles is expressed by the inequality

$$v^2 > \frac{2F_{ad}}{m} \cdot \frac{1-k^2}{k} \qquad (V.10)$$

where m is the mass of a particle, and k is the ratio of the vertical components of particle velocity before and after impact, i.e., v_2/v_1 (see Fig. V.3).

Putting the value of F_{ad} from (V.8) into Eq. (V.10) and taking $k = 0.8$, $H = 5 \cdot 10^{-8}$ cm, and $\rho = 2$ g/cm^3, Jordan obtained a relationship between the velocity at which particle adhesion was possible and the particle diameter (d_p):

$$v < \frac{30}{d_p} \tag{V.11}$$

On analyzing Gillespie's results, we may conclude that the kinetic energy of the impact exceeds the adhesion energy for particles 2 μ in diameter when the velocity exceeds 10 cm/sec. According to Eq. (V.11) and Jordan's results, quartz particles 2 μ in diameter may stick to a smooth quartz surface for particle velocities of under 15 cm/sec.

Thus, Jordan [128] and Gillespie [208, 209] give almost identical results.

It follows from Eq. (V.7) that the energy of adhesion of moving particles to paint coatings depends on the radius of the contact area (a), which in turn depends on the elastic properties of the substrate. In order to reduce the adhesion of dust particles to a painted surface, very hard coatings must be chosen.

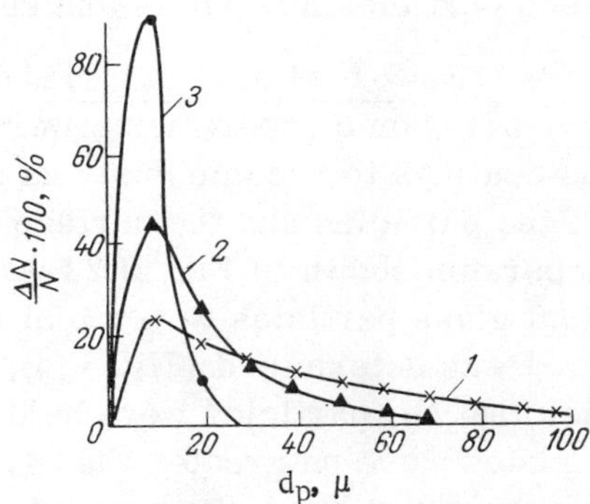

Fig. V.4. Differential particle size distribution curves. 1) In the original dust; 2,3) in dust adhering to plates moving at 1.4 and 2.8 m/sec, respectively.

There is a standard method of estimating the hardness of coating materials [210]. Tests are usually made on the basis of the pendulum method (impact tester) relative to the elastic properties of a glass surface, the hardiness of which is taken as unity. Below we present some results of hardness tests based on the pendulum method for various paints and enamels [211]:

Pentaphthalic PF-115	0.15 to 0.2
Lacquer No. 67	0.1 to 0.3
Nitro-glyptal furniture lacquer 754	0.28
Chemically resistant enamels (KhSE, KhSL)	0.3 to 0.4
Nitro-enamels for freight vehicles, Nos. 507, 508, 230, 907	0.35 to 0.45
Perchlorvinyl enamels KhV-124, KhV-125	0.35 to 0.4
Nitro-enamels for light motor vehicles	0.34 to 0.61
Butylmetacrylate-base water emulsion, ASML-23	0.5
Black lacquer Ch-2	0.60
Epoxy enamel OEP-107 A1	0.92 to 0.98

Under otherwise equal conditions, we may expect that the adhesive force between dust from an air flow and pentaphthalic lacquer will be greater than that relating to epoxy enamel. Naturally, the hardness of a coating is not constant; it varies with the temperature of the surrounding air (particularly on raising the temperature) and the period elapsed since coating. However, the data presented may be used as a preliminary estimate for the adhesive forces acting between particles and paints and enamels.

Characteristics of the Deposition of Dust on Certain Surfaces. Some experiments were made on the tendency of various coatings to become dusty as a function of the relative velocity of the particles and the surface on contact. For this purpose the apparatus shown in Fig. II.23 was employed. The adhesion of spherical glass particles to vertical plates coated with perchlorvinyl enamel was determined (Fig. V.4). The experiments led to the conclusion that the particles were held on the surface if the velocity of the plates were no greater than 4.2 m/sec. As the velocity of the surfaces diminished, the proportion of large particles among the total number of adhering particles increased, in agreement with formula (V.11). Hence, in dry weather, only small dust particles will stick to the surface of motor vehicles.

The results obtained for the deposition of dust from an air flow on plates 1 × 2 cm in size cannot be mechanically transferred to large objects such as automobiles and locomotives. Special features in the air flow around macroscopic objects as well as the existence of shaped surfaces, grooves, and projections, produce local regions of severe dust deposition. In addition to this, the horizontal and vertical parts of moving surfaces will gather dust to different extents. In order to determine this difference, horizontal and vertical plates were both tested. The following gives the dust density (number of particles per square meter for spherical glass particles 40-60μ in diameter on polyurethane-enamel coated plates* as a function of the velocity of the latter:

Velocity of the plates, m/sec.	1.4	4.2	8.4	12.6	16.8
Dust density for plates placed:					
vertically	$1.07 \cdot 10^6$	$8.8 \cdot 10^6$	$1.6 \cdot 10^7$	$1.8 \cdot 10^7$	$2.1 \cdot 10^7$
horizontally	$2.06 \cdot 10^6$	$9.1 \cdot 10^5$	$8.5 \cdot 10^5$	$7.8 \cdot 10^5$	$6.9 \cdot 10^5$

We see that with increasing velocity of the plates the amount of dust deposited on the vertical plates increased and that on the horizontal diminished.

Let the rate of free settling of the dust particles (v_{free}) be 1 m/min, i.e., dust from a rectangular parallelepiped 1 m high with a base of 1 cm^2 falls on a horizontal plane 1 cm^2 in area in 1 min. If now the same plate is set vertically and caused to move at a rate of 10 m/min (v_m), then in the same period it will gather dust from a parallelepiped with the same base but 10 m long. Since the dust concentration and surface tackiness are the same in each case, while the height (length) of the parallelepiped has increased by a factor of 10, the dustiness of the vertical surface will be 10 times as great as that of the horizontal (without taking account of the air-flow conditions).

Thus, on these assumptions, the number of particles held on vertical surfaces (N_v) may be expressed in terms of the number adhering to a horizontal surface (N_h), thus:

*In order to improve the dust catching and eliminate the rebounding of particles from the surface a layer of oil 0.5 mg/cm^2 in density was placed on the plates.

Table V.3. Adhesion Numbers of Spherical Glass Particles (diameter 40 ± 5 μ) in a Water Drop (diameter 550 μ) to a Glass Surface as Functions of the Composition of the Drop

Drop composition	Hydrophilic surface		Ordinary glass		Hydrophobic surface		Oiled with Avtol*	
	θ,°	γ_F, %	θ,°	γ_F, %	θ,°	γ_F, %	θ,°	γ_F, %
Distilled water	18	89	30	24	65	21	55	96
Conduit water.	16	73	28	33	60	10	53	95
0.1 M solution of NaCl.	15	69	17	37	51	8	53	85
0.1% solution of Sulfonol	10	22	19	8	22	2	19	27

*Oiling density 0.3 mg/cm^2.

$$N_v = K_{ch} N_h \frac{v_m}{v_{free}} \qquad (V.12)$$

where v_m is the velocity at which the vertical plates are moving, v_{free} is the rate of free settling of the particles, and K_{ch} is a coefficient allowing for the air-flow characteristics of the surface. If $K_{ch} \sim 1$, then for $v_m \geq v_{free}$, we have $N_v > N_h$ and for $v_m < v_{free}$ we have $N_v < N_h$.

The dust in the atmosphere may stick to the surface of aircraft. The number of adhering particles may easily be determined if the particles are radioactive.

The radioactive dust formed as a result of nuclear explosions charges the atmosphere and stratosphere and may be carried six to eight thousand kilometers in 24 h. In the troposphere radioactive aerosols last for 2-3 weeks and in the stratosphere several years [213, 214]. The diameter of the dust particles is no greater than 5 μ [215]. According to BOAC data [216], when a Comet 4 passenger airliner passed through a radioactive cloud formed as a result of the global propagation of radioactive dust particles, the maximum surface activity was no greater than 10^{-3} Ci/cm^2. If we take the activity of 1 g of dust as equal to 0.1 Ci [217], then 1 cm^2 will hold 10^{-5} g of dust, or a few hundred particles 1 μ in diameter. The impact of cosmic dust particles fractions of a micron in size

[218] at high speeds on the surface of spaceships or rockets may lead to the formation of craters and the retention of these particles on the surface of such vehicles [219].

Adhesion of Particles Together with Drops of Water to a Solid Surface. In periods of rain, drops of water may capture dust particles suspended in the air. Such processes are employed in water curtains [220, 221] for dust suppression. The particles here fall on the surface at the same time as the drops of water. There is also another possibility in which drops of pure water fall on a dust-laden surface, capture adhering particles, and remain fixed to the surface together with these.

In experiments carried out by the centrifuge method [67], the adhesion of particles attached to a surface with drops of water was studied. The effect of the state and properties of the original surface, the dimensions of the particles and water drops, the direction of the detaching force, and the presence of surface-active substances in the water on the adhesion characteristics was determined.

We see from the experimental results (Table V.3) relating to the removal of spherical glass particles in a drop of water by a force directed normally to the dust-laden surface (centrifuge turning at 2200 rpm) that as the wetting angle increases, i.e., with increasing hydrophobic properties of the surface, the adhesion number falls (for a drop of specific composition). This may be explained by the fact that on a hydrophobic surface the particles lie entirely in the water drop; on a hydrophilic surface the drop may spread and the particles may remain on the surface outside the liquid.

The detachment of particles inside a drop is affected by the composition of the aqueous medium, as in all cases of liquid adhesion (see Chapter IV). Thus, if instead of distilled water we use solutions of certain salts (conduit water, 0.1 M NaCl solution), adhesion diminishes. Particle adhesion is still more reduced by solutions of surface-active substances (0.1% Sulfonol solution). The reduction in adhesion number resulting from the effects of SAS solutions as compared with other media is greater for hydrophobic than hydrophilic surfaces.

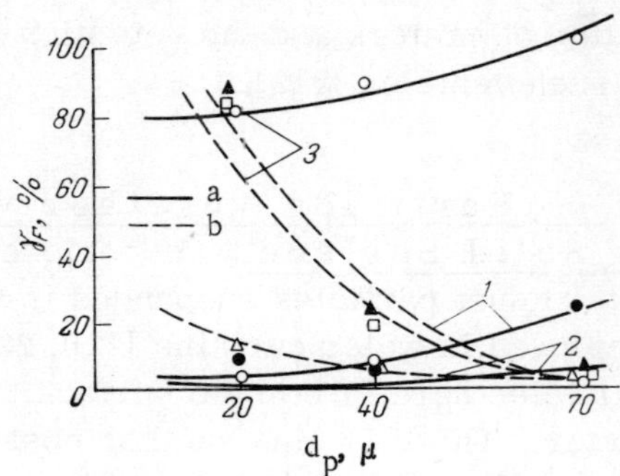

Fig. V.5. Adhesion number of spherical glass particles in a drop of water 550 μ in diameter as a function of the particle diameter for a force directed normally (curves a) and tangentially (curves b) to a glass surface. 1) Ordinary; 2) hydrophobic; 3) hydrophilic.

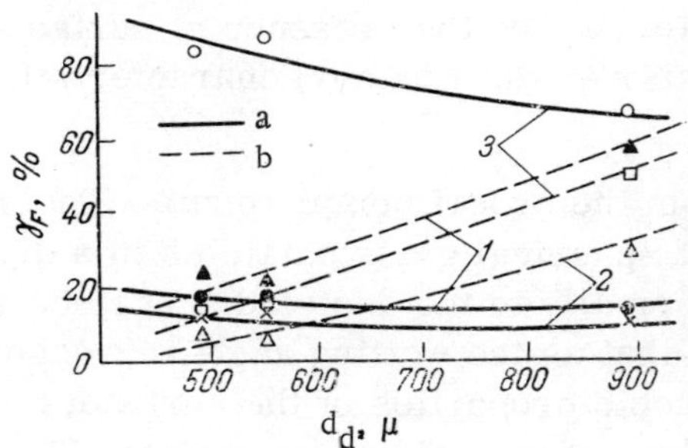

Fig. V.6. Adhesion number of spherical glass particles 40 ± 5 μ in diameter in a drop of water as a function of the size of the drop for a force directed normally (curves a) and tangentially (curves b) to a glass surface. 1) Ordinary; 2) hydrophobic; 3) hydrophilic.

The conclusions drawn on the basis of laboratory experiments are borne out in practice. Hydrophobic coatings do in fact become less contaminated in rain [222], and it is easier to clean them [223].

It is always more difficult to remove particles in water drops from oiled surfaces, but even here the desired result may be achieved if the water contains salt, or more particularly surface-active substances.

As shown by the experiments mentioned earlier [67], the adhesion numbers of particles in water drops depend both on the particle and drop size, and also the direction of the detaching force (Figs. V.5 and V.6).

If the detaching force is directed normally to the surface (curves a), the removal of the particles depends chiefly on the properties of the surfaces; it is more difficult from a hydrophilic surface. The particle size affects the results less than the properties of the surface. The adhesion number in fact varies very little with particle (Fig. V.5) and drop (Fig. V.6) size for any particular surface.

The centrifugal force (see §6) acts both on the particles (F_c') and on the drop of liquid (F_c''). The total detaching force acting on the particle is $F_{det} = F_c' + F_c''$. If the centrifugal force is directed normally to the surface, then the value of F_c'' will be constant for the same drop size, while F_c' and F_{det} will be directly proportional to the particle size. At the same time the adhesive forces will be directly proportional to particle size (see §24). Hence, for the same F_c'' the adhesion number for relatively small particles will be smaller. This tendency is confirmed experimentally (Fig. V.5, curves a).

For the same particle dimensions F_c' is constant, while F_c'' and F_{det} are directly proportional to the drop size, and the adhesion number is inversely proportional to the drop size. This tendency is also confirmed experimentally (Fig. V.6, curves a).

If the detaching force is directed tangentially, then the particle and drop sizes have a greater effect on the adhesion number. This is because a tangentially acting centrifugal force causes the drop to spread. Here the force acts only on the particles, and this makes it more difficult to remove the smaller particles for the same size of drop (Fig. V.5, curves b), since for the same number of revolutions the detaching force is proportional to the volume of the particles, i.e., to r^3 [formula (II.5)], while the adhesive force is proportional to their radius [224] (see §24), i.e., to r. However, under the influence of a tangential force, the large drops spread more easily; hence particles of the same size are less easily detached as drop size increases (Fig. V.6, curves b).

The more hydrophilic the surface, the more favorable are the conditions for spreading of the drop and the more difficult is it to remove the particles. On an ordinary surface as distinct from a hydrophilic one, there is no drop spreading before centrifuging (wetting angle equals 30°). Drop spreading starts for a centrifuge speed of 2200 rpm, i.e., before the instant of particle detachment. The centrifugal force only acts on the adhering particles, which means that the conditions for the removal of particles from ordinary and hydrophilic surfaces are almost the same. This is confirmed by experimental results (curves b1 and 3 in Figs. V.5 and V.6 are analogous).

Adhesion of Particles After the Evaporation of the Water Drop. Particles situated in a drop of liquid remain on the surface after the evaporation of the liquid; however, the adhesion of the particles will now differ from the adhesion of the same type of dust in air.

Figure V.7 shows the variation in the adhesion of spherical glass particles 50 ± 5 μ in diameter to a glass surface (ordinary and hydrophobic) as a function of the conditions of particle settling: in a drop of liquid (water, acetone, or alcohol) with subsequent evaporation, or by free settling in air. In the first case (curves 1-3) the adhesion is greater than in the second (curves 4).

Similar data were obtained for the adhesion of spherical glass particles 70 ± 2 μ in diameter to painted surfaces. For example, with a detaching force of $2.2 \cdot 10^{-1}$ dyn, the adhesion number for particles after free settling on ordinary and hydrophobic surfaces respectively equalled 26 and 5%, whereas, for settling in a drop of conduit water with subsequent evaporation, the values were 100 and 95%, or in drops of alcohol, acetone, or carbon tetrachloride, 95-98 and 75-86%.

The original surface has an effect on the adhesion of the particles: on hydrophobic surfaces the adhesive forces are in all cases smaller than on ordinary surfaces. Evidently on evaporation of the liquid a "crust" of salts and other contaminating material forms in the particle—surface contact zone and this causes the greater adhesion.

In order to verify this assumption, the adhesion of dust was studied after evaporation of drops of NaCl solution of various con-

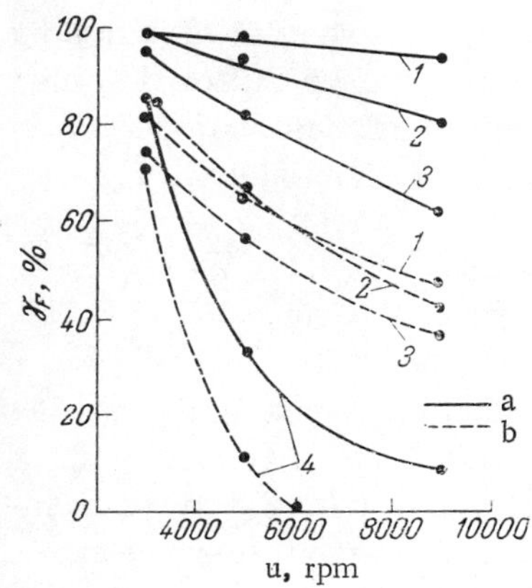

Fig. V.7. Adhesion number of spherical glass particles 50 ± 5 μ in diameter on interaction with ordinary (a) and hydrophobic (b) glass as a function of the centrifuge speed for various conditions of particle settling. 1,2,3) In drops of water, alcohol, and acetone, respectively; 4) free air settling.

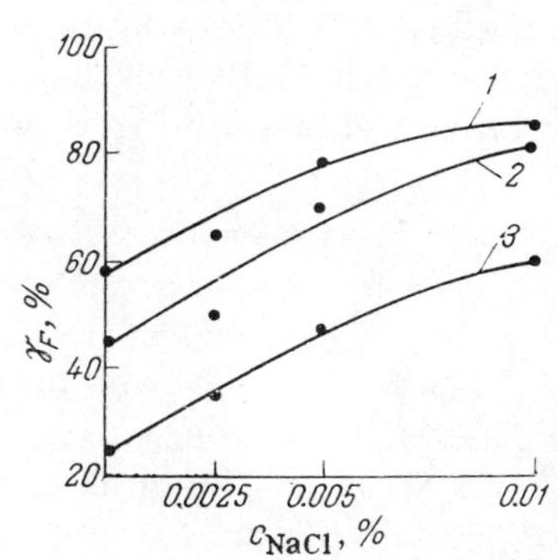

Fig. V.8. Adhesion number of spherical glass particles adhering to various surfaces after the evaporation of the drop, as a function of NaCl concentration in the liquid. 1) Surface coated with perchlorvinyl enamel; 2) ordinary glass; 3) painted and hydrophobized glass.

centrations (Fig. V.8). The adhesion rose considerably with increasing NaCl concentration, confirming the assumptions made earlier.

Similar processes may take place after the evaporation of a liquid film lying between the particle and the surface [225].

Adhesion of Particles to an Oily Surface. Usually, atmospheric dust contains not only solid particles but also oil contaminations [166] which settle on surfaces, making them oily. In addition to this the surface of objects such as motor vehicles may become oily in the course of use. Thus it is found that 17% of the particles adhering to the motor vehicles of Moscow contain such oil layers. Similar results are obtained on analyzing the contaminations [226] adhering to the outer surfaces of railroad rolling stock. For example, on locomotives, 23% of the surfaces in contact are oily, and in trucks, 19%.

Oiliness makes the surfaces tacky and increases the adhesion of particles (see §14).

Oiliness of coatings changes the adhesion of particles in liquids as well as in air. Let us consider some results obtained by the method of immersing in a liquid medium (0.1% Sulfonol solution) for the detachment of particles from surfaces coated with perchlorvinyl enamel for various degrees of oiliness:

Oil density, mg/cm^2 . . .	0.05	0.25	0.5	1	2
γ_F, %	32	48	41	49	10

Thus, as in the case of air, the adhesive force of the particles first increases with increasing oiliness as a result of the fact that the particles sink in the oil film and are held more firmly on the sample surface, and then falls as the particles are removed together with the film of oil. The maximum adhesion occurs for an oiliness factor of 0.25-1 mg/cm^2.

The effect of water temperature on the adhesion of particles to an oily surface has also been studied. Below we present some data obtained by the immersion method at various temperatures for the detachment of spherical glass particles less than 60 μ in diameter from a surface coated with PKhV enamel with an oiliness of 0.5 mg/m^2:

t, °C	2	10	20	40	60	80
γ_F, %:						
in 0.1% Sulfonol	71	56	47	38	32	29
in 0.25% DB	49	38	30	22	20	18
in 0.25% OP-10	76	74	70	60	56	52
γ_F for a glass surface in 0.25% OP-10, % . . .	87	85	79	74	68	62

The reduction in the adhesion of particles to oily surfaces with increasing temperature is due to the fact that the viscosity of the oil falls, and hence the adhesive forces between the oil film and the particles do likewise.

§ 27. Dependence of the Adhesion of Particles on the Physicochemical Properties of Paint and Varnish Coatings

Adhesion of Particles and Wetting Capacity of the Coating. Of the fair number of physicochemical properties influencing the adhesive forces between particles and coatings, let us consider the hydrophoby and hydrophily of the coating, and also its continuity (affected by the destruction of the coating material) and tackiness (resulting from film-forming substances and solvents).

It was pointed out earlier (see §§4, 18) that one of the factors determining adhesion was the molecular force. The molecular interaction of contiguous solids is due to the properties of the surface functional groups of molecules. For paints and lacquers [205] the molecular interaction between dust particles and the surface may be associated with the polarity of the polymers. The molecular component of the adhesive forces depends on the polarity of both particles and coatings. The following illustrates the possible influence of the functional groups of polymers on the adhesion of polar and nonpolar particles*:

Molecular groups†	$-NH_2$	$-OH$	$-Cl$	$-CH_3$	$-CH=CH_2$	$-C_6H_5$
Change‡ in adhesion:						
of polar particles. . . .	+	+ +	−	− −	+ −	−
of nonpolar particles .	−	− −	+	+	+ −	+

These data relating to the effect of the functional groups of polymers on the adhesion of particles agree with the empirical De Bruyne rule (see §4): the adhesive forces are at a maximum when the polarity of the contiguous surfaces is the same, i.e., hydrophobic to hydrophobic and hydrophilic to hydrophilic.

As a rule, the functional groups of molecules on the surface are accompanied by other groups, which affect the surface proper-

*A measure of polarity is the wetting angle of the surface with respect to water, i.e., the hydrophobic or hydrophilic quality of the coating.

†Here as earlier (see §10) we must allow for the effect of the numbers of these groups, not merely their presence, on adhesion.

‡Arbitrary notation, see page 158.

Table V.4. Wetting Angles of Certain Coatings for Water and Adhesion Numbers of Spherical Glass Particles 40-60 μ in Diameter to These Surfaces (F_{det} = 1.4 · 10^{-1} dyn)

Coating	θ,°	γ_F, %
Hydrophobic enamel KhS-127-2	100	4
Perchlorvinyl resin PSKh-S	76	57
Epoxy resin E-49 hardened with DGU ..	70	83
Polyurethane lacquer No. 930.......	64	85

ties. It is therefore better to classify the adhesive properties of the surface by reference to the wetting angle rather than the properties of the functional groups. Experimental results [11] (Table V.4) show that the adhesion of hydrophilic glass particles increases as the surface becomes less hydrophobic, i.e., in this case the De Bruyne rule is adequately satisfied.

Jordan [128] also observed a fall in the adhesive forces of glass dust as the surface became more hydrophobic. By hydrophobizing the surface, the adhesion of particles, films, and solid surfaces may be reduced, as well as the sticking of ice [129, 227]. In addition to this, adhesion relative to the same coating will vary in accordance with the properties of the particles under consideration.

Thouzeau and Taylor [24] measured the adhesion of inert lime dust of two types (nonwetting and ordinary) by the method of inclining a dust-laden hydrophilic surface and found that the adhesive force of the ordinary dust was 0.53 g/cm^2 and that of the nonwetting (hydrophobic) material 0.33 g/cm^2.

Hence, by choosing a coating with hydrophobic or hydrophilic properties, we may reduce or increase the adhesive force of dust. It should be remembered that changing the adhesive force in this way is only possible if the adhesion is determined by molecular forces, while the electrical component of the adhesive force is negligible and humidity of the air exerts no influence on the adhesion. This is valid for microscopic particles.

Unfortunately, there are no systematic data regarding the hydrophobic and hydrophilic properties of coating materials in present use. Myagkov [228] determined the wetting angle of certain coatings with respect to a 3% aqueous solution of KCl:

	θ,°		θ,°
Drying oil + 80% Pb_3O_4 . .	55	Unpigmented ethynol lac . . .	38
Drying oil + 50% Fe_2O_3 . .	33	Coal lac + 40% ZnO.	50
Unpigmented drying oil . .	23	Coal lac without pigments. . .	48
Ethynol lac + 30% Fe_2O_3. .	42		

A considerable proportion of paints and coatings, such as perchlorvinyl and polyurethane enamels, nitro-enamels, etc., have a wetting angle broadly within the limits indicated above.

Steudel [229] verified the dust-holding capacity of various coatings by reference to the increment in the weight of contaminants on plates fixed to the sides of railroad trucks passing through industrial, agricultural, steppe, and alpine zones in Diesel, electric, and steam trains. The coatings thus studied were placed in the following order with respect to degree of contamination, and hence falling adhesion: enamels of the epoxy, polyurethane, and alkyd types, nitro-alkyd enamels with inorganic and organic pigments, transparent protective lacquers for stainless steel [229]. The adhesion of contaminations to these coatings does not depend on preliminary treatment of the coating with a detergent solution. The lower the adhesion of water to the coating, the more weakly does the latter hold contaminant particles.

Thus, hydrophobization of paint coatings reduces the adhesion of dust particles. In view of this we must consider methods of hydrophobizing coatings.

Hydrophobization of Coatings. At the present time wide use is being made of hydrophobic silico-organic resins and lacquers based on these [230, 231]. The hydrophobization of paints and lacquers may be effected by introducing silico-organic compounds (in particular, the liquids GKZh-10, -11, or -94) into the original paint or enamel, or by treating painted surfaces with these liquids [232].

The process of hydrophobization is based on the interaction of silanes with hydroxyl groups, always existing in macroscopic molecules, and on the polycondensation of silanes with the formation of a polysiloxane film. Such additives ensure [233] resistance of the coating to mechanical damage.

In order to reduce tackiness and increase the period of service of the coating, silicone wax polishing compositions are used (0.5-2% wax, 1.5-3% dimethylsilicone, solvent turpentine) [231].

Hydrophobization may be effected by the addition of carbon tetrafluoride [227], the introduction of mica [234], modification with film-forming agents [235, 236] and pigments [237], the use of surface-active substances for dispersing the pigments [238, 239], and by the uniform distribution of specially introduced silico-organic resins [240]. Treatment of a glass surface with a 0.05% aqueous solution of ethyl alcohol derivatives also imparts water-repelling qualities [241].

In introducing various additives to paint and lacquer coatings, one must take account of their compatibility with other components of the coating materials [242]. Otherwise, the service qualities of the coatings may be impaired.

Hydrophobization of a paint coating in order to impart heat and moisture resistance is in practice carried out for structures consisting of large blocks or panels, such as concrete and reinforced concrete buildings [232, 243, 244]. Thus, in silicate paints used for treating concrete panels, sodium ethyl and methyl siliconates are introduced as hydrophobizing additives; in cement paints used for painting plaster (stucco) foundations, sodium stearate is used. In polyvinyl acetate paints used for internal finishing work no additives are introduced; the coatings are covered with lacquer, which gives them water-repellant properties. It is noteworthy that hydrophobized coatings are contaminated less, and adhering particles are removed from them more easily, than in the case of ordinary coatings. Considering that a considerable proportion of the dust has hydrophilic properties, we may regard these results as satisfactory confirmation of the De Bruyne rule.

<u>Changes in the Dust-Holding Capacity of a Coating Resulting from the Properties of the Film-Forming Materials.</u> The dust-holding capacity of paint coatings depends on the time elapsed since the moment of painting; this is evidently a result of processes taking place in the drying and aging of the coatings [245]. The following results show the variation in the adhesion of spherical glass particles 40-60 μ in diameter with the time elapsed after coating the surface with perchlorvinyl enamel [11]:

F_{det}, dyn	$2.8 \cdot 10^{-4}$	$5.7 \cdot 10^{-4}$	$9.0 \cdot 10^{-4}$	$2.6 \cdot 10^{-3}$	$2.6 \cdot 10^{-2}$	$1.1 \cdot 10^{-1}$
γ_F, %:						
after 8-10 days	76.7	69.7	68.0	66.7	60.5	60.0
25-30 days	68.9	62.5	60.7	60.0	52.3	47.5
10-12 months	–	–	–	–	75.3	68.0

Eight to ten days after coating, the adhesion of the glass particles to the perchlorvinyl coatings was slightly greater than after 25-30 days. After 10-12 months the adhesion increased again as a result of the breakup of the surface film.

The dust-holding capacity of oil paint also gradually falls with time; however, the adhesion increases again after only 2-2.5 months.

The dust-holding capacity of coatings depends greatly on the properties of the solvents and film-forming substances contained in the paints or enamels. Thus it is noteworthy that oil coatings prepared on a natural drying oil base hold contaminants much more strongly than other coatings, since the drying oil imparts a property of tackiness [203]. The adhesion of glass particles to surfaces coated with oil paints is greater than their adhesion to perchlorvinyl coatings [11], and 25-30 days after painting the values are as follows:

F_{det}, dyn	$2.8 \cdot 10^{-4}$	$2.6 \cdot 10^{-3}$	$2.6 \cdot 10^{-2}$	$1.1 \cdot 10^{-1}$
γ_F, %:				
for oil paint	100	98.3	–	92.0
for perchlorvinyl enamel, clean	68.9	60.0	52.3	47.5
covered with Avtol	100	100	98.7	98.7

These results [11] were obtained after careful treatment of the test surfaces with distilled water. For the surface covered with Avtol (calculated surface density 1.4 mg/cm^2), the dust-holding capacity of the coating may be affected not only by the properties of the contaminants, but also by the adhesion of the Avtol to the coating (if the adhesion of the particles to the Avtol layer is greater than that of the Avtol to the coating).

The tackiness of paint coatings may be reduced by using quick-drying film-forming materials [246] or painting in a fluidized bed [247], thus eliminating the need for solvents.

Adhesion in the Case of Painting in a Fluidized Bed. In the aerosol method of painting, the initial process is the adhesion of thermoplastic-polymer and low-molecular-resin powder particles suspended in the air to the surface in question. In order to make the particles adhering to the treated surface form a continuous layer, painting is carried out either with heating of the particles in the suspended state or else with heating of the original surface.

Powders used for fluidized-bed painting [248, 249] include low- and high-pressure polyethylene, polyvinyl butyral (Butvar), polystyrene emulsion, polyamide resin type AK-7, polyoxymethylene, polyvinyl chloride, a copolymer of styrene with methylmetacrylate, epoxy resin type E-41, phenolformaldehyde resin No. 108. In addition to the main components, pigments and dyes imparting heat resistance and other qualities to the coating are introduced. In particular, graphite is introduced in order to increase the adhesion of polyethylene powders to the surface under treatment, while, in order to reduce the adhesion of dust and fluff to the coating, compounds of the

$$RCON\begin{cases}(C_2H_4O)_nH\\(C_2H_4O)_mH\end{cases}$$

type [250] are introduced into the original powder, where n + m = 2-20, and R is an alkyl radical with 4 to 22 carbon atoms.

In order to increase the adhesion [248] of polyethylene and polyvinyl butyral powders to steel, polystyrene is added to the first of these powders and chromium oxide to the second. The adhesion of powders prepared from polyamide resins to surfaces of aluminum and aluminum alloys or copper and copper alloys (brass, bronze) may be increased by introducing ultramarine and aluminum powders into the original material.

One finds that substances of similar chemical composition to polyvinyl chloride adhere most strongly to films of this material [251]. In the present case the De Bruyne rule also finds practical confirmation: maximum adhesion to surfaces is exhibited by films with analogous properties.

In order to create a uniform coating, it is very important that the particle size of the powder should be between 50 and 400 μ,

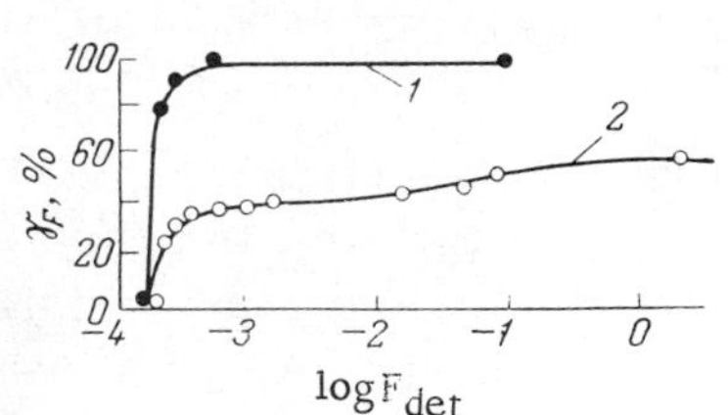

Fig. V.9. Adhesion of dust in ionized (2) and un-ionized air (1).

the size of the particles in the fractions differing by no more than 100 μ. Otherwise, the particles will be unequally heated, and the adhesive forces will vary, which in the long run will have an adverse effect on the quality of the coating.

In order to fix particles to plastic objects, one uses a solvent imparting the property of tackiness to the surface [252].

Coulomb forces (see §12) may be used in order to achieve firm adhesion of the powder forming a coating. For this purpose, for example, dry thermoplastic (phenol resin not completely hardened) and silicone powder is sprayed through a mesh raised to a potential of 90 kV. The negatively charged particles settle on the object to be painted, forming a dense, uniform layer which remains stable under vibration. Various pigments may be added to the powder, provided that a bulk resistance of 10^{-5} Ω · cm is maintained. In order to obtain a continuous coating, one either fuses the adhering layer or heats the object first; alternatively these operations may be combined [253].

§28. Change in the Dust-Holding Capacity of Paint and Varnish Coatings Under the Influence of Electric Forces

By varying the electrical component of the adhesive forces, we may considerably reduce the adhesion of dust to a surface or even prevent the latter from becoming dusty at all.

We considered the possibility of changing the electrical component of the forces of particle adhesion by changing the properties of the surface earlier (§11). The adhesive forces may also be reduced by an amount equal to their electrical component (or, at any rate, by an amount proportional to this) by ionization of the air surrounding the dust-laden surface. Thus the adhesion of spherical glass particles 40-60 μ in diameter is smaller after ionization of the air (using a TsNIIShellak ionizer) than in un-ionized air (Fig. V.9) [254]. Ionization of the air surrounding the dust-laden surface leads to partial neutralization of the charges in the double layer, and hence reduces adhesion.

The electrical forces may also act on dust particles before these come into contact with the surface (see § 12), i.e., in principle, there may be such a thing as a dust-repellant coating, chiefly in relation to the dust particles floating in the air. Alkyd-styrene construction paints with a low binder content are an example of this [255]. A similar result may be achieved by treating the object with a 1% aqueous solution of a cation-active substance, such as Arkvad 18 which, according to the American firm of Armour [256], prevents dust accumulating on objects for several months. In this case, the dust-holding power is reduced by the removal of electric charges from plastic articles in the same way as this is done by the use of antistatic preparations [254].

There are also some dust-collecting paints and enamels which create their own electric field.

Like dust-repellant properties, the dust-collecting properties of coatings are determined by the charge on the particles and the surface (coatings). Let us consider the mechanism underlying the repulsion (dust-repellant coatings) and attraction (dust-collecting coatings) of dust particles in the air on approaching a painted surface, and also the adhesion of the particles after contact with the surface. Work of this kind was carried out by Frederick [257] when studying the electrical properties of dust in order to select material for filter cloth. In order to classify the possible particle–substrate relationships arising from electrical forces, Frederick divided dust into *three classes*. We shall modify Frederick's classification slightly, allowing for dust repulsion and dust deposition.

Class I includes electrically active dusts, on deposition of which the adhesive forces exceed the autohesive, so that there is no aggregation of the particles. Such dusts include roasted zinc silicate and oxide, ready-made cement, converter soot, zinc oxide, corn starch, lump clay, and diatomaceous earth after thermal alkali treatment.

Class II includes electrically active dusts which discharge on adhesion and may form strong aggregates. These include fumes (smoke) of the zinc oxide type, raw cement, discharges from nickel-melting furnaces, magnesites, chromium ore, wheat starch, powdered sugar, and molybdenum oxide.

The separation of dust into classes is a little arbitrary, since the contact potential difference and rate of charge leakage depend not only on the properties of the dust, but also on those of the paint coating.

Class III includes electrically inactive dusts; these include soot, silicon dioxide, kaolin clay, and granulated oats. Electrically inactive dust in the air is not subject to electrical effects on the part of the substrate. We shall therefore not consider it.

Dust particles are attracted to the substrate if the charges on the dust and substrate are opposite and the forces of electrical interaction exceed the characteristic weight of the particles, taken together with the inertial and diffusion forces [6, 7], i.e., if $F_e > (P + F_{in} + F_d)$.

Below we present the conditions for the adhesion and nonadhesion of electrically active dusts of Classes I and II:

	Class I	Class II
Potential difference on contact:		
for adhesion	max.	max.
for nonadhesion.	min.	max.
Rate of charge leakage in contact zone:		
for adhesion	min.	min.
for nonadhesion.	max.	max.

Thus, in order to ensure the maximum adhesion of Class I dust (i.e., dust not susceptible to aggregation) and prolonged action of the charged substrate, it is essential that the contact potential difference between the surface of the particles and substrate should be a maximum and the rate of charge leakage on contact a minimum.

The adhesion process is rather different for dust capable of aggregation. On contact with the surface such particles are discharged. The greater the potential difference between the particle and substrate and the faster the charge leakage, the more intensively does discharge and the aggregation of particles take place. All this tends to reduce the adhesion of dust particles, and sometimes even prevents dust from settling (see §11). However, it must be remembered that under such conditions an additional force of attraction may be created as a result of image effects (see § 12).

Paint and varnish coatings possess either dust-repellant or dust-holding properties in accordance with their ability to conduct an electric current, i.e., in accordance with their electrical resistance.

Conducting coatings may receive charges in various ways: from an external source, as a result of deformation arising from the drying of paint coatings [258], as a result of the tribo-effect [146], as a result of air flowing around the coatings [232], etc. Thus, the deformation of smoked sheet vulcanized with sulfur produces electrification of the samples as a result of a change in the orientation of the polymer molecules [259], and this in turn is determined by the composition. The charges resulting from the deformation of such rubber products are illustrated below:

Number of parts by weight of sulfur per 100 parts of smoked sheet.	5	10	20	50
Maximum surface charge density ($\times 10^9$), C/cm^2	1.1	1.3	2.3	0.7

By using these data we may calculate the forces of attraction or repulsion. Thus, if we suppose that for particles 10 μ in diameter an image force (see § 12) acts on the projections of the particles, then, allowing for the resultant charges, the force of repulsion (or attraction, depending on the sign of the charge on the particles) required to overcome the weight of the particles should be not less than $1.3 \cdot 10^{-6}$ dyn and should act at a distance of $2.3 \cdot 10^{-3}$ cm from the contact zone.

Charge leakage, and hence the discharge of the dust particles, are determined by the electrical resistance of the film of coating material (§ 12), and also the electrical conductivity of the medium in the contact zone.

Vasserman et al. [260] studied the dependence of the electrical resistance of film-forming materials with roughly the same resistivity on the amount of moisture in the film after immersing the sample in a 0.5 N solution of NaCl and holding in the solution for 24 h. The results of these measurements were as follows:

Film.	Perchlorvinyl	Nitrocellulose	Butylmetacrylate	Ethylcellulose
Resistance · 10^{-7} in Ω·cm				
on immersion	32.0	4.4	1.6	1.8
after 24 h. . .	5.5	1.1	0.7	0.5

It follows from these results that the electrical resistance of a perchlorvinyl film, which has a low vapor permeability and swells badly in water, greatly exceeds that of the other films after immersion in the solution. This is because the dust holding of the swollen perchlorvinyl film is greater than that of the other materials.

§ 29. Reducing the Adhesion of Particles by Insulating the Original Surface

In order to reduce the adhesion of dust particles and powders to a painted surface we may insulate it by depositing some kind of solid or liquid insulating film, or by using multilayer coatings. Then the adhesion of microscopic particles will be determined by the nature of the top film rather than the properties of the paint.

On using a thin liquid film, the adhesion is in effect no longer of the gas type but of the liquid, and this tends to reduce the adhesive forces (see Chapter IV).

A thin surface layer may be created by imposing an electric field. This moves the positive charges on the particles to the cathode (substrate) and simultaneously attracts water; the resultant water film on the cathode plays the part of a wetting interlayer and reduces adhesion and friction.

This method may be used in practice. For example, an electrode may be mounted in the body of a dump truck. As a result of electro-osmosis [261], the water starts moving in the finely porous hydrophilic medium to the bottom (cathode) and forms a film on this, which reduces the sticking of soil to the lining of the dump-truck body. This method can of course only be used in the transport of fine-grained soils (clay and loam). The electrical system is [261] rather cumbersome and its installation in the truck increases the weight and complexity of the latter. It is therefore best to use this method on stationary or semistationary equipment.

Electrokinetic processes are used for preventing the overgrowth of the bottoms of vessels (ships) and at the same time for increasing the resistance of these to sea water [262].

Solid films have found application for treating the inner surfaces of oil pipes in order to reduce the adhesion of the paraffin components of the oil to these. The film-forming substances include amides of phosphoric acid ester [263], which are dissolved in gasoline, benzene, or Diesel oil and deposited on the surface. After evaporation of the solvent a thin film remains.

The surfaces of trucks, machines, and various types of industrial apparatus and equipment may be protected by an emulsion of wax in water, particularly the protective wax Expotect [264]. After drying, the emulsion coats the surface, providing reliable insulation. The wax layer is not wiped off by hand. Moisture, dust, soot, and other contaminants are kept away from the original surface. The layer of wax is easily removed together with the contaminants adhering to it.

Multilayer coatings are used in the atomic industry as a protection for easily accessible, simply shaped surfaces (walls, floors, partitions) from radioactive substances [265]. The outer layer of the coatings subjected to radioactive contamination is easily removed or washed off. Special plastics are used in the atomic industry as insulating layers [266].

The change in the adhesion of particles after depositing aqueous solutions of certain surface-active substances (SAS) of various types (nonionogenic, cation- and anion-active) on a painted surface was studied in [267].

The following represents the results of adhesion measurements (γ_F) made on spherical glass particles 80-100 μ in diameter adhering to a surface painted with polyurethane enamel U-21* and treated with various SAS†

*Similar results are obtained by using perchlorvinyl coatings; coatings based on oil paints give slightly different results.

†The surface-active substances are deposited on the surface in the form of a 1% aqueous solution at the calculated rate of about 0.075 liter/m^2; after drying, a thin film of the SAS is formed. Tests were made with a detaching force of $8.4 \cdot 10^{-2}$ dyn.

	γ_F, %		γ_F, %
Arkvad S*	64	Gardinol (a "Novost'" preparation)	8
Arkvad T	68	Étomid NT/25*	46
Arkvad 18	78	OP-10	58
Arkvad 12	82	Armid 0*	73
Arkvad 2	84	Sulfonate	99
Arkvad 2NT	89	Standard surface†	79

We see from these results that the majority of the SAS studied change the dust-holding capacity of the original coatings very little. The most acceptable SAS for appreciably reducing the adhesion of particles is Gardinol (a "Novost'" preparation) and to some extent Étomid NT/25. The former is more accessible and is widely manufactured in the Soviet Union. For this reason only Gardinol has been used for studying the dependence of dust-holding power on the amount of SAS in the original solution. Two methods are used for depositing a film of Gardinol on the surface. In the first method the solution is deposited and dried, and a white film is left on the surface. In the second method the solution is deposited and left for 30 min, after which the surface is washed with distilled water. In practice this is analogous to the rain washing of the deposited film.

Figure V.10 shows the variation in the adhesion of spherical glass particles 80–100 μ in diameter to a surface painted with polyurethane enamel as a function of the concentration of the aqueous solution of Gardinol taken for treating the surface. The detachment of the particles was carried out by the vibration method after drying the film. For between 0.1 and 1% of Gardinol in aqueous solution, the dust holding of the coatings was much lower than that of the original surface. For under 0.01% Gardinol in solution the adhesive force increased and differed little from that corresponding to particles on the clean surface (see Fig. V.10).

For a detaching force similar to the weight of the particles, almost all the particles were removed from a surface treated with

*The SAS Arkvad, Armid, and Étomid are made in the Moscow Branch of the All-Union Scientific-Research Institute of Oils and Fats (VNIIZh); the other commercial products are made by Soviet industry.

† The "standard" is a painted surface washed with distilled water.

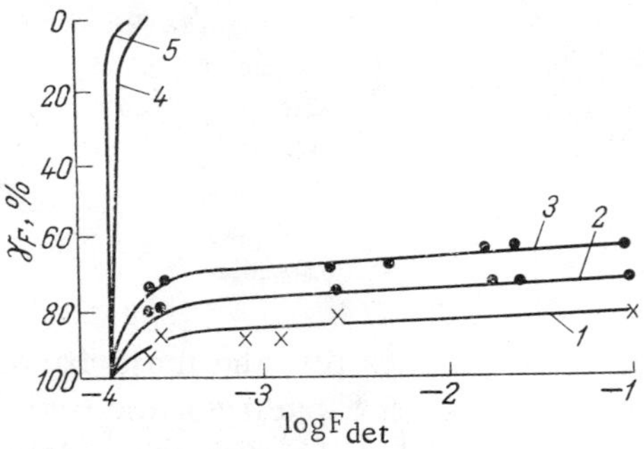

Fig. V.10. Adhesion of spherical glass particles to a surface painted with polyurethane enamel and treated with Gardinol solutions of various concentrations. 1) Untreated surface; 2) treated with a 0.001% solution; 3) 0.01%; 4) 0.1%; 5) 1%.

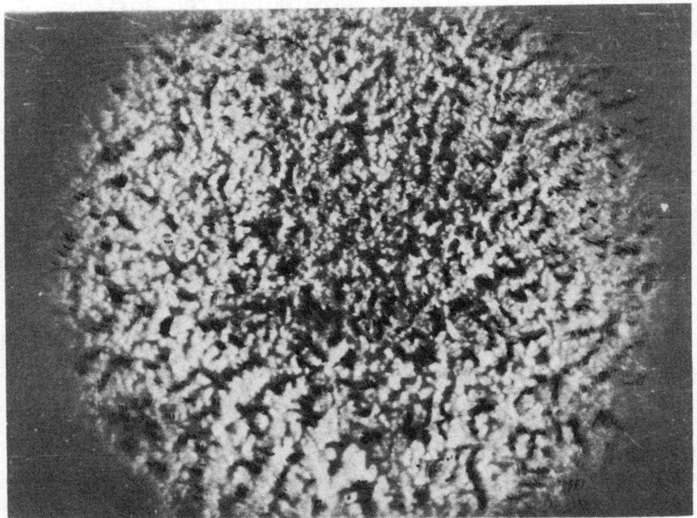

Fig. V.11. Appearance of a Gradinol film after drying a 1% aqueous solution (photographed with a microscope attachment).

a Gardinol solution, whereas, for an ordinary surface only 15% of the particles were removed for a detaching force of 10^{-1} dyn, i.e., almost three orders higher than the weight of the particles.

The reason for the fall in adhesion is the fact that the Gardinol film formed (Fig. V.11) has a crystal structure, so that the surface acquires a relief; this tends to reduce the true contact area between particles and substrate (see §14).

On using other SAS no crystalline film is formed.

In addition to this, a Gardinol film sharply reduces the charge on the particles in the contact zone, i.e., it screens the original surface and hence reduces the electrical component of the adhesive forces. By way of example, the following results represent the charge on spherical glass particles measured by the method developed earlier [11] for the detachment of particles from ordinary painted surfaces and those having a Gardinol film:

d_p, μ	40-60	20-30	10-20
$q \cdot 10^{15}$ (referred to one particle, C):			
on a painted surface	6.7	2.7	4.0
on a surface with a Gardinol film	0.3	0	0

For a relative air humidity of over 90% the effect of the Gardinol film weakens, but even under these conditions the adhesion remains less than that of a sample without any Gardinol film. The advantages of a Gardinol film as compared with an ordinary coating become manifest when the dust-laden surface is acted upon by an air flow. In order to make a detailed study of these phenomena, dust-laden surfaces (with and without Gardinol films) were placed in an aerodynamic tube; the air speed was 8 m/sec. The results of experiments on the detachment of particles 40-60 μ in diameter for various directions of the air flow relative to the surface (ordinary surface and one treated with a 1% aqueous solution of Gardinol) are given below:

Direction of flow relative to the surface	Perpendicular	Inclined	Parallel
γ_F, %:			
ordinary surface	98	98	97
Gardinol surface	34	6	2

These results indicate that, first, the particles are held worse on a Gardinol film than on an ordinary surface and, secondly, detachment of the particles takes place to a smaller extent when the air flow is directed perpendicularly to the dust-laden surface.

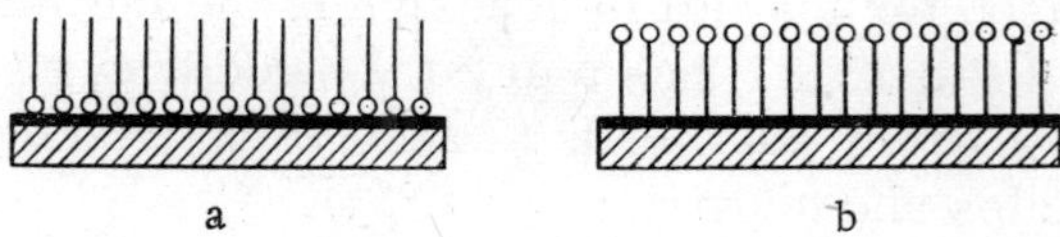

Fig. V.12. Orientation of SAS molecules on unpainted (a) and painted metal surfaces (b).

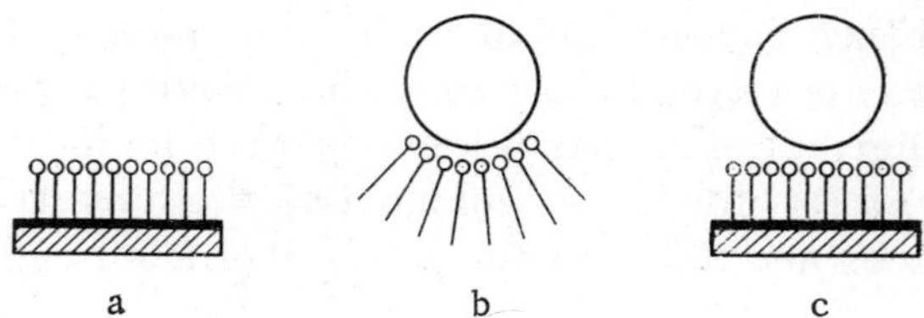

Fig. V.13. Orientation of SAS molecules on a painted surface (a), on a particle (b), and in the space between these (c).

With the second method of depositing the Gardinol film the excess of surface-active substances is washed away and only layers or even a monolayer of SAS molecules remain on the substrate. We may suppose that the orientation of the SAS molecules will be different on different surfaces. On unpainted steel surfaces with a greater affinity toward polar groups, the hydrophilic part of the SAS molecules will be oriented toward the substrate; on painted surfaces the reverse will apply (Fig. V.12). Thus, if the orientation of the SAS molecules on the surfaces of the particles and substrate is opposite, treatment of the particles with a solution of SAS is equivalent to depositing a monolayer of the solution on a painted surface.

For example, on depositing a SAS on a painted surface (Fig. V.13a) and on a glass particle (Fig. V.13b) and bringing their surfaces close together, a layer of specific orientation will develop (Fig. V.13c). The adhesion of spherical glass particles to a painted surface was studied as a function of the preliminary treatment of the particles and surface with SAS solutions of various concentrations and of the time taken to dry the film before washing with distilled water.

The following adhesion numbers were obtained:

1) For an ordinary surface, the particles not having been treated with SAS:

d_p, μ	80-100	40-60
F_{det}, dyn.	$8.4 \cdot 10^{-3}$	$4.1 \cdot 10^{-2}$
γ_F, %	50	44

2) For an ordinary surface, the particles having been treated in a solution of Gardinol and washed with distilled water after careful drying:

d_p, μ	80-100	40-60
F_{det}, dyn.	$8.4 \cdot 10^{-3}$	$4.1 \cdot 10^{-2}$
γ_F, % for SAS concentration		
0.1%	28	29
1.0%	19	-

3) For untreated particles, the surface having been treated with a Gardinol solution, held for 10 min, and washed with distilled water, for the same particle sizes and detaching forces:

γ_F, % for SAS concentration:		
0.1%	18	10
1.0%	20	16

4) For the same particles and surface, but after holding the latter for 24 h before washing in water:

γ_F, % for SAS concentration:		
1.0%	14	8
0.1%	18	20

It follows from these results that treatment with an aqueous solution of Gardinol, even after washing the deposited film with water, reduces the dust-holding capacity of painted surfaces by a factor of 2-4 times or even more, the effect being almost the same for SAS-treated surfaces and SAS-treated particles. As the time of holding the SAS-treated surface before washing increases (from 10 min to 24 h), adhesion remains unaltered.

The reduction in the adhesion of particles after SAS treatment indirectly confirms the idea expressed in Fig. V.13 regarding the effect of the orientation of SAS monolayers and its relation to the dust-holding capacity of a coating.

§30. Methods of Combating the Adhesion of Paraffin and the Overgrowth of the Underwater Parts of Vessels

Method of Combating the Adhesion of Paraffin. Crystalline paraffin enters into the composition of petroleum. When petroleum moves through oil pipes, the crystalline paraffin may adhere to the inner surfaces of the latter [268]. First of all the paraffin particles line up on rough and uneven parts of the surface. The adhering particles create an additional resistance to the flow and tend to catch subsequent crystalline paraffin particles. Hence, local narrowing of the flow cross section takes place, and this greatly reduces the transmission of the pipeline, sometimes stopping it up entirely. The adhesion of paraffin takes place still more strongly when the petroleum is saturated with gas. A gaseous medium promotes the adhesion of paraffin.

The adhesion of paraffin is affected by surface-active substances accompanying the petroleum [268].

The deposition of paraffin in oil pipes may be combated in the following ways: the actual removal of blockages already formed, a reduction in the roughness of the paint coating, the choice or development of paints preventing or at any rate reducing the adhesion of paraffin.

The first of these methods is of a passive nature. Blockages formed by adhering paraffin may be removed by mechanical or ultrasonic means. One may also heat the blocked parts and melt the paraffin blockages, or use steam or solvents. Methods of eliminating adhering layers of paraffin are set out in treatments by Galonskii [269] and by Lyushin et al. [270].

Of the various methods indicated, the most efficient is the third, i.e., the use of paint coatings preventing the adhesion of paraffin. Even in 1948, Boltyshev, as indicated in Galonskii's book [269], noted that paraffin would not stick to glossy surfaces. Later [270], therefore, the inner surfaces of oil pipes began to be treated with gasoline-resistant, quick-drying paint coatings, creating a smooth, glossy surface. The main coating materials preventing the deposition of paraffin [270] include:

Priming KhSG-26 (GOST 7313-55) and priming VKhGM (TUMKhP 4204-55); solvent R-4; cover in one coat.

Lacquer KhSL-1 (GOST 7313-55) and lacquer VKhL-4000 (TUMKhP 2647-55); solvent R-4; cover in 5 coats.

Lacquer based on É-40 resin; composition É-40 resin (VTU and KU 44-56), plasticizer tricresylphosphate (TUMKhP 723-44) and hardener No. 1 (TU MLKZ 3-55); solvent a mixture of toluene and ethyl Cellusolve; cover in 3 and 6 coats.

Bakelite lacquer for coatings (GOST 901-56); solvent ethyl alcohol (technical); cover in 3 and 6 coats.

Lacquer No. 976 based on polyurethane resin; solvent cyclohexane; cover in 3 and 4 coats.

It follows from experimental results [271] that the adhesion of paraffin depends not only on the smoothness of the coating but also on the nature of the surface. Thus it was found that on holding cylindrical rods of copper, wood, bronze, Teflon, Duralumin, cast iron, Plexiglas, brass, and steel in a gushing oil well for 24 h the greatest amount of paraffin was deposited on the Plexiglas cylinder, although it had a smooth surface. Unfortunately, the experimental data obtained in [271] did not lead the authors to any conclusions regarding the effect of roughness on the adhesion of paraffin, since the class of finish characterizing the various metal surfaces tested was unknown.

As in the case of surface hydrophobization (see §§4, 10, 18), the "oleophobization" of a surface reduces the adhesion of paraffin particles [272, 273].

The oleophoby of a substance is proportional to its dielectric constant (ε). The adhesion of paraffin, which has a constant of $\varepsilon = 2$, is a minimum for a Bakelite film ($\varepsilon = 6.9$) and a maximum for a perchlorvinyl film ($\varepsilon = 3.5$), particularly if the latter contains aluminum powder [274]. The adhesion to a Bakelite film may be further reduced by adding epoxy and phenolformaldehyde resins to Bakelite lacquer [275].

It must not be considered that the oleophobization of coatings is a universal method for preventing the adhesion of paraffin. It is well known that surface-active substances enter into the composition of petroleum, and that these are capable of collecting on the paraffin particles [268], imparting oleophilic properties to these. In this case, the oleophobization of the coatings will not lead to the desired results.

In view of the fact that the properties of petroleum and paraffin emanating from different fields are different, coating materials should only be used selectively. Only research and experiment (at present unfortunately very limited) can give a true answer to questions regarding the suitability of various coatings for particular conditions.

Paint Coatings Preventing the Overgrowth of Ships. The layer adhering to the hull of a ship increases its resistance to motion and reduces the speed of the ship; hence, such layers must periodically be removed, demanding considerable expenditure of labor and enforced docking.

Overgrowth takes place after the adhesion of seaweed and marine organisms, capable of multiplying rapidly, to the hull of a ship. Hence, the first act of overgrowth is the adhesion of microorganisms. This is further indicated by the fact that overgrowth only starts while the ship is stationary. When the ship is moving, the micro-organisms cannot adhere to it [276]. This fact is supported by laboratory experiments which showed that for a water speed of the order of 1 m/sec, particles of the order of 2 μ in size were detached from the surface, being no longer able to stick [224].

Methods of combating overgrowth usually reduce to the mechanical removal of the adhering layer, active insulation of the coating with subsequent disinfection, and the use of paint coatings preventing overgrowth.

Mechanical cleaning may be carried out, in particular, by an ultrasonic method [277]. However, mechanical cleaning requires docking, and is thus not very promising.

An example of the method of active insulation is the "Toksikon" system [278]. A solution consisting of kerosene and toxin (poison) is supplied from a tube placed in the upper submerged part of the ship. This solution covers the submerged part of the vessel with a thin layer, preventing the adhesion of seaweed and marine organisms; the kerosene cleans the surface and the toxin kills the micro-organisms.

The mechanism underlying the overgrowth-counteracting method is based on the gradual dissolution of the film-forming base [279]; this gives rise to flow of toxic substances passing out of the coating. Thus colophony, which constitutes a base for compositions used in treating the submerged parts of vessels, dissolves in weakly alkaline sea water; this releases toxins (cuprous oxide with added mercury and zinc oxides) present in the composition. The use of organic poisons (DDT, hexachloran) yields no positive results. A review of the use of fungicides for preventing the overgrowth of vessels has been given in [280]. Laboratory testing methods have been developed for testing the suitability of paint coatings.

Of the various paint coatings tested, the most acceptable include a coating of KhV-53 on a base of SPS resin with the addition of linseed oil, and a coating of KR-29 on a polyisobutylene base. Below we present some data relating to the overgrowth of these coatings as compared with conventional coatings of the NIVK-2 type used on seagoing vessels in southern latitudes [279, 282]:

Coating	KR-29	KhV-53	NIVK-2
Service period, months	22	16	15
Degree of overgrowth	20	5	70

A vessel coated with KhV-53 enamel may be used without docking for a year and a half; with ordinary coatings this is reduced to 10 months.

The absence of roughness from the coating partly prevents the gathering of micro-organisms [276].

Chapter VI

Adhesion of Dust in an Air Flow

§31. Detachment of a Monolayer

Forces Acting on Adhering Particles. On subjecting a dust-laden horizontal surface to a flow of air, the forces acting on the adhering particles include the adhesive forces F_{ad}, the weight of the particle P, the head or frontal pressure* F_f, and the lift F_l. The conditions for the detachment of the particle may be expressed by the inequality

$$F_f \geqslant \mu (F_{ad} + P - F_l) \qquad \text{(VI.1)}$$

where μ is the coefficient of friction.

For particles of smallish dimensions, when $F_{ad} \gg P$ and $F_{ad} \gg F_l$, inequality (VI.1) simplifies: $F_f \geq \mu F_{ad}$. For $F_{ad} = 0$, from (VI.1) we obtain $F_f \geq \mu(P - F_l)$. If $F_l \sim 0$, then $F_f \geq \mu P$.

The frontal pressure associated with a flow of air around a particle may be calculated from the formula

$$F_f = c_x \cdot \rho S \frac{v^2}{2} \qquad \text{(VI.2)}$$

where c_x is the resistance coefficient of the particles; ρ is the density of the air; S is the middle cross section of a particle; and v is the air speed.†

---

*It would be more accurate to call this quantity the frontal force exerted by the flow on the adhering particle.

† The adhesive force is inversely proportional to the particle size (see §16). From the condition $F_f = \mu F_{ad}$ and Eq. (VI.2), it follows that

$$v \sim \frac{1}{r^{3/2}} \qquad \text{(VI.2a)}$$

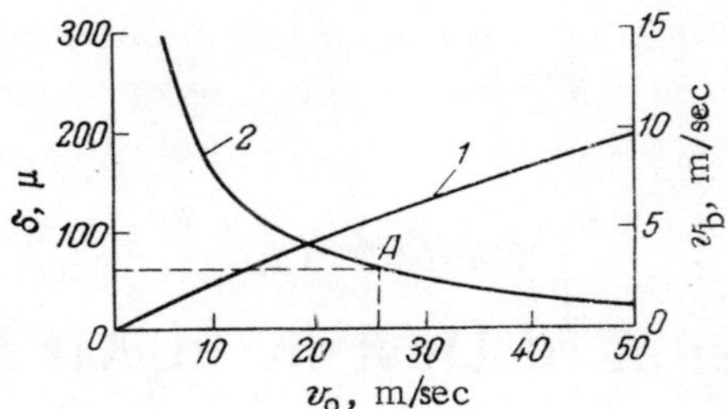

Fig. VI.1. Velocity in the boundary layer (1) and boundary layer thickness (2) as functions of the axial velocity of the flow in a tube 200 mm in diameter.

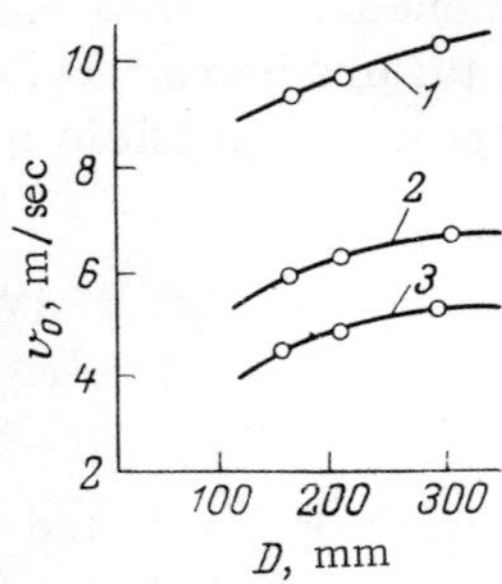

Fig. VI.2. Axial flow velocity corresponding to the detachment of particles 2.65 g/cm^3 in density as a function of pipe diameter for different diameters of the particle. 1) d_p = 21; 2) 58; 3) 89 μ.

However, the conditions governing the air flow around particles situated right in the flow and attached to the walls respectively are not exactly the same [283].

The air speed in the main part of the flow is distributed more or less uniformly. On passing around adhering particles lying in the boundary layer, the speed of the flow changes from a specific value to zero. The change in the velocity and the indeterminacy of the coefficient c_x [$c_x = f$ (Re); Re = ψ (v)] in this case makes calculations based on (VI.2) difficult.

The frontal pressure (F_f) depends on the thickness of the boundary layer. For a linear distribution of velocity and a laminar boundary layer [284], the force of interaction between the flow and the particles is given by Stokes' law:

$$F_f = \frac{3\pi\eta v_b \cdot d_p^2}{2\delta} \tag{VI.3}$$

where η is the viscosity, v_b is the velocity in the boundary layer, and δ is the thickness of this layer.

Equation (VI.3) is valid for particles of relatively small dimensions, when the thickness of the boundary layer exceeds the diameter of the particles. In practice, such a case is rarely met.

Figure VI.1 gives the results of Schlichting [284] for the variation in the thickness of the boundary layer and the velocity at its boundary as functions of the axial velocity of the flow (v_0). We see from Fig. VI.1 under what conditions the particles may "sink" and in what cases the particles will project outside the boundary of the laminar layer, i.e., experience laminar-turbulent interaction with the flow.

For example, for an axial velocity less than 25 m/sec (point A in Fig. VI.1), particles of diameter not exceeding 50 μ will reside within the laminary layer. For higher velocities the diameter of such particles exceeds the thickness of the laminar layer and they will project beyond the boundaries of the latter.

For considerable air-flow velocities (over 100 m/sec) the laminar boundary layer transforms into a turbulent one, but the turbulent boundary layer has a laminar sublayer. The thickness of the laminary sublayer is much smaller than that of the laminar boundary layer, and when a flow with a velocity of 150-500 m/sec passes around the plate, the thickness fluctuates between 9 and 2 μ. In this case also we may distinguish the laminar and laminar-turbulent action of the flow on adhering particles; the difference from the case of the laminar boundary layer is simply that the minimum size of particles only experiencing laminar effects is appreciably smaller.

We cannot characterize the effect of the flow on the adhering particles by the axial velocity only, since the effect also depends on the diameter of the pipe. We must therefore relate the effect to the Reynolds number (Re). Thus, for pipes 100, 250, and 400 mm in diameter the thickness of the laminar boundary layer for $Re = 5.6 \cdot 10^4$ is, respectively, 1.52, 1.31, and 2.1 mm, and for $Re = 4.7 \cdot 10^6$ it equals 0.01, 0.026, and 0.042 mm [285], i.e., it may be smaller than the diameter of the adhering particles.

For constant axial velocity and diminishing pipe diameter [286], the Reynolds number, and hence the coefficient c_x and the frontal pressure, also diminish, making the detachment of the particles more difficult [see (VI.2)]. With increasing pipe diameter

(Fig. VI.2) a greater velocity is required along the axis of the pipe in order to detach adhering particles.

It is best to define the character of the interaction between the flow and the adhering particle not simply by reference to the Re number, but also by reference to the particle diameter (Rumpf [287]):

$$\frac{d_p v_{\text{det}}}{\nu} < 5 \text{ — laminar action}$$

$$5 < \frac{d_p v_{\text{det}}}{\nu} < 70 \text{ — laminar-turbulent action} \qquad \text{(VI.4)}$$

$$\frac{d_p v_{\text{det}}}{\nu} > 70 \text{ — turbulent action}$$

where d_p is the particle diameter or the distance from the wall of the tube in a direction perpendicular to its axis, v_{det} is the velocity at which detachment of the adhering particles takes place (or the velocity at a height equal to the radius of the particle), and ν is the kinematic viscosity of air.

The action of an air flow on an adhering particle may also be taken into account by means of the coefficient of aerodynamic resistance α. The conditions for the detachment of the particles will in this case be given by the expression [288]:

$$v_{\text{det}} \sqrt{\alpha} = \text{const} \qquad \text{(VI.5)}$$

where v_{det} is the average air speed at which the particles slip and become detached, while α is the coefficient of aerodynamic resistance (used for characterizing air conduits with cross-sectional areas of a few square meters), expressed in $\text{kg/sec}^2 \cdot \text{m}^4$.

With increasing turbulence of the flow, i.e., increasing α, the detachment velocity (v_{det}) diminishes. Condition (VI.5) takes no account of adhesive forces.

According to experimental results, the removal of carbon particles more than 75 μ in diameter [the adhesive force of such particles is negligible, and condition (VI.5) may be used] with $\alpha = (1.8\text{–}4.0) \cdot 10^{-3}\ \text{kg/sec}^2 \cdot \text{m}^4$ takes place at flow velocities of 2.5–1.8 m/sec.

With increasing relative humidity of the air, there is a rise in the adhesive forces, and since the flow velocity at which dust

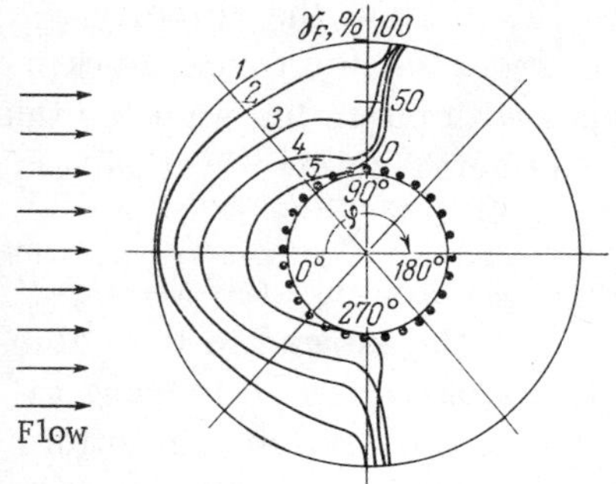

Fig. VI.3. Adhesion number associated with the detachment of loess particles 40-100 μ in diameter by an air flow from a cylindrical porcelain surface arranged vertically in an aerodynamic tube, expressed as a function of the angle φ at which the flow meets the surface. Flow velocities: 1) v = 0; 2) 5; 3) 7; 4) 10; 5) 15 m/sec.

particles are removed depends on F_{ad} [see (VI.1)] the value of v_{det} also increases.

In order to detach spherical particles 20 μ in diameter for 40% relative humidity of the air, a flow velocity of 10 m/sec is required; for 80% humidity the figure is 14 m/sec.

An increase in surface roughness only raises v_{det} for small particles [16], 20 μ in diameter; it has no effect on v_{det} for larger particles (over 50 μ), in accordance with the concepts of micro- and macro-roughness (see § 14).

Macroscopic roughnesses may "screen" adhering particles from the flow or change the character of the interaction between the flow and these particles.

The frontal pressure and the adhesive force depend on the size of the particles. Hence, we must expect the velocity at which the particles become detached to change in accordance with their sizes. The following represents some experimentally determined air speeds in an aerodynamic tube 10 cm in diameter for which corundum particles lying on an iron wall suffer detachment [286]:

d_p, μ	70	100	160	200	400	1000
v_{det}, m/sec..	11.4	10.6	10.8	10.9	12.7	16.3

For particles larger than 100 μ in diameter the adhesive force is negligible (i.e., $F_{ad} \ll P$). In this case, the conditions for the removal of the particles are given by the inequality $F_f \geq \mu P$. In accordance with Eq. (VI.3), $F_f \sim r^2$, while the weight of the particles $R \sim r^3$. Thus, for the same c_x, the air speed causing the removal of the particles should rise with increasing particle size; this is confirmed experimentally.

As the particle size falls from 100 to 70 μ, the velocity v_{det} also rises slightly owing to the increased adhesive force (Syrkin confined attention [286] to experiments with particles smaller than 70 μ in diameter, i.e., precisely the diameter below which adhesion becomes particularly significant).

The velocities at which particles of constant size are detached vary over a certain range owing to the spread in the values of adhesive force (see §17). Thus, loess particles of average diameter 12 $\pm$ 7 μ adhering to a glass surface placed at the bottom of a horizontal tube of square cross section 15 $\times$ 15 cm^2,* are detached at a mean air-flow velocity [83] of 7-14 m/sec.

The detachment of adhering particles from spherical and cylindrical surfaces has a number of special features [83].

It follows from Fig. VI.3 that the detachment of the particles depends not only on the air-flow velocity, but also on the position of the surface relative to the axis of the flow, i.e., the angle at which the flow meets the surface. Maximum detachment (minimum γ_F) occurs for φ values of 90 and 270°; only a small number of particles are detached from the front (φ = 0°) and none at all from the back at the velocities in question. With increasing flow velocity (curves 2-5), the adhesion number diminishes. However, even under these conditions more particles come away from surfaces placed parallel to the flow (φ = 90 and 270°) than frontally. This fact is of particular importance in the filtration of aerosols (see §43), in the flow of furnace gases around a group of pipes (see §48), and in certain other cases.

Estimating the Degree of Cleaning of a Surface. The extent to which particles have been removed from a surface may be characterized by the coefficient K_N (a quantity inverse to the adhesion number) or by the coefficients K_S and K_m; these indicate the extent to which the number of particles (K_N), the area occupied by adhering particles (K_S), or the mass of adhering particles (K_m) have been reduced as a result of an air flow or vibration:

$$K_N = \frac{N_i}{N_f} \quad \text{(VI.6)}$$

*We are considering the detachment of a layer of adhering particles.

$$K_S = \frac{S_i}{S_f} \qquad \text{(VI.7)}$$

$$K_m = \frac{m_i}{m_f} \qquad \text{(VI.8)}$$

where N_i is the initial number of particles adhering to the surface before being acted upon by the flow, N_f is the final number of particles remaining on the plate after subjection to the flow, while S_i and S_f represent the corresponding areas occupied by the adhering particles, and m_i and m_f represent the masses of the adhering particles before and after being acted upon by the flow respectively.

For dust containing a wide range of particle sizes, from d_1 to d_n, $S=\sum_1^n \frac{\pi}{4} d^2$, $m=\sum_1^n \frac{\pi}{6} d^3$.

The following represents some typical values of the coefficient K_N determined by removing spherical glass particles with a side spread of particle sizes from horizontal plane surfaces in an air flow:

v, m/sec	2.8	5.6	11.2
K_N:			
for d_p = 100-150 μ	15	33	54
for 50 < d_p < 100 μ.	1.1	1.2	1.3
for d_p < 50 μ	1.05	1.05	1.10

Small particles (those under 100 μ or more, particularly those under 50 μ in diameter), the adhesive force of which greatly exceeds the particle weight, are not removed very easily by an air flow. Large particles more than 100 μ in diameter, which act on the surface with a force smaller than their own weight, are readily removed from horizontal plates. Even for a low air speed (2.8 m/sec), the number of adhering particles falls by a factor of 15.

Thus, large particles are quite easily removed from dry surfaces, whereas an air flow of 2.8-11.2 m/sec will only remove a small proportion of the smaller particles.

In order to improve the efficiency with which the adhering particles are removed, one must either raise the air-flow velocity or intensify some auxiliary particle-removing process. For ex-

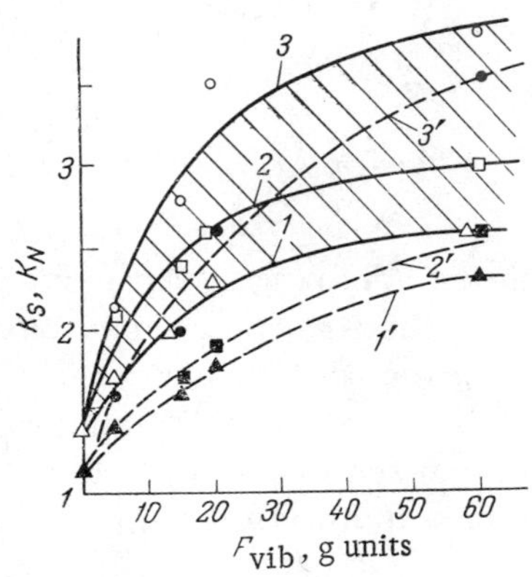

Fig. VI.4. Value of the coefficients K_S (1,2,3) and K_N (1',2',3') as a function of the vibrational forces for particles smaller than 100 μ in diameter (surface inclined at φ = 90°) for various air speeds: 1,1') v = 2.8; 2,2') 5.6; 3,3') 11.2 m/sec.

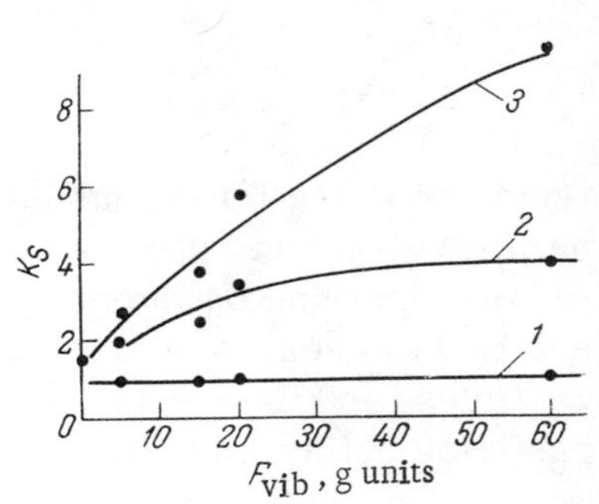

Fig. VI.5. Coefficient K_S (rate of air flow 11.2 m/sec; φ = 90°) as a function of the vibrational forces. 1) For an oily surface (oil density 0.03-0.07 mg/cm^2); 2) for a clean surface; 3) after holding the surface in the open air for 90 h.

ample, a combination of an air blast and mechanical vibration may be used to remove the particles.

The value of K_S is presented as a function of the vibrational forces for various air speeds in Fig. VI.4. As we should expect, the coefficient K_S is always larger than K_N (other conditions being equal). The value of K_S for particles less than 100 μ in diameter varies over the range shown shaded in Fig. VI.4.

We may consider that under these conditions, K_S = 2-4 and K_N = 1.5-3. The smaller particles (less than 50 μ in diameter) are removed less easily:

F_{vib}, g units.	5	10	15	20	60
K_S for air speed:					
5.6 m/sec*.	1.15/1.1	1.20/1.13	1.25/1.16	1.45/1.22	1.65
11.2 m/sec	1.25	1.40	1.50	1.75	2.15

*The value of K_S in the numerator relates to φ = 90°, and that in the denominator to φ = 0°.

For small particles (smaller than 50 μ), K_S fluctuates between 1 and 2. Particles are removed rather more easily from vertical than horizontal surfaces.

The removal of particles depends on the state and properties of the surface. By way of example, the following represents the values of K_S for particles smaller than 100 μ in diameter, as determined by removing the particles from perchlorvinyl—enamel-coated and uncoated tinplate surfaces in an air flow of 5.6 m/sec:

F_{vib}, g units	0	20	60
K_S for surfaces:			
unpainted	1.8	4.6	5.8
painted	1.4	2.6	3.0

In view of the fact that the adhesion of the particles to the unpainted surfaces is lower than to the painted surfaces, it is easier to clean the former.

The way in which K_S depends on the state of the painted surface is shown in Fig. VI.5. As a result of holding the sample in air (curve 3), atmospheric dust adheres to their surfaces; the original surface is thus screened, and the value of K_S may raise by a factor of 2-2.5. Spherical glass particles cannot be removed at all from oily surfaces in an air flow (curve 1).

If the particles are deposited on the surface in a drop of water, then on subsequent drying of the drop the adhesion of such particles increases (see §26) and it becomes more difficult to remove the particles (diameter under 100 μ) in an air flow (air speed 11.2 m/sec, angle φ = 90°):

F_{vib}, g units	0	5	15	20	60
K_S for settling:					
free	1.3	2.1	2.8	3.5	4.0
in a water drop, with subsequent drying	1.0	1.1	1.1	1.0	1.1

A surface receiving dust under these conditions is hardly freed from dust at all in an air flow of 2.8-11.2 m/sec.

By using the experimental results obtained, we may derive an equation characterizing the removal of adhering particles in an

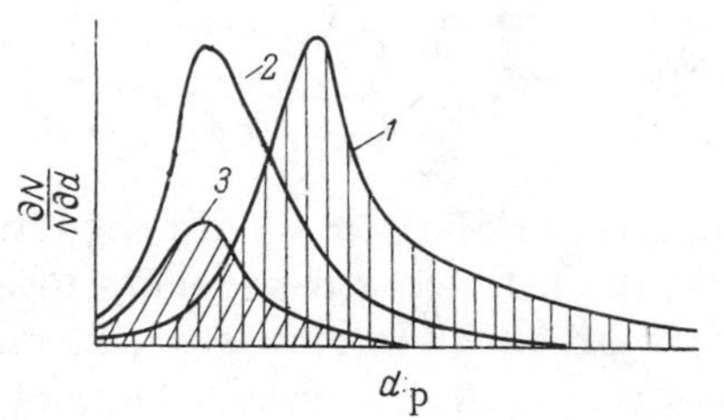

Fig. VI.6. Particle size distribution. 1) Initial; 2) final; 3) weighted with respect to the initial.

air flow, i.e., develop a method for calculating the adhesion number or the coefficient K_N.

Figure VI.6 shows the particle-size distribution. The original distribution (curve 1) changes under the influence of the air flow or vibration (or both together). This change is expressed by curve 2 in terms of the number of particles remaining, or by curve 3 in terms of the original distribution (an analogous situation holds for the distribution of particles expressed in terms of the area which they occupy, and also for the mass distribution).

The ratio between the areas of the figures formed by the curves 1 (S_1) and 3 (S_3) equals K_N, K_S, or K_m, depending on which distribution the curves represent.

In general form the value of K_N may be expressed by the formula

$$K_N = \frac{N_i}{N_f} = \frac{S_1}{S_3} = \frac{\int_{d_1}^{d_2} F_{N,i}(d)\,\partial d}{\int_{d_1}^{d_2} F_{N,f}(d)\,\partial d} = \frac{\int_{d_1}^{d_2} F_{N,i}(d)\,\partial d}{\int_{d_1}^{d_2} F_{N,i}(d)\gamma_F(d)\,\partial d} \qquad \text{(VI.9)}$$

where $F_{N,i}$ (d) and $F_{N,f}$ (d) are functions characterizing the initial and final size distributions of the particles.

The function F_f(d) should allow for the change in the final distribution on the scale of the original, i.e.,

$$F_{N,f}(d) = F_{N,i}(d)\,\gamma_F(d)$$

The function γ_F(d) characterizes the change in the adhesion number as a function of particle dimensions. The adhesion number for a given particle diameter is numerically equal to the ratio of the ordinates of the final (Fig. VI.6, curve 3) and initial (curve 1) particle-size distributions. Whereas, usually the adhesion number

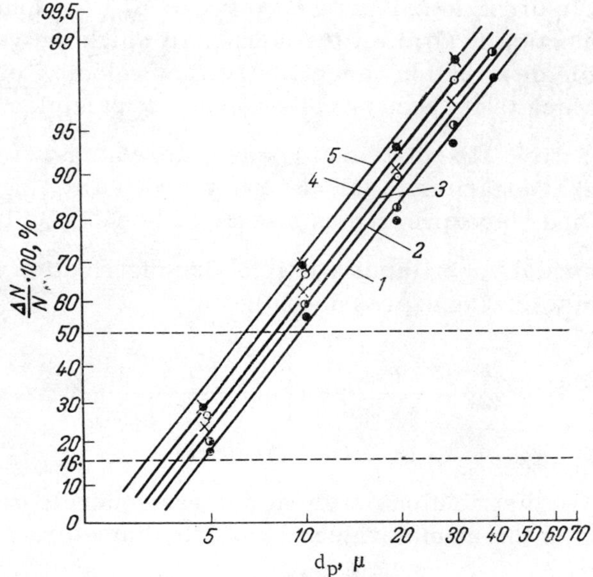

Fig. VI.7. Size distribution of spherical glass particles before and after being acted upon by an air flow for surfaces placed vertically and perpendicularly to the axis of the flow. 1) Initial; 2-5) after interaction with flows of 5.6, 8.4, 11.2, and 14 m/sec, respectively.

is quoted in percents, in calculating K_N from formula (VI.9) the value of γ_F must be given in relative units.

By analogy with formula (VI.9) we may write for the coefficient K_m

$$K_m = \frac{m_i}{m_f} = \frac{\int_{d_1}^{d_2} F_{m,i}(d)\,\partial d}{\int_{d_1}^{d_2} F_{m,f}(d)\,\partial d} = \frac{\int_{d_1}^{d_2} F_{m,i}(d)\,\partial d}{\int_{d_1}^{d_2} F_{m,i}(d)\,\gamma_F(d)\,\partial d} \tag{VI.10}$$

and for the coefficient K_S

$$K_S = \frac{S_i}{S_f} = \frac{\int_{d_1}^{d_2} F_{S,i}(d)\,\partial d}{\int_{d_1}^{d_2} F_{S,f}(d)\,\partial d} = \frac{\int_{d_1}^{d_2} F_{S,i}(d)\,\partial d}{\int_{d_1}^{d_2} F_{S,i}(d)\,\gamma_F(d)\,\partial d} \tag{VI.11}$$

Thus, in order to calculate K_N, K_S, or K_m, we must determine: the function $F_f(d)$, i.e., the manner in which the parameters of the particle distribution vary with the flow velocity, or the function $\gamma_F(d)$, which takes account of the number of particles removed.

The particle size distribution encountered most frequently obeys a normal logarithmic law [6] and in probability logarithmic coordinates may be expressed as a straight line (Fig. VI.7, line 1).

The normal-logarithmic particle size distribution is approximated by the following expression [6]:

$$\frac{\partial N}{N\partial d} = \frac{0,43}{\sigma\sqrt{2\pi}\,d}\exp\left[-\frac{(\log d - \log \overline{d})^2}{2\sigma^2}\right] \qquad \text{(VI.12)}$$

where σ is the distribution parameter (mean square): $\sigma = \log \overline{d} - \log d_{16}$; and $\overline{d}$ is the median value of particle diameter.

The values of $\log d$ and $\log d_{16}$ are obtained from the particle size distribution (see Fig. VI.7).* By integrating (VI.12) we find the probability distribution of particles of a given size, from d_1 to d_2.

If $d_1 = 0$ and $d_2 = \infty$, then $\int\limits_{d_1}^{d_2} F_i(d)\,\partial d = 1$, i.e., the numerator in formula (VI.9) equals unity. For finite values of d_1 and d_2 this integral differs from unity.

It follows from Fig. VI.7 that the size distribution of the particles remaining on the surface after interaction with an air flow obeys the same law and is expressed by a straight line.

Analogous results are obtained for the particle size distribution before and after the action of various vibrational forces on the dust-laden surface (Fig. VI.8).

The following results represent the values of the parameters $\overline{d}$, $\log \overline{d}$, $\log d_{16}$, and σ for calculating the particle size distribution before (initial) and after being acted upon by air flows of various velocities and vibrational forces:

---

*$\log \overline{d}$ corresponds to the value $(\Delta N/N) \cdot 100 = 50\%$, and $\log d_{16}$ to 16%.

v, m/sec.	0	5.6	8.5	11.2
$\bar{d}$	10.0	9.0	8.2	7.6
$\log \bar{d}$	1.0	0.954	0.914	0.881
$\log d_{16}$.	0.672	0.643	0.602	0.568
σ.	0.328	0.311	0.312	0.313

F_{vib}, g units.	10	15	20
$\bar{d}$	9.8	9.2	8.4
$\log \bar{d}$	0.990	0.963	0.924
$\log d_{16}$.	0.662	0.643	0.602
σ.	0.328	0.320	0.322

The following empirical equations may be used to express the relation between the distribution parameters and the air-flow velocity:

$$\bar{d} = 10 - 0.017\,v \quad \text{for} \quad \sigma = 0.312 \tag{VI.13}$$

or vibrational forces

$$\bar{d} = 10 - 0.07\,F_{\text{vib}} \quad \text{for} \quad \sigma = 0.325 \tag{VI.14}$$

where v is the air-flow velocity in m/sec and F_{vib} is the vibrational force in g units.

The quantity σ equals the tangent of the slope of the straight lines characterizing the particle size distribution (Figs. VI.7 and VI.8). The straight lines 2, 3, 4, and 5, which determine the size distribution of the remaining particles, are parallel, i.e., σ = const. The deviations of the quantity σ from the mean value reflect the accuracy of the graphical constructions.

In general, $\sigma = f(\text{v})$.

The manner in which particle-adhesion forces vary with particle diameter [i.e., the function $\gamma_F(d_p)$] for various flow velocities is indicated in Fig. VI.9. An analogous relationship between the adhesion number and particle size is obtained for various vibrational forces (10, 15, and 20 g units).

The relation between adhesion number and particle size, expressed in the form of logarithmic coordinates in Fig. VI.9, may be represented by the following formula:

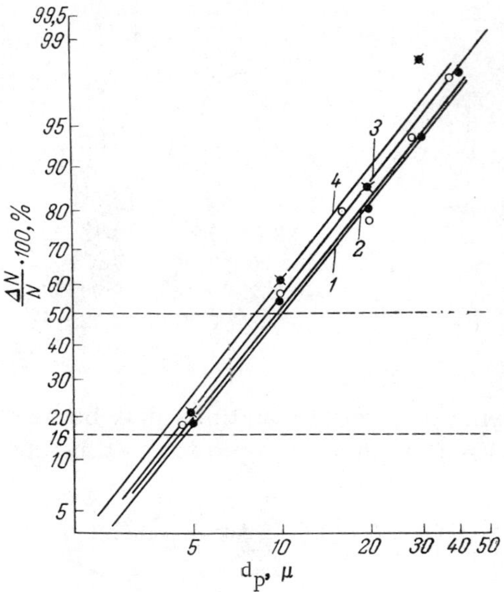

Fig. VI.8. Size distribution of spherical glass particles before and after being acted upon by vibrational forces for surfaces placed vertically. 1) Original; 2-4) for vibrational forces of 10, 15, and 20 g units, respectively.

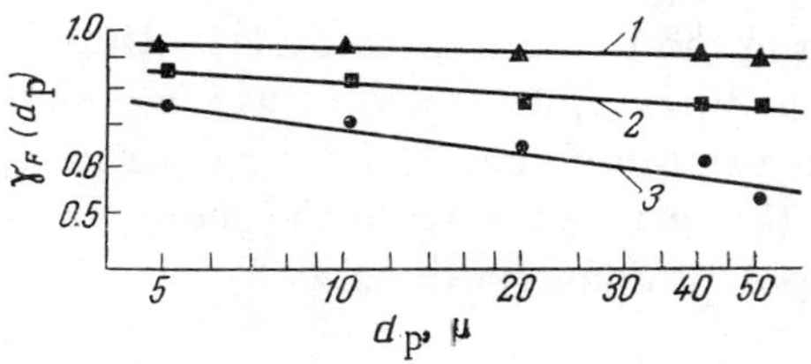

Fig. VI.9. Adhesion numbers as a function of particle size after interaction with air flows of various velocities. 1) v = 5.6; 2) 8.4; 3) 11.2 m/sec.

$$\gamma_F(d_p) = ad_p^b \tag{VI.15}$$

The constants a and b are determined from Fig. VI.9:

v, m/sec	5.6	8.4	11.2
a	0.98	0.96	0.90
b	$-1.75\cdot10^{-2}$	$-5.24\cdot10^{-2}$	-0.1

The value of the constant b for vibrational forces equal to 10, 15, and 20 g units is numerically equal to the value of b given above for various flow velocities; the constant a has the values 0.98, 0.96, and 0.94, respectively.

Allowing for possible errors in carrying out the graphical work, we take $a = 1$.

Like the distribution parameters σ and d, the constants a and b may be expressed [see formulas (VI.13) and (VI.14)] in terms of quantities on which the detachment of the particles depends, i.e., v and F_{vib}. In order to transform the denominator of formula (VI.9), we put

$$d^b = e^{2.3\, b \log d} \tag{VI.16}$$

Allowing for Eqs. (VI.12), (VI.15), and (VI.16), we transform the denominator in (VI.9):

$$N_f = \int\limits_{d_1}^{d_2} \frac{0.43\, a}{\sqrt{2\pi}\,\sigma d} \exp\left[\frac{-(\log d - \log\bar{d})^2 + 4.6\, b\sigma^2 \log d}{2\sigma^2}\right] \partial d \tag{VI.17}$$

Let us now transform the argument of the exponential in Eq. (VI.17):

$$\frac{-(\log d)^2 + 2\log d \cdot \log\bar{d} - (\log\bar{d})^2 + 4.6\, b\sigma^2 \log d}{2\sigma^2} =$$

$$= \frac{-(\log d)^2 + 2\log d\,(\log\bar{d} + 2.3\, b\sigma^2) - (\log\bar{d})^2 \pm (\log d + 2.3\, b\sigma^2)^2}{2\sigma^2} =$$

$$= -\frac{[\log d - (\log\bar{d} + 2.3\, b\sigma^2)]^2}{2\sigma^2} + \frac{(\log\bar{d} + 2.3\, b\sigma^2)^2 - (\log\bar{d})^2}{2\sigma^2} \tag{VI.18}$$

Thus, the argument of the exponential decomposes into two terms. The first of these enables us to reduce the distribution obtained to standard form, while the second is a constant quantity (for a given distribution and d are constant). Let us denote

$$\frac{(\log\bar{d} + 2.3\, b\sigma^2)^2 - (\log\bar{d})^2}{2\sigma^2} = B \tag{VI.19}$$

allowing for (VI.18) and (VI.19), Eq. (VI.17) may be written

$$N_f = \int_{d_1}^{d_2} \frac{0.43\, a \cdot e^B}{\sqrt{2\pi}\, \sigma d} \exp\left\{ - \frac{[\log d - (\log \bar{d} + 2.3\, b\sigma^2)]^2}{2\sigma^2} \right\} \partial d \qquad \text{(VI.20)}$$

Formula (VI.20) may be transformed by regarding d_1 as zero.

Changing variables, we denote

$$\frac{\log d - (\log \bar{d} + 2.3\, b\sigma^2)}{\sigma} = u \qquad \text{(VI.21)}$$

After differentiating Eq. (VI.21), we obtain $(1/d)\partial d = (\sigma/0.43)\,\partial u$. Then,

$$N_f(0, d_2) = \frac{a \cdot e^B}{\sqrt{2\pi}} \int_0^{\frac{\log d_2 - (\log \bar{d} + 2.3\, b\, \sigma^2)}{\sigma}} \exp\left(-\frac{u^2}{2}\right) \partial u \qquad \text{(VI.22)}$$

The tabulated probability integral $\Phi(u)$ equals

$$\Phi(u) = \frac{2}{\sqrt{2\pi}} \int_0^u \exp\left(-\frac{u}{2}\right) \partial u \qquad \text{(VI.23)}$$

Finally we obtain

$$N_f(0,\ d_2) = \frac{a \cdot e^B}{2}\ \Phi\left[\frac{\log d_2 - (\log \bar{d} + 2.3\, b\sigma^2)}{\sigma}\right] \cdot 100 \qquad \text{(VI.24)}$$

We may transform the numerator and the denominator of Eqs. (VI.9)-(VI.11) in the same way.

Finally, Eq. (VI.9) may be put in the form

$$K_N = \frac{\Phi_1 \left| \frac{\log d_2 - \log \bar{d}}{\sigma} \right|}{a \cdot e^B \cdot \Phi_2 \left[\frac{\log d_2 - (\log \bar{d} + 2.3\, b\sigma^2)}{\sigma} \right]} \qquad \text{(VI.25)}$$

In view of the fact that the particle-size distribution functions [see (VI.9)] are not normalized, we may define their moments of the n-th order as:

$$\alpha_{\mathrm{i}}^{n}=\int_{d_1}^{d_2} d^n F_{N,\mathrm{i}}(d)\,\partial d$$

$$\alpha_{\mathrm{f}}^{n}=\int_{d_1}^{d_2} d^n F_{N,\mathrm{f}}(d)\,\partial d \tag{VI.26}$$

Hence, the zero-order moments will be equal to

$$\alpha_{\mathrm{i}}^{0}=\int_{d_1}^{d_2} F_{N,\mathrm{i}}(d)\,\partial d \qquad \alpha_{\mathrm{f}}^{0}=\int_{d_1}^{d_2} F_{N,\mathrm{f}}(d)\,\partial d \tag{VI.27}$$

In accordance with (VI.9) we obtain from (VI.27)

$$\frac{1}{\gamma_F}=K_N=\frac{\alpha_{\mathrm{i}}^{0}}{\alpha_{\mathrm{f}}^{0}} \tag{VI.28}$$

The value of the n-th order moment for the normal logarithmic distribution is given by the formula

$$\alpha_{\mathrm{i}}^{n}=\int_{d_1}^{d_2} d^n \frac{0.43}{\sigma\sqrt{2\pi}\,d}\exp\left[-\frac{(\log d-\log\bar{d})^2}{2\sigma^2}\right]\partial d=$$

$$=\int_{d_1}^{d_2}\frac{0.43}{\sigma\sqrt{2\pi}}\exp\left[\frac{-(\log d-\log\bar{d})^2+4.6\,(n-1)\log d\sigma^2}{2\sigma^2}\right]\partial d \tag{VI.29}$$

For n = 0, formula (VI.29) transforms into (VI.12). Thus, in order to calculate the coefficients K_N, K_m, and K_S, we must know: the initial particle size distribution, i.e., d_2, $\bar{d}$, and σ, and the variation in the adhesion number as a function of particle size, i.e., the quantities a and b [Eq. (VI.15)].

If one has to determine the degree of cleaning of a surface which has earlier been treated in an air flow or by vibrations, the original values of d and may be determined from Eqs. (VI.13) and (VI.14).

Table VI.1 gives the calculated values of the coefficients representing the removal of dust by an air flow, by vibrational forces, and also by both of these acting togehter: $K_S = (K_S)_v \cdot (K_S)_{F_{vib}}$

Table VI.1. Calculated and Experimental Coefficients* K_S and K_N for Various Air-Flow Velocities and Vibrational Forces and K_S for a Combination of an Air Flow and Vibration

Air-flow velocity, m/sec	K_S for a vibrational force of				K_N for a vibrational force of			
	0	10 g	15 g	20 g	0	10 g	15 g	20 g
0	–	1.09	1.24	1.48	–	1.03/1.04	1.10/1.10	1.21/1.14
5.6	1.07	1.16/1.14	1.33/1.25	1.58/1.46	1.02/1.05	–	–	–
8.4	1.30	1.42	1.61	1.92	1.08/–	–	–	–
11.2	1.70	1.85	2.11	2.52	1.18/1.10	–	–	–

* Denominator gives experimental values.

In addition to this, calculated and experimental values are given for the coefficient K_N (for an air flow and vibrational forces acting separately).

It follows from Table VI.1 that the calculated and experimental values of the coefficients agree quite satisfactorily. In the present case, K_S/K_N varies from 1.05 to 1.45, this ratio increasing with an increasing absolute value of K_S.

The results of calculations based on formulas (VI.9)-(VI.29) will depend on the properties of the surface and of the particles. Hence, despite the generality of the computing method, the values of the coefficients characterizing the removal of the particles will differ for different specific cases.

In order to estimate the influence of the adhesive properties of the particles and their size distribution on the degree of dust removal, a wider range of experimental data must be secured; so far, such data are very limited.

§ 32. Detachment of a Layer

Denudation and Erosion. In the detachment of an adhering dust layer by an air flow, the following processes may occur: the removal of the top particles, i.e., the overcoming of autohesion, the detachment of a layer of dust, i.e., the overcoming of the adhesive forces in the layer, and the detachment of individual particles remaining after the removal of the layer.

The removal of the top particles is possible when $F_{ad} > F_{aut}$. In this case, the dust is raised to a comparatively short distance above the original surface (Fig. VI.10a). The autohesive process of dust-layer detachment is called *erosion* [289].

For stronger forces of autohesion (appreciably exceeding the adhesive forces), detachment occurs at the boundary between the surface and the dust layer. In this case it is the adhesive forces which are overcome (Fig. VI.10b, c). This process is called *denudation* [62]. In denudation, detachment of the dust starts at the leading edge of the dust deposit and a dust cloud rapidly fills the whole channel.

There is a certain class of dusts for which $F_{ad} \gg F_{aut}$. In dusts of this kind there is no denudation. This class includes room dust, shale, some kinds of gypsum, carbonates, etc.

An adhering dust layer of regular shape (with plane boundaries) is detached more rapidly than an irregular one. For an air speed of 30 m/sec, only 0.25 sec from the start of the air flow we find that 60% of the particles in an adhering dust layer of regular shape are removed, while for an irregular shape the figure is 20%. The rate of denudation may be calculated thus *

$$v_d = K_1 (F_{aut} \cdot \rho)^{1/2} + K_2 \quad \text{(VI.30)}$$

where v_d is the air-flow velocity for which denudation occurs, K_1 and K_2 are coefficients, F_{aut} is the autohesive force between the particles, and ρ is the particle density.

---

*If we consider that $F_{aut} \sim 1/r^2$ (see § 17), then

$$v_d \sim \frac{1}{r} \quad \text{(VI.30a)}$$

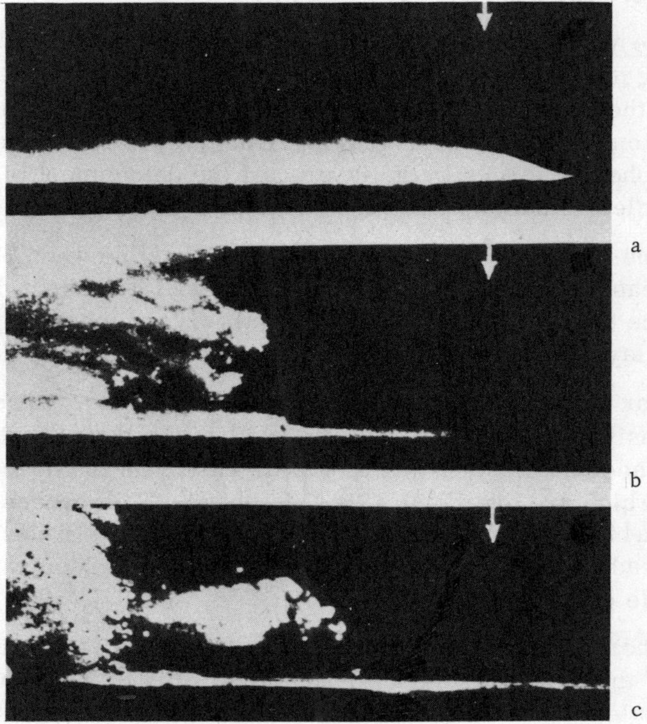

Fig. VI.10. Detachment of an adhering layer of irregular (a,b) and regular shape (c). a) Erosion; b,c) denudation.

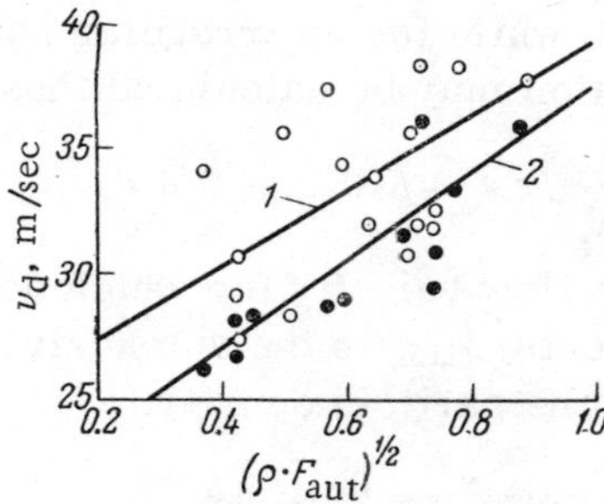

Fig. VI.11. Denudation velocity of dust deposits from a surface covered with emery paper of zero grain size as a function of the parameter $(\rho F_{aut})^{1/2}$ for an adhering layer of regular (2) and irregular shape (1) (dimensions of dust deposit 17.8 × 5.1 × 0.6 cm).

For an adhering layer of regular shape, $K_1 = 17.6$, $K_2 = 21.8$; for dust deposits of irregular form, $K_1 = 16.6$, $K_2 = 26.6$.

The velocity of denudation is shown as a function of the parameter $(F_{aut} \cdot \rho)^{1/2}$ in Fig. VI.11. In formula (VI.30) only the autohesive force and the density of the particles are allowed for, the adhesive force not being taken into account, although Davies [62] noted that dust could be detached more easily from polished brass surfaces than from surfaces covered with emery paper of zero grain size.

In denudation, all dust deposits are removed in about 0.5 sec. Hence, the denudation velocity is the main parameter determining this process. If $F_{ad} \sim 0$ and $F_{aut} \sim 0$, the removal of large particles (2-4 mm in diameter) depends on the air-flow velocity only and occurs along the boundary between the dust layer and the surface [290], since when particles move along the surface of similar particles the coefficient of friction is greater than when they move along a solid (hard) surface.

In erosion, a considerable amount of adhering dust remains even 18 sec after the start of an air flow at a velocity of 25 m/sec (Fig. VI.12). Hence, erosion depends not only on the velocity of the air flow but also on the period for which it acts on the adhering dust. Thus, the erosion process may be estimated [62] by reference to some arbitrary parameter E indicating the amount of dust (g/sec) removed by an air flow with a velocity of 25 m/sec in 5 sec. For this air velocity, no adhesion-type detachment of the dust layer takes place over a period of 4-6 sec (Fig. VI.12); the area of the remaining layer of adhering dust remains equal to the original. The parameter E may be expressed in terms of the density of the particle material and the autohesive force of the dust layer in the following form:

$$E = \frac{a\rho^2}{F_{aut}} - b \qquad \text{(VI.31)}$$

For an adhering layer of regular shape, $a = 0.37$, $b = 0.25$ and for one of irregular shape $a = 0.27$, $b = 0.10$.

Figure VI.13 shows the erosion parameter E as a function of ρ^2/F_{aut}.

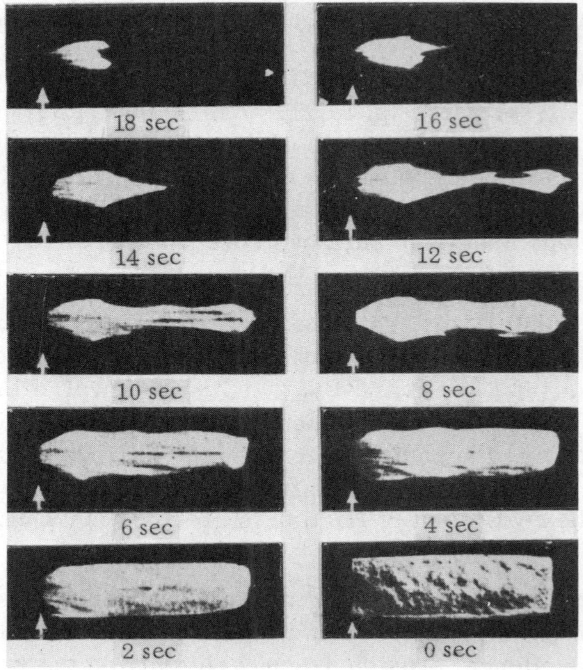

Fig. VI.12. Erosion of gypsum particles (autohesive force of the layer 300 dyn; flow rate 25 m/sec) for various periods of action of the flow.

The parameter E constitutes a relative characteristic of the erosion process, since the velocity of removing the adhering layer (25 m/sec) was selected arbitrarily.

Erosion may also be estimated by reference to the material loss, i.e., the reduction in the mass of the adhering dust layer, from 1 m^2 of surface in 1 sec.

We see from Fig. VI.14 that the relation between the material loss and the air-flow velocity for a given dust fraction is of a power nature.

For magnetite dust, with particles less than 10 μ in diameter, adhering to the bottom of a rectangular channel, this relationship may be expressed by the semi-empirical formula:

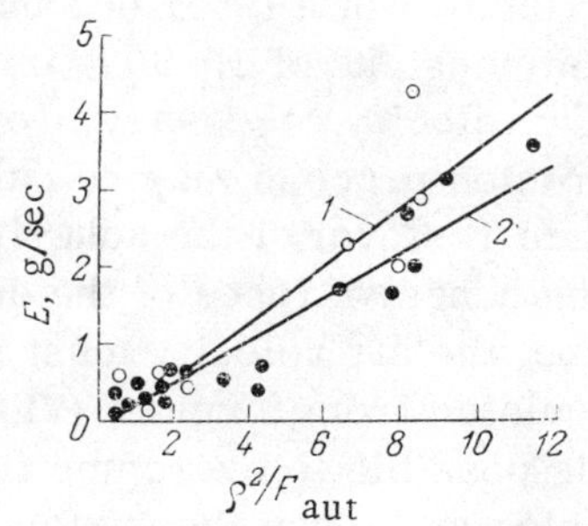

Fig. VI.13. Erosion of dust deposits as a function of the parameter ρ^2/F_{aut} for an adhering layer of regular (1) and irregular shape (2).

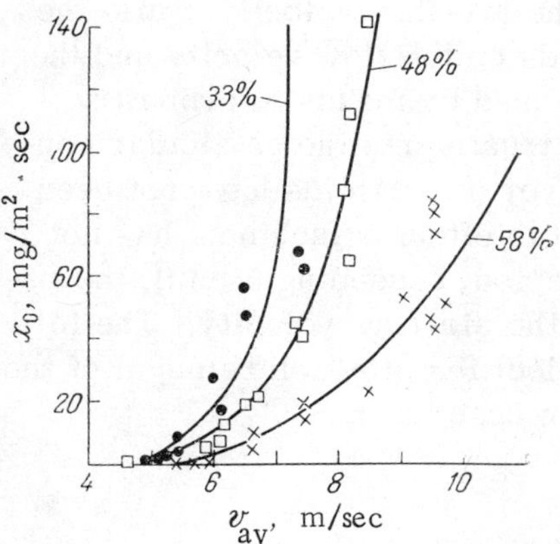

Fig. VI.14. Loss of quartz dust containing various numbers of particles with $d_p < 30$ μ (figures on curves) as a function of the average air-flow velocity.

$$x_0 = a\,(v_{av} - v_{det})^b \qquad \text{(VI.32)}$$

where x_0 is the loss of dust in g/(m² · sec); v_{av} is the average air-flow velocity in m/sec; v_{det} is the flow velocity causing detachment of the dust layer in m/sec; and a and b are experimental coefficients. For a metallic surface, a = 0.7, b = 4.0; for magnetite, a = 1.2, b = 2.5.

When using formula (VI.32) to calculate the loss of particles, one must know not only the values of the experimental coefficients (a and b) but also the values of v_{av} and v_{det}, which vary with the diameter of the air pipe (see §31). According to formula (VI.32) a uniform removal of the adhering dust layer is assumed (loss rate constant). This assumption is valid over particular intervals of time in the erosion process, but it cannot be extended to the whole process; this limits the possibility of calculating the loss of material from formula (VI.32).

For calculating the air-flow velocity at which the detachment of the dust particles begins (v_{det}, m/sec), we may use the empirical formula [293]:

$$v_{det} = \frac{50(\rho D)^{1/6}}{\sqrt{d_p}} \qquad \text{(VI.33)}$$

where ρ is the density of the dust in g/cm³; D is the pipe diameter in m, and d_p is the diameter of the dust particles in μ.

Formula (VI.33) is valid for the removal of a layer of sand and coal 0.5-1 mm thick with particle dimensions of 15-90 μ in pipes 100-400 mm in diameter. The air velocity necessary to overcome the forces of autohesion in the erosion process may be calculated from this formula. For complete removal of the adhering particles, i.e., in order to overcome the adhesive force of the dust layer to the inner surface of the pipeline, the air velocity must be considerably higher than the value calculated from formula (VI.33). On increasing the air-flow velocity it is possible to overcome the adhesive forces of the remaining particles and clean the surface from the adhering dust layer. Hence, for $F_{ad} > F_{aut}$, we must distinguish two air-flow velocities. The first characterizes the conditions under which the forces of autohesion are overcome, while the second relates to adhesive forces. The first of these velocities is always smaller than the second.

Thus, the adhesion-type detachment of a layer of adhering particles (denudation) depends on the air-flow velocity, while the autohesion process (erosion) depends on the flow velocity and the period of its operation, being expressed by means of arbitrary quantities. Unfortunately, in many treatments the particular conditions of removing the adhering layer of particles have not been distinguished and the time required to effect detachment has not been given (this is usually a long period, exceeding 5 min), the results being estimated simply from the air-flow velocity. The following results relate to the flow velocities at which removal of the layer of adhering particles has been observed:

	v, m/sec
Coking coal, anthracite dust [294]	3-4.5
Magnetite dust [291] ($d_p < 10\ \mu$)	4.0
Quartz dust containing 98% silica [295] ($d_p < 30\ \mu$)	7-10
Fine-grained coal dust [296]*	5.0
Coal dust [297]:	
$d_p \approx 21\ \mu$	10.0
$d_p \approx 58\ \mu$	6.2
$d_p \approx 87\ \mu$	5.0

*In all cases except this the diameter of the pipe was no greater than 300 mm.

On the basis of the properties and dimensions of the particles forming the adhering layer, we may suppose that the investigations in question related to the autohesive detachment of particles, i.e., erosion, the velocities quoted simply being sufficient to overcome the forces of autohesion, since a monolayer of adhering particles with diameters smaller than 100 μ is very little affected by velocities of 3-10 m/sec (see the results presented on page 203). In order to remove a monolayer of adhering particles, air-flow velocities exceeding 100 m/sec are required.

The action of an air flow on an adhering layer may also be expressed in terms of the frontal pressure, i.e., the pressure of the air flow on unit area of cross section of the adhering layer (this is usually expressed in g/cm^2).

The frontal pressure acts either on the end surface or on projections in the adhering layer of particles. The value of the frontal pressure increases with increasing flow velocity and increasing area of interaction between the flow and the attached layer, as in the case of a monolayer of particles with an air flow passing around it [see Eq. (VI.2)].

Thus, the frontal pressure of an air flow in a tube of diameter 200 mm will be $6 \cdot 10^{-3}$ and $5 \cdot 10^{-2}$ g/cm^2 for air-flow velocities of 15 and 30 m/sec, respectively [287]. In the presence of undulating deposits, this pressure rises as a result of the greater area over which the air flow acts on the dust layer. Thus, if there is a projection 3 mm high in a tube 200 mm in diameter, then for an average flow velocity of 15 m/sec, the frontal pressure is $6 \cdot 10^{-1}$ g/cm^2, i.e., 100 times greater than for an even dust layer [287].

If the dust-laden plates are placed at an angle to the flow, the rate of detachment of the upper layers of magnetite dust held by autohesive forces may be determined [291, 292] from the empirical formula*

$$v_{\varphi} = v_{det} \cdot k_a \cdot \sin \varphi \qquad (VI.34)$$

where v_{φ}, v_{det} are the dust-layer detachment velocities for flow—surface angles of φ and 0°, respectively, in m/sec, and k_a is a coefficient depending on the arodynamic resistance of the channel (α): for $\alpha = 3 \cdot 10^{-4}$, $k_a = 1.7$; and for $\alpha = 14 \cdot 10^{-4}$, $k_a = 1.1$.

*The formula is valid for φ = 60-80°.

We see from formula (VI.34) that, on increasing the angle between the air flow and the dust-laden surface (from 60–80°), the dust is detached with greater efficiency [291, 292].

Detachment by Means of a Dust-Laden Air Flow. The air flow may itself contain solid particles. The detachment of an adhering dust layer will then be due partly to the effects of the flow of air moving at a specific velocity and partly to collisions between the airborne and static (adhering) particles. When these two forces are added together, the detaching force will increase, and detachment may take place in a less rapid air flow. The more particles the air flow contains, the more will the detachment velocity be reduced.

Thus, the velocity required to detach shale particles from the surface of a glass tube by an air flow containing particles 250–475 μ in diameter falls from 9.8 to 8.5 m/sec as the number of these particles increases [290].

In contrast to the case of a sphere striking a plane [298, 301], when an airborne particle collides with an adhering particle the change in momentum equals the momentum associated with the total force employed in deforming the contact zone and removing the adhering particle (F_1 and F_{det}, respectively), i.e.,

$$m\Delta v = F_m t \text{ and } F_m = F_1 + F_{det} \qquad \text{(VI.35)}$$

where F_m is the force with which the moving particles acts on the adhering particle.

The value of F_1 may be calculated from the Hertz formula (see §14):

$$F_1 = \int_0^x F_x\,dx \quad F_1 = F_m \cdot \sin\beta$$

where F_x is the force arising for central impact between the bodies, x is the deformation of the contact zone, and β is the angle at which the particle meets the surface.

Before becoming detached, the particle may slide on the surface. The conditions under which the particle may slip are

$$F_{det} \geqslant \mu F_{ad}, \text{ where } F_{det} = F_m \cos\beta$$

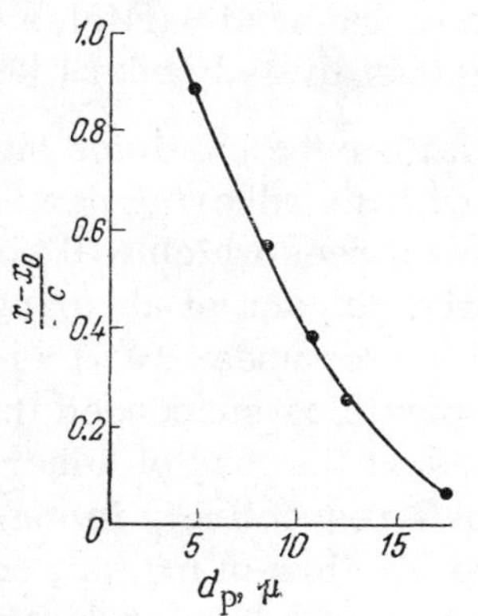

Fig. VI.15. Loss of adhering particles as a function of the diameter of the particles suspended in the air flow.

The coefficient of friction (μ) and the adhesive force (F_{ad}) are determined experimentally [13, 21].

The particle moving at velocity v_1 strikes an adhering particle and rebounds at a velocity v_2. In order to be specific, let us suppose that v_2 equals the velocity of free settling of the particles. If not, the particle may again reach the surface and not be caught up in the air flow.

For calculations based on formula (VI.35), we must determine the duration of the impact. For this purpose we make use of the experimental results of Malyshev [207] (see Chapter V).

After the substitution we obtain the following from Eq. (VI.35):

$$mv_1^{1/5}(v_1 - v_2) = 6.11 \cdot 10^{-4} (F_m \sin\beta + \mu F_{ad} \cdot \cos\beta) K \quad \text{(VI.36)}$$

where K is a coefficient accounting for the energy losses associated with the deformation of the contact zone.

From Eq. (VI.36) we may determine the velocity which the particles moving in the air flow should have in order to detach the layer of adhering particles.

If there is a large number of particles present in the air flow, the effect of the flow on the adhering dust particles will increase. If 1 kg of air contains 1 kg of sand particles [287] 2 mm or over in diameter, the shearing force of the flow will increase by five or six times as compared with a flow of pure air. The following represents the values of frontal pressure created by an air flow in a smooth tube 200 mm in diameter:

v_{av}, m/sec	15	50
Re	$2.1 \cdot 10^5$	$7 \cdot 10^5$
F_f, g/cm^2:		
air flow	$6 \cdot 10^{-3}$	$5 \cdot 10^{-2}$
flow containing 1 kg sand in 1 kg air	$3 \cdot 10^{-2}$	$3 \cdot 10^{-1}$

In the present case the dimensions of the sand particles in the air flow exceed those of the adhering dust by 1-2 orders [302].

We see from Eq. (VI.35) that, the larger the particles suspended in the flow, the more efficiently will the adhering particles be detached, i.e., the greater the adhesive forces which will be overcome. Figure VI.15 shows the relative amount of adhering magnetite dust carried away (particle diameter under 10 μ) as a function of the diameter of similar dust particles suspended in an air flow [291, 292]. Here, x and x_0 represent the loss of adhering particles in dust-laden and dust-free flows respectively in mg/(m^2 · sec), c is the concentration of dust in the air flow in mg/m^3, and the quantity $(x - x_0)/c$ characterizes the loss of adhering dust due to the kinetic energy of the particles in the incident flow referred to unit dust concentration (mass constant).

The relationship between the amount of dust blown away in unit time from unit area of the dust-laden surface and the number of particles in the flow capable of ejecting the adhering particles may be expressed as follows:

$$x = a\,(v_{\mathrm{av}} - v_{\mathrm{det}})^b + k\,(v_{\mathrm{av}} - v_{\mathrm{det}})\cdot c \qquad \text{(VI.37)}$$

where a, b, and k are coefficients, and c is the dust content of the air flow in mg/m^3. The remaining notation is the same as in Eq. (VI.32).

The first term on the right-hand side of Eq. (VI.37) characterizes the removal of dust particles by the air flow [see Eq. (VI.32)], and the second represents the removal of adhering particles by the dust particles contained in the air. Under the conditions of iron-ore mines with $v_{\mathrm{det}} = 1.8$ m/sec, we have $a = 1.85$, $b = 2$, $k = 0.083$.

§33. Adhesion of Particles to the Inner Surfaces of Air Conduits

The adhesion of particles borne in an air flow to the surface of an air conduit may take place if forces preventing the removal of the particles from the surface occur (see Chapter III); however, a necessary condition of adhesion is that the particles should be brought up to the surface in the first place. As in the earlier sections of this chapter, in this section we shall not pay attention to

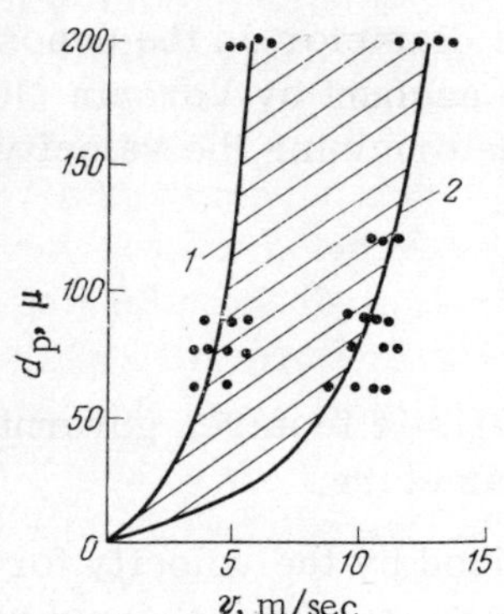

Fig. VI.16. Diameter of the particles carried away by a flow as a function of the flow velocity. 1) Boundary corresponding to no deposition; 2) boundary corresponding to no detachment of the adhering layer.

the actual motion of the particles on the surface but simply select the conditions facilitating or preventing contact and adhesion between the particles and the substrate.

Adhesion of Particles to the Bottom of an Air Pipe. Dust particles never settle at the bottom of a channel (and this of course excludes the question of adhesion) if the vertical pulsating velocity (v_v) of the air flow exceeds the velocity of free settling of the dust particles in the air, i.e., $v_v > v_{free}$. Knowing v_v and its relation to the flow velocity, we may calculate the velocity of the air flow for which there will be no settling of dust. Ryzhenko [303, 304] found that for particles less than 10 μ in size in moving air, the permissible velocities in air pipes of round (v_r), rectangular (v_{rect}), and trapezoidal (v_t) cross section were expressed by the formula

$$v_i \geqslant \frac{a_i \cdot 10^3 \rho}{\sqrt{\alpha}} \tag{VI.38}$$

where v_i is the average velocity in the air pipe for which dust particles less than 10 μ in diameter fail to settle, in m/sec (i = r, rect, or t), a_i is a coefficient ($a_r = 9.6$, $a_{rect} = 6.8$, $a_t = 7.8$); ρ is the density of the dust material in g/cm³, and α is the aerodynamic resistance factor of the tube.

Formula (VI.38) is valid for an air pipe with a round cross section for $\alpha = (5.25\text{–}18.7) \cdot 10^{-4}$ and Re = 80,000–186,000; for a rectangular air pipe with $\alpha = (4.2\text{–}8.05) \cdot 10^{-4}$ and Re = 90,000–250,000; and for a trapezoidal air pipe with $\alpha = (3.66\text{–}15.1) \cdot 10^{-4}$ and Re = 140,000–280,000.

According to Ryzhenko's results [304], for $\alpha = (12\text{–}24) \cdot 10^{-4}$ and $\rho = 2.65$ g/cm³ the minimum velocities are, respectively, $v_r = 0.5\text{–}0.7$, $v_{rect} = 0.37\text{–}0.5$, and $v_t = 0.42\text{–}0.6$ m/sec.

Formula (VI.38) takes no account of the effect of the cross-sectional area of the channel and particle diameter on the deposition process. This aspect was taken into account by Voronin [305] who proposed the following formula for determining the velocity in the channel:

$$v = \frac{v_{\text{free}}}{\sqrt{\alpha}}$$

The quantity v_{free} allows for the characteristic features governing the deposition of particles of different diameters.

Thus an air flow may be characterized by the velocity for which particles are not deposited on the bottom of the channel and by the velocity resulting in the detachment of adhering dust (see §§31, 32). The relation between these velocities for particles of mud, sand, and coal dust (particle diameter under 100 μ) in tubes of diameter 125 mm and length 2000 mm made of Plexiglas, steel, and cast iron was determined by Herning [306].

Figure VI.16 shows the diameter of the particles carried away by a flow as a function of the flow velocity above which no particles are deposited (curve 1) and the flow velocity above which detachment of the adhering particle layer takes place (curve 2). In order to determine the velocity in the first case, the experiment was carried out under conditions in which the dust particles moved together with the air flow, and in the second case the particles were spread on the bottom of the tube and an air flow was then directed over them. We see from the resultant data that the detachment of adhering particles demands greater air-flow velocities than are required to prevent the deposition of the same particles from the air flow. In the shaded region only some of the adhering particles will be removed. For particles less than 60 μ in diameter an air speed of over 10 m/sec will not only prevent the deposition of dust, but also remove any adhering layer. Unfortunately, Herning [306] gave no indication of the thickness of the dust layer or the amount of dust introduced into the flow.

Adhesion of Particles to the Walls (Sides) of an Air Conduit. Adhesion to the vertical walls (sides) takes place as a result of the normal component of the velocity of the dust-laden air flow. This component arises from turbulent pulsations of the flow in a direction perpendicular to the surface of the air conduit [302]. The validity of this was confirmed by Ryzhenko and Shcherbina [307], who showed that the amount of dust adhering to Duralumin plates 80×80 mm^2 in size placed along the perimeter of the ventilation drift of the "Kochergarka" mine was roughly the same on the side walls and the roof.

Contact of dust particles may take place when these line up on uneven parts of the surface (adhesion) or on already deposited particles (autohesion). A band of adhering particles first forms on the surface and then steadily increases with time until the whole is covered.

Microphotographs show that particles are primarily caught and held on surface irregularities. Not all the particles come into contact with the surface and adhere to it. Thus the dimensions of the adhering particles are principally of the order of 2-3 μ for the fine fraction, although the particles suspended in the air are up to 12 μ in size.

Averbukh [308] studied the adhesion of alumina particles to the dry and wet bottom (moistened with water in the latter case) of a horizontal channel of rectangular cross section 35×75 mm^2 and 1000 mm long; he found that for a certain air-flow velocity the deposition of the dust on the dry and wet surfaces was identical (Fig. VI.17). On increasing the flow velocity the deposition coefficient* for the dry surface became smaller than for the wet. Moreover, the difference in the dust-catching power of the wet and dry surfaces is less appreciable for fine than for coarse dust (Fig. VI.17). This is due to the following circumstance. Coarse dust, having a considerable kinetic energy, in general rebounds from a dry surface but sticks to a wet one. A wet surface captures coarse particles (having a greater kinetic energy than fine ones) more easily.

The sticking fraction of the total number of particles touching a surface may be characterized [309] by the adhesion probability:

$$w = A \cdot v^n$$

*The deposition coefficient or factor (K_{dep}) is the ratio of the number of adhering particles touching the surface (see § 34).

where A is a coefficient depending on the form of the dust, the properties of the surface, and the humidity of the air as well as the dampness of the surface, v is the average velocity of the air in the air conduit, and n is a power index.

The variation in dust concentration associated with adhesion to the sides of the air conduit may be put in the form

$$c = c_0 \cdot e^{kL}$$

where c_0 and c are the dust concentrations before and after passing through the air conduit in mg/m^3; k is a coefficient varying (for particles above 1 μ in diameter) from $8.4 \cdot 10^{-3}$ to $4.6 \cdot 10^{-3}$ as the mean flow velocity rises from 0.25 to 0.85 m/sec; and L is the length of the air conduit in m.

This equation is valid for smallish flow velocities at which no detachment of the layer of adhering particles takes place and for smallish particles (diameter up to 10 μ), the rate of free settling of which is insignificant. The reduction in the concentration of such particles on passing along a mine drift (Fig. (VI.18) takes place as a result of their adhesion and not free settling [309]. The greater the flow velocity, the less will be the difference in dust concentration on passing away from the source of dust formation (Fig. VI.17), since, as velocity increases, the layer of adhering particles may be detached and the dust converted into the suspended state. In the present case this phenomenon is undesirable, since adhesion is responsible for a peculiar kind of filtration of the air flow.

The adhesion of dust particles moving in a flow is possible if $F_{ad} > F_{det}$. The class of detaching forces (F_{det}) in general includes forces determined by the elastic properties of the bodies in contact*

$$F_{elas} = K_{elas} r^2 \cdot v_p^{6/5} \qquad \text{(VI.39)}$$

where K_{elas} is a constant depending on the elastic properties of the material, r is the radius of the particles, and v_p is the particle velocity.

*The dependence of adhesion on the elastic properties of a paint coating was considered earlier (see § 26).

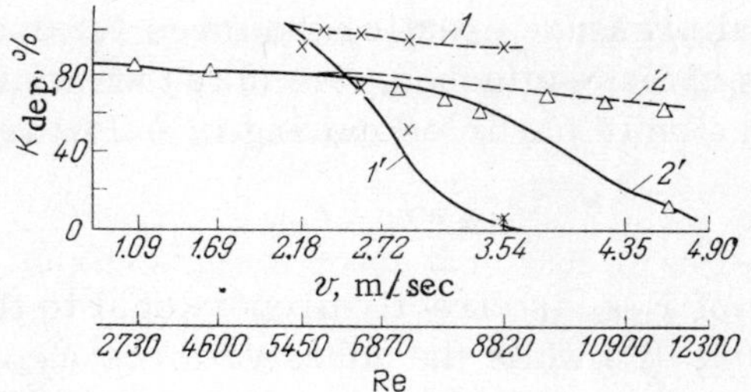

Fig. VI.17. Deposition coefficient of alumina particles on wet (1,2) and dry (1',2') surfaces as a function of the air-flow velocity. 1,1') d_p = 50-60; 2,2') 20-30 μ.

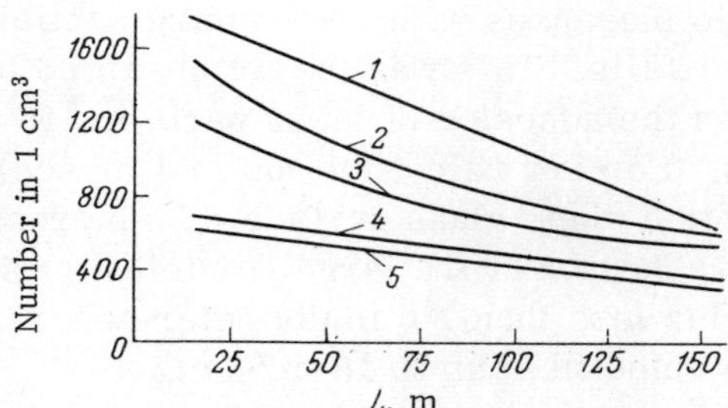

Fig. VI.18. Variation in the concentration (number of particles in 1 cm^3 of dust more than 5 μ in diameter along a mine drift 6.7 m^2 in cross section for air-flow velocities of: 1) v = 0.25; 2) 0.35; 3) 0.6; 4) 0.7; 5) 0.85 m/sec.

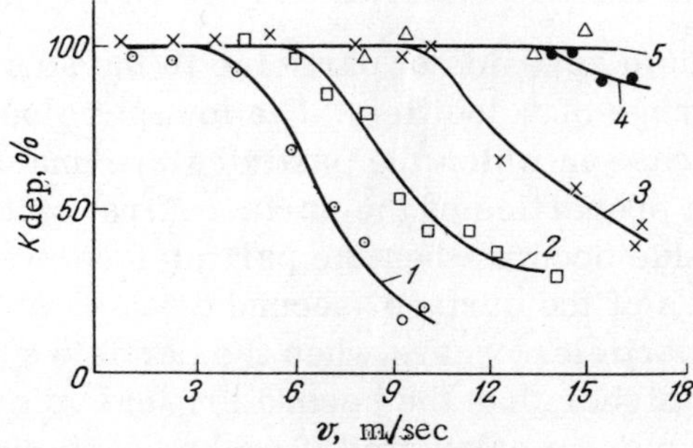

Fig. VI.19. Coefficient of deposition as a function of flow velocity for particles of diameter: 1) d_p = 30-40; 2) 20-30; 3) 5-20; 4) 1-5; 5) less than 1 μ.

The frontal pressure creating the force F_f also tends to tear off the particles already attached. We may thus write the condition for the adhesion of particles moving in a flow as follows:

$$F_{elas} + F_f \leqslant F_{ad} \qquad \text{(VI.40)}$$

The value of F_{elas} is directly proportional to the square of the particle radius (r^2) while the adhesive force (see §16) is proportional to $1/r$. Hence, the ratio $F_{ad}/F_{elas} \sim 1/r^3$.

The coefficient of deposition (K_{dep}) depends both on the conditions governing the flow around obstacles and on the elastic properties of the surface. For the same flow conditions the value of K_{dep} is directly proportional to the ratio F_{ad}/F_{elas}. Since the value of this ratio increases with diminishing particle size, K_{dep} will increase as r falls. This was in fact observed by Tekenov [83] in experiments on the adhesion of loess particles to a glass surface (Fig. VI.19). For low velocities (up to 4 m/sec) particles of all sizes adhere to a plane glass surface. On increasing the airflow velocity the adhesion of the large particles diminishes (curves 1 and 2). Particles less than 1 μ in diameter adhere even for relatively large flow velocities (up to 15 m/sec).

Thus we may assert that for certain particle sizes there is a certain critical velocity above which the particles will rebound from the surface, i.e., adhesion is impossible (see §26). On further raising the particle velocity to several hundreds of meters per second (in Fig. VI.10 these are not shown), the dust particles penetrate into the surface material and are firmly fixed.

The minimum adhesion of particles to the surface will occur over a certain range of velocities. The lowest velocity value corresponds to the case in which the particles are unable to overcome the elastic properties of the surface (first critical velocity), and the upper value occurs when the particles do overcome the elastic properties of the surface (second critical velocity). Thus, adhesion of the particles occurs when the particle velocity is lower than the first or higher than the second critical velocity. The first critical velocity may be calculated if we know the elastic properties of the surfaces in contact. These calculations have been verified experimentally (see §26). The second critical velocity is only to be determined experimentally, and the number of experiments is very limited.

Deposition of Particles from a Heated Flow. The mechanism underlying the deposition of aerosol particles from a hot flow onto a cold surface is based on the motion of particles situated in a nonuniformly heated medium in a direction opposite to the temperature gradient, i.e., from the high- to the low-temperature zone [310]. Under the influence of the thermophoretic force there is a radial component of velocity from the center of the heated flow to the colder wall, giving rise to the further possibility of contact between the particle and the surface. The deposition of dust is characterized by a deposition coefficient K_{dep} equal in the present case to the ratio (in percents) of the mass of the particles settling on 1-cm length of tube to the mass of all the particles passing through the section of tube in question. Since all the settling particles are fixed to the wall of the tube, the coefficient K_{dep} determines the adhesion of the particles for the thermophoretic deposition of particles from the flow:

$$K_{dep} = c\frac{\nu \Delta T}{Q_0 \cdot T^2} \tag{VI.41}$$

where c is an experimental coefficient [for a $PbCl_2$ aerosol, $c = (3.3 \pm 0.4) \cdot 10^5$ in the cgs system); ν is the kinematic viscosity in cm^2/sec; ΔT is the temperature head in the tube in degrees; Q_0 is the flow of air reduced to normal conditions in cm^3/sec; and T is the temperature of the flow in °K.

A study of the thermophoretic deposition of $PbCl_2$ particles for a concentration of 5-70 g/cm^2 on the walls of horizontal and vertical tubes 4, 7, 14, and 24 mm in diameter with water-cooling of the flow in the Reynolds number range 100-300 (air flow 33.3-133.2 cm^3/sec) showed the validity of Eq. (VI.41).

The coefficient of deposition in the present case is independent of the tube diameter (between 4 and 24 mm). The change in the coefficient along the tube (maximum at the initial cross section) is due to the fall in flow temperature. A study of the deposit on the walls of the tube by means of an electron microscope showed that the deposited $PbCl_2$ particles had dimensions between 0.5 and 3 μ.

§ 34. Adhesion of Dust to Obstacles Situated in an Air Flow

Adhesion of Dust to Cylindrical and Spherical Surfaces. The number of particles of a given size (N) settling on an obstacle may be calculated from

$$N = K_{dep} \cdot v\, Snt \tag{VI.42}$$

where K_{dep} is the coefficient of deposition; v is the flow velocity in cm/sec; S is the middle section of the obstacle in cm^2; n is the dust-carrying factor of the flow, or the number of particles in 1 cm^3; and t is the time in sec.

In the present case, K_{dep} is the ratio of the number of adhering particles to the total number of particles which have passed through the middle section of the obstacle. The amount of adhering dust and the value of the deposition coefficient depend on the conditions governing the flow of the dust-laden air stream around the obstacles, the possible rebounding of particles from the surface, and also the adhesive forces capable of holding these particles. The value of the deposition coefficient is usually less than unity.

The possibility of controlling the deposition of dust on obstacles situated in a flow is important in connection with the solution of certain practical problems. For example, the sticking of dust to the surfaces of high-voltage insulators greatly reduces their insulating properties [311].

The conditions governing the deposition of dust particles on obstacles lying in an air flow were studied experimentally in [312-316].

The deposition and adhesion of dust particles on cylindrical and spherical surfaces take place in a nonuniform manner. The number and diameter of deposited loess dust particles [291] are presented as functions of the angle of incidence on a cylindrical surface for various flow velocities in Fig. VI.20. The number and maximum size of the deposited particles fall as the angle rises from 0 to 90°. For a φ close to zero, the flow velocity is minimal, so that the detachment of particles as a result of aerodynamic forces will be negligible. For φ close to 90°, the number of adhering particles falls sharply, since the oblique impact communicates a rotatory motion to the particles.

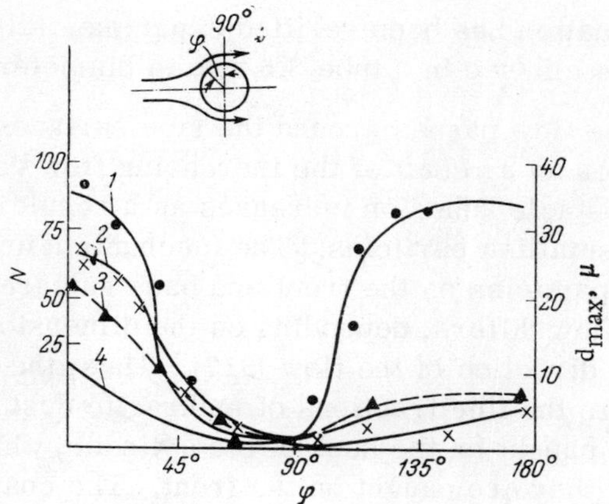

Fig. VI.20. Numbers (curves 1 and 2) and dimensions (curves 3 and 4) of the particles deposited on a cylinder as functions of the incident angle for various flow velocities: 1) v = 5.0; 2) 14.2; 3) 6.0; 4) 16 m/sec.

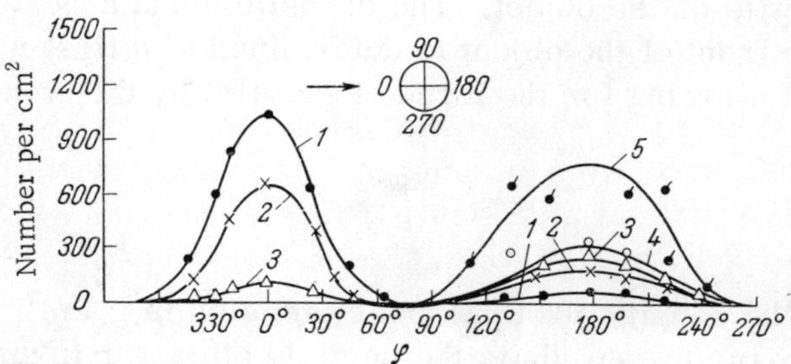

Fig. VI.21. Number of drops of transformer oil deposited on 1 cm² of the surface of a cylinder 2.5 cm in diameter as a function of the angle of rotation of the cylinder for a flow velocity of about 10 m/sec and various drop sizes: 1) d_d = 60; 2) 30; 3) 15; 4) 7.5; 5) 2.5 μ.

The number of paraffin particles 1.8 μ in radius adhering to a quartz cylinder 300 μ in diameter may be expressed as a function of the angle φ (Fig. VI.20) by the equation

$$N = N_0 \cdot \cos \varphi \tag{VI.43}$$

where N_0 is the number of particles adhering for $\varphi = 0$, i.e., along the axis of the flow.

This equation has been verified experimentally for air speeds of 5.8 and 20.4 cm/sec in a tube 8.5 mm in diameter.

When the flow passes around the side surfaces, the detaching force increases as a result of the increasing flow velocity. At the back of the obstacle adhesion increases as a result of eddies, particularly for smallish particles. The mechanism underlying the deposition of particles on the front and back surfaces of the obstacle in the flow differs, depending on the dimensions of the particles and the direction of the flow [317]. Thus, the greater part of the particles in the fine fractions of anthracite dust in a rising vertical flow are caught on the back of the cylinder, while for a descending flow they are caught on the front. The coarse fractions (particle diameter 238.5 μ) in both cases settle on the front; however, the amount of adhering dust is smaller for a rising flow than for a descending one [318].

When flow takes place around an object the trajectories of the particles suspended in the flow deviate from the flow lines on account of inertia. Hence particles pass through the boundary layer and settle on the object. The deposition and adhesion of particles on the front of the object is determined to a first approximation (without allowing for the force of gravity) by the Stokes criterion:

$$\mathrm{St} = \frac{\rho v d_p^2}{18\eta D} \qquad \text{(VI.44)}$$

where ρ is the density of the aerosol particles in g/cm^3, v is the flow velocity in cm/sec, d_p is the particle diameter in cm, η is the dynamic viscosity of the air in P, and D is the diameter of the obstacle in cm.

Theoretically there is a certain critical value of the Stokes criterion; for values of St below this critical value there is no inertial deposition of the particles [6]. However, in practice, particularly for very large flow velocities, where turbulence is involved, this situation frequently fails to hold, and it is therefore difficult to calculate the number of particles deposited.

Figure VI.21 shows the number of oil drops deposited on a cylinder as a function of the angle of rotation of the latter for drops of various sizes [317]. The use of drops eliminates the question

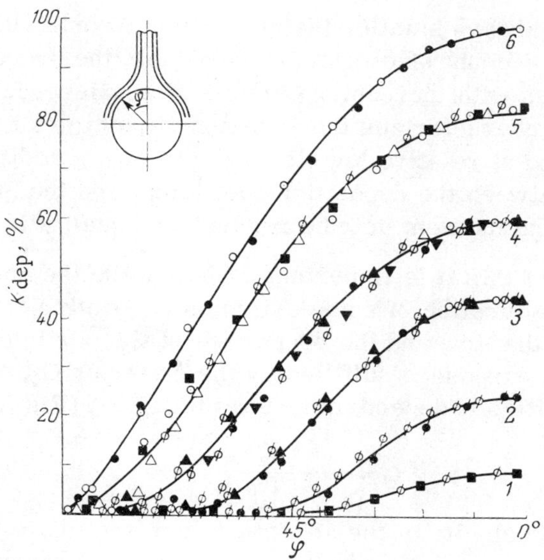

Fig. VI.22. Deposition coefficient of dust on the surface of a cylinder as a function of the angle of rotation of the cylinder for various values of $\log C_1$. 1) 1.73; 2) 2.53; 3) 3.28; 4) 3.90; 5) 4.84; 6) 6.07.

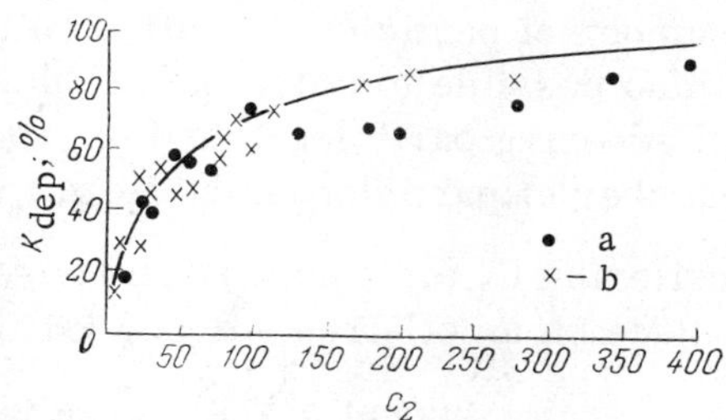

Fig. VI.23. Dependence of K_{dep} on the criterion C_2 for the surfaces of a sphere (a) and a cylinder (b).

of rebounding so that adhesion may be observed in pure form. In accordance with the value of the Stokes criterion, in the present case deposition of the particles occurs on the front. Analysis of the experimental results shows that when flows pass around cylindrical and spherical objects, the number of solid particles held on the surface is always smaller than the number contained in the incident flow on account of rebounding particles, the probability of which increases with increasing particle size. Hence, the Stokes criterion only characterizes the adhesion of particles to the front of an object and at relative low flow velocities. In addition to this, the relation between the deposition coefficient and the Stokes criterion can at the moment only be regarded as qualitative.

In view of this it is important to determine the coefficient of deposition as a function of some criterion accounting for the characteristics of the flow and the properties of the particles and the surface. This purpose is fulfilled by the criterion C_1, on which the deposition coefficient depends quite unambiguously (Fig. VI.22) [318]:

$$C_1 = k \frac{v d_p^5 \rho_a^{2.5} \cdot g \rho_p^{0.5}}{D \eta^3} \qquad \text{(VI.45)}$$

where v is the velocity of the air flow in cm/sec, d_p is the particle diameter in μ, ρ_a and ρ_p are the densities of the air and the particles in g/cm^3, D is the diameter of the cylinder in mm, and η is the dynamic viscosity of the air in P.

Knowing the characteristics of the flow and the obstacle, we may calculate the criterion C_1 and determine K_{dep}. Using the deposition coefficient and the number of particles in the flow, we may calculate the number of particles deposited on the obstacle [see (VI.42)]. It is also possible to solve the problem in reverse. From the number of adhering particles N and the value of K_{dep} we may calculate the number of particles in the flow.

We take the criterion C_1 for a horizontal flow with deposition of particles taking place on an oil-free surface [318].

For a vertical flow the amount of potassium bichromate dust of various fractions settling (Fig. VI.23) on spherical and cylindrical surfaces covered with a layer of vaseline was determined experimentally in [117] and the dimensionless criterion C_2 was calculated:

$$C_2 = \rho \frac{(v \pm v_{free}) d_p^2}{\eta D} \qquad \text{(VI.46)}$$

where ρ is the density of the particle substance in g/cm^3, v is the velocity of the incident flow in cm/sec, v_{free} is the velocity of free settling of the particles in cm/sec, η is the dynamic viscosity of the air under the experimental conditions in P, and D is the diameter of the sphere or cylinder in cm.

This equation is valid for a vertical flow. The plus sign in front of v_{free} holds when the deposition of the particles coincides in direction with the velocity of the dust-laden flow; otherwise it is replaced by minus.

By using the value of C_2 we may determine the coefficient of deposition and then continue the calculations by means of Eq. (VI.42).

The calculated value of n [see (VI.42)] obtained in terms of C_2 and K_{dep} differs from the experimental value by 10% [117].

We take the criterion C_2 for deposition on a surface covered with an oil film [117], i.e., under conditions in which the rebounding of the particles is excluded.

The criteria C_1 and C_2 are valid for specific conditions [117, 318] (specified dust, surface, and air conduit) and as yet there are no grounds for extending their significance to other cases of dust deposition. In addition to this, the distribution of the dust particles in a flow depends on their dimensions and it is in practice difficult to ensure a uniform concentration of dust in a flow. This fact limits the possibility of calculating K_{dep} by reference to the criteria C_1 and C_2 in practice. However, the calculation of the deposition coefficient as a function of the properties of the flow, surface, and dust, and also the number of particles in the flow, deserves attention and further development.

Cylindrical surfaces may arise not only in the center of the flow but also near the wall of the air conduit. Thus, for fixing the walls of certain air conduits, for example in mine drifts, it is customary to use fixing devices ordinarily constituting cylindrical surfaces. The air flow passes around these surfaces. The particular characteristics governing the way in which air flows around cylindrical surfaces in contact with a plane determine the deposition and adhesion of the aerosol particles.

The reduction in the dust concentration arising from adhesion when a dust-laden flow passes successive fixing devices may be de-

Table VI.2. Coefficient of Deposition of Magnetite Dust Less Than 10 μ in Diameter as a Function of the Velocity of the Air Flow

Velocity of air flow, m/sec	K_{dep}, %								
	Angle of plate to the air-flow axis								
	0°	20°	40°	60°	90°	−20°	−40°	−60°	−90°
0.2	–	–	–	1.20	0.95	1.80	1.05	0.35	0.40
0.6	5.67	5.80	8.0	1.30	0.75	2.70	1.25	0.70	0.75
1.0	4.80	5.14	7.20	4.0	1.55	2.80	1.80	1.05	0.95
1.5	3.67	4.60	6.30	7.40	1.85	3.00	2.00	1.10	1.25
2.0	3.20	3.80	5.60	10.0	2.25	3.20	2.40	1.60	1.57
2.5	2.06	3.50	5.00	10.40	2.50	3.40	2.60	1.90	2.00
3.0	1.74	3.30	4.70	11.20	3.17	3.70	2.90	2.10	2.20

termined from the formula [309]

$$c = c_0 \cdot e^A;\ A = 0.002\, w \frac{PL}{SR_g} \sqrt{\frac{\rho l d_p^2}{SD_{fix}}} \qquad \text{(VI.47)}$$

where c_0 and c are the dust concentrations before and after the passage of the air flow in mg/m^3, w is the adhesion probability $w < 1$, P and S are the perimeter and area of cross section of the flow in m and m^2, respectively, L is the length of the working in m, R_g is the gauge of the fixing system ($R_g = l/D_{fix}$), ρ is the density of the dust material in g/cm^3, l is the distance between the fixing devices in cm, and D_{fix} is the diameter of the fixing device in cm.

This equation holds for specific conditions. The distance between the fixing devices is no greater than five or six times their diameter, the flow cross section is a few square meters, the flow velocity is no greater than 1 m/sec, and the particle diameter has a maximum of 10 μ.

Adhesion to Plates. The number of dust particles adhering to plates placed in a flow is determined by the concentration in the flow, the properties of the dust and of the surface, and also the arrangement of the surface relative to the axis of the flow. In view of the absence of any general review on research on this subject, we shall confine attention to experimental data relating to the adhesion of magnetite dust [319] of diameter smaller than 10 μ and a horizontal plate 40 cm^2 in size situated in a horizontal, rectangular air conduit of cross section 40 × 40 cm^2 (see Table VI.2).

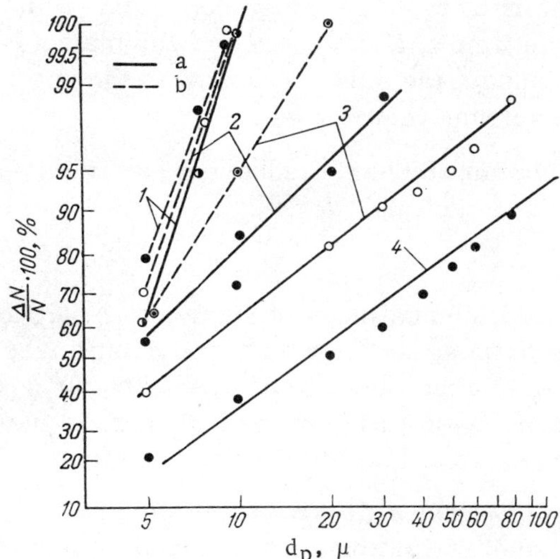

Fig. VI.24. Size distribution of spherical glass particles adhering to an oily (oil density 0.5 mg/cm^2) surface as a function of the inclination of the plate to the flow axis. a) Flow velocity 5 m/sec; b) flow velocity 25 m/sec. 1) 0°; 2) 45°; 3) 90°; 4) distribution of dust in the flow.

The value of K_{dep} rises as the angle between the plate and the flow axis increases from 0 to 40° (for flow velocities, up to 1.0 m/sec) and from 0 to 60° (for velocities of 1.5-3.0 m/sec).

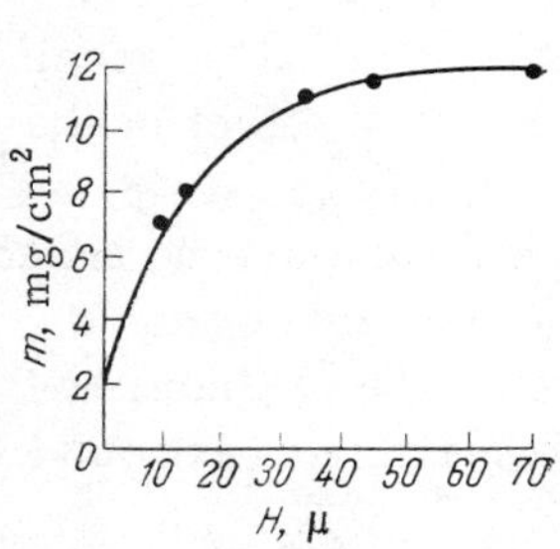

Fig. VI.25. Number of dust particles settling on 1 cm^2 as a function of the thickness of the oil layer.

The presence of an oil film on the surface facilitates adhesion. Rosinski, Nagamoto, and Ungar [320], generalizing their experiments on the deposition of fluorescent powders more than 10 μ in diameter on a sticky surface, proposed the following formula for determining the number of adhering particles:

$$N_t = \left[2S\,\frac{\sin\varphi}{\pi d_p^2}\right]\left[1 - \exp\left(-\frac{\pi d^2}{2S\sin\varphi}\right) n_1 t\right] \qquad \text{(VI.48)}$$

where N_t is the number of particles deposited on the sticky surface from the flow in time t, S is the area of the plate, φ is the angle at which the flow meets the surface, and n_1 is the number of particles carried to the surface in unit time.

The maximum number of adhering particles equals

$$N_{\infty} = -\frac{2S \sin\varphi}{\pi d_{\mathrm{p}}^2}$$

We verified the dustiness of surfaces in air flows experimentally. For this purpose we used an aerodynamic tube containing plates 1 × 2 cm in area clamped at 0, 45, and 90° to the air flow. The original dust (with a wide spread of particle size) was held in a funnel and introduced into the flow at velocities of 5 and 25 m/sec.

Figure VI.24 shows (in probability—logarithmic coordinates) the integral size-distribution curves for particles adhering to surfaces painted with polyurethane enamel and made oily with Avtol, placed at various angles to the air-flow axis, for various velocities of the air flow.

As the angle between the surface and the axis of the air flow diminishes (straight lines 3, 2, and 1), the proportion of small particles increases, and particles of diameter smaller than 10 μ are deposited from a flow directed tangentially to the surface (straight lines 1). This agrees with the results of calculations based on (VI.42) except for the case in which $\varphi = 0$.

On increasing the velocity of the air flow from 5 to 25 m/sec there is also a rise in the proportion of small particles fixed on the surface (straight lines a and b); this is evidently due to the special characteristics of the flow around the obstacle. The effect of the velocity of the air flow is only considered indirectly in (VI.48) by the number n_1. Only at the initial instant, when there are no adhering particles on the surface, is the number of particles settling proportional to the number striking. Thereafter the probability that incident particles will strike those already adhering and rebound from the latter increases.

The number of adhering particles depends on the thickness of the oil film as well. Figure VI.25 shows the amount of settling dust (101 to 165 μ) as a function of the thickness of the oil film for a mean velocity of the dust-laden air blast equal to 26 m/sec. With increasing thickness of the oil film on the surface, the amount of adhering dust increases and reaches a maximum value for an oil thickness of 40-50 μ. In the present case the thickness of the oil film ensuring efficient capture of the dust particles roughly equals half the particle diameter (50-80 μ). However, there are no grounds for extending this rule to particles of all sizes, as suggested by Dergachev [117] (see § 14).

From Eq. (VI.48) we may determine only the number and not the adhesive force of the particles; the force depends on the velocity of the air flow. It has been found experimentally (see Chapter III) that for higher air-flow velocities the particles penetrate more deeply into the oil film (owing to the heavier impact) and are harder to remove from the oily surface.

The number of adhering particles may be increased as a result of electric forces (see § 12). If a steady voltage of 12 kV is applied to a Plexiglas cylinder 1.5 cm in diameter and 7 cm long with built-in copper electrodes, we find that for a flow speed of 3 m/sec a larger number of particles is deposited on the front surface of the sample than would otherwise be the case. If the number of adhering particles in the absence of the electric field is taken as unity, then under the influence of the electric field the deposition of particles 1-5 μ in size increases by a factor of 3 and that for particles 10-20 μ in size by a factor of 1.5. This rise in the number of adhering particles is due to image forces. For particles larger than 40 μ no rise in the adhesive force occurs in the present case when the electric field is present [83].

§ 35. Adhesion when Particles Touch the Sides of Air Conduits. Triboelectricity

Adhesion and Thermal Processes. When particles touch the sides of a conduit, in addition to the forces already considered in §§ 11-15, we observe other forces, associated with the melting of the particles at the contact zone and the triboeffect, which promotes firm adhesion of the particles to the surface.

As regards thermal stability or heat resistance, all particles may be arbitrarily divided into two groups, namely, those respectively sensitive and insensitive to the rise of temperature caused by friction.

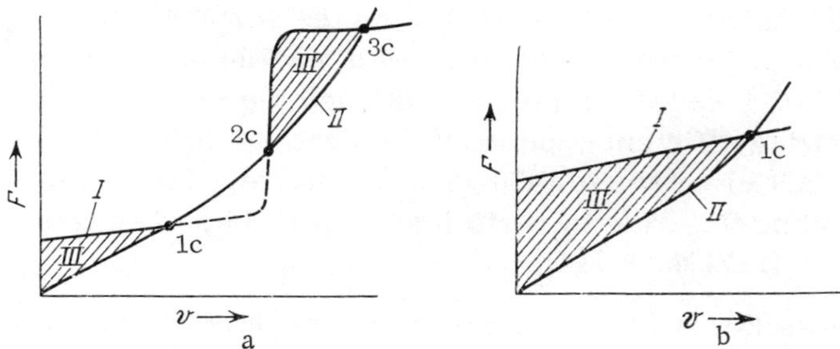

Fig. VI.26. Effect of air-flow velocity on the adhesive force of different classes of dust. a) Temperature-insensitive; b) temperature-sensitive. I) Adhesive forces; II) detaching forces; III) regions of adhesion. 1c,2c,3c) Critical points.

The first group includes particles of substances with low melting and softening points, such as wax, thermoplastics, sulfur, organic dyes, insecticides, and starch sugar. The small amount of heat developing when these undergo friction with a surface is sufficient to effect surface melting or softening. In this case, adhesion arises as a result of tackiness in the contact zone and intensifies on cooling (as a result of thermal processes). Dust particles containing oil, fats, or fatty and oily shales are also capable of being held on a surface as a result of tackiness.

The second group of substances has no such property.

The adhesion characteristics of these two groups of substances were considered by Rumpf [278]. For non-temperature-sensitive particles coming into contact at low velocities (Fig. VI.26b) the adhesive force predominates over the elastic detaching force. As the flow velocity increases, however, rebounding of the particles may in fact take place.

The rebounding of spherical glass particles up to 30 μ in diameter from a steel surface arranged perpendicularly to the dust-laden flow starts at about 13 m/sec.

For temperature-sensitive particles (Fig. VI.26a) the initial adhesion is due to the same processes as in the case of elastic particles, except that the first section (up to the point 1c) is restricted to narrow limits. A rise in the flow velocity leads to a rise in the impact energy and to the melting of the particles in the contact zone, which appears in the form of increased adhesion (point 2c). As the velocity increases further, the amount of heat evolved remains unaltered, and this gives rise to the third critical point (3c). Experiments show that, for all forms of dust, adhesion fails to occur at low velocities exceeding a certain value, 3c for the first, and 1c for the second group of particles (Fig. VI.26).

If a particle sticks in the molten state and then solidifies, the adhesive force of this particle increases. Thus, in the case of iron fragments formed in the braking of railroad trucks as a result of friction between the wheels and the brake shoes and tires [321] the true area of contact with the surface increases as a result of the ductility of the material, and when the particles cool they become welded to the original surface.

The temperature resulting from friction in the contact zone may be determined (assuming that the particles are pressed to the surface by a force F and that all the work of friction is transformed into heat) from the equation [322]:

$$T = T_0 \cdot \left(A \frac{\sqrt{\pi}}{2} - 1 \right); \qquad A = \frac{F \mu v \sqrt{at}}{kT_0} \tag{VI.49}$$

where T is the temperature in the contact zone, T_0 is the melting point, μ is the coefficient of friction between the particles and the wall, v is the velocity of the particles relative to the wall, a and k are the coefficients of thermal diffusivity and thermal conductivity of the particle, and t is the time of contact. For $A < A_{cr} = 2/\sqrt{\pi}$, melting is impossible, since the temperature in the contact zone will be lower than the melting point; for $A \geq A_{cr}$, the particles do melt in the contact zone and the adhesion thus increases on cooling.

Adhesion and Triboelectricity. When dust particles move in an air flow limited by walls, the adhesive force may be increased as a result of the electric charges arising when the particles come into contact with the solid surface. The magnitude and sign of the charge associated with contact between two bodies under static conditions can only be calculated for clearn surfaces (see §11). However, such calculations are difficult to carry out for real systems involving the motion of particles. Hence we shall have to confine our attention meanwhile to qualitative results [323].

Table VI.3. Charge on the Particles of Various Minerals* as a Function of Particle Size

Mineral	$q \cdot 10^{-9}$, C/cm² for particle sizes:					
	5-10 μ	10-55 μ	50-100 μ	100-250 μ	250-500 μ	500-1000 μ
Quartz	−1.01	**−4.63**	−4.34 / −3.08	−2.59 / −1.34	−2.17 / −0.55	−0.37 / −0.21
Microcline	+1.62	+1.97	+2.71 / +0.95	**+12.90** / +0.82	+4.03 / −0.24	+0.31 / −0.22
Calcite	+0.51	+1.48	+6.29 / +3.78	**+11.88** / +3.83	5.18 / 1.91	+4.49 / +1.70
Gypsum	–	–	+14.67	**+20.53** / +3.09	11.52 / 2.76	+7.86 / +3.07
Hornblende	−0.62	**−4.50**	−3.97	−1.54 / +1.62	−0.55 / −0.81	−0.14 / −0.31
Muscovite	–	+0.53	+42.27	+89.52 / −1.59	**134.3** / −2.53	+79.0 / +4.64
Biotite	−1.09	−10.01	**−25.25**	−20.32	−4.48 / −16.45	+0 / −21.06

*Maximum charge values indicated by darker print. The numerator refers to the results obtained on blowing the particles through the tube and the denominator to those obtained by pouring them through.

When quartz, talc, or starch particles come into contact with a Pyrex surface, symmetrical charging occurs, i.e., the number of positively and negatively charged particles is approximately equal [324]. In some cases the particle charge of one sign prevails over the particle charge of the other sign. Particles coming into contact with a layer of polycrystalline sulfur receive a negative charge. However, sulfur particles coming into contact with a quartz surface are charged positively.

The sign of the charge cannot affect the adhesion associated with image forces (see §12). In the case of symmetrical charging, when a layer of adhering particles is formed, there may be a discharge, and this will reduce the autohesive forces between the particles in the layer and the adhesion of the layer to the surface by the equivalent of the Coulomb constituent.

Runge [325] found the greatest charge on cocoa powder and maize flour in a copper tube. On blowing sugar powder through a narrow glass tube a potential difference of the order of 20 kV may develop. On blowing coal dust through a glass tube it becomes charged negatively and adheres to the inner walls of the tube.

Thomas [326] and Walther and Franke [327] passed coal dust (sifted through a sieve and oxidized in air at 350°C) through an iron drum so that the dust struck the metal surface several times. The measured charge on the surface was $6.6 \cdot 10^{10}$ unit charges per gram of dust, or 10^3-10^4 unit charge on each particle. With the very small area of contact involved, the charge density in the contact zone reaches values sufficient to have an appreciable effect on the adhesive force (see §11).

Chao Tse-san [328, 329] studied the electrification of powders on blowing through a copper tube 40 mm in diameter and 0.46 m long. The most widespread minerals were taken for study: quartz, microcline, calcite, muscovite, biotite, gypsum, and hornblende. The charge was determined by reference to the specific surface of 1 g of powder calculated by a geometrical method. In one series of investigations [328] the charge density of the particles was determined by blowing them through a copper tube (with $\varphi = 0°$) in an air flow with an average velocity of 6 m/sec (Table VI.3, see the results presented in the numerator); in the other series [329], the particles were poured through the same tube inclined at an angle of $\varphi = 60°$ (results in the denominator). It was found that on blowing particles (of any of the materials) through the tube, the charge first rose with increasing diameter, reached a maximum, and then fell. On comparing the magnitudes and signs of the charges obtained by pouring (with $\varphi = 60°$) and blowing (with $\varphi = 0°$) the particles, we see that in practice these were not the same (except for hornblende, and to some extent, quartz). This was because, first, in an air flow traveling at 6 m/sec the particles jumped along the surface rather than moved smoothly and, secondly, the magnitude and sign of the charge were random quantities and may have varied considerably in the course of the experiments.

We notice that the charges on the particles may also vary with temperature. Figure VI.27 shows this relationship for quartz and microcline particles moving along a quartz tube [329]. We see that the charge on the heated quartz particles is greater than that

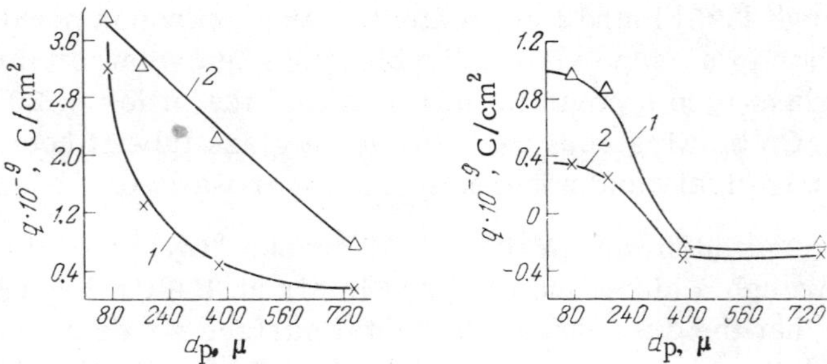

Fig. VI.27. Charge on quartz (a) and microcline (b) particles formed by friction on a quartz tube as functions of the dimensions of the particles for various temperatures of the medium. 1) Particles dried in air; b) heated to 105°C.

on the unheated ones. For microcline particles (Fig. VI.27b) the reverse is true. In the range of particle sizes considered, the charge increases with increasing size of the particles. Heating the quartz particles to 105°C removes condensed moisture and reduces the specific resistance of the material, which falls from 10^6 to 10^3 M [329]; this promotes the transfer of electrons and raises the contact potential difference and hence the adhesion (see §11). A rise in temperature causes internal ionization, which affects the double layer in different ways and may even change its sign, reducing the original charge on the particles; this occurs for microcline (Fig. VI.27b).

§36. Detachment and Adhesion of Particles of Different Sizes

It follows from the foregoing arguments that, when a dust-laden surface is subjected to an air flow, two processes may take place: the detachment of previously adhering particles, and the deposition of particles carried along by the flow.

The degree of removal of a monolayer of particles will be affected by the adhesive force, the velocity of the flow, and the structure of the boundary layer at the surface around which the air is flowing. If we know the change in the size distribution of the adhering particles arising as a result of the air flow, we may calculate the extent to which the surface is freed from adhering particles and express this as a function of the flow velocity.

The removal of the layer of particles depends on the ratio of the forces of adhesion and autohesion. The adhesion-type detachment of the adhering layer (denudation) is determined by the velocity of the air flow and the adhesive force. Autohesion-type detachment (erosion) depends not only on the autohesive force and air speed but also on the time for which the air flow operates. Hence the removal of either a layer or a monolayer of adhering particles is determined, other conditions being equal, by the velocity of the air flow. The flow velocity required for the detachment of the particles in turn depends on the dimensions of these particles.

The results presented below indicate how the rate of air flow is related to the radius (r) of the detached particles for various forms of detachment:

Form of detachment. . .	Detachment of a monolayer	Denudation	Erosion
Dependence on r	$\sim 1/r^{3/2}$	$\sim 1/r$	$\sim 1/r^{1/2}$
Formula	(VI.2a)	(VI.30a)	(VI.33)

It follows from these results that, as the size of the adhering particles diminishes, the greatest increase in velocity should correspond to the removal of a monolayer and the smallest to the autohesive removal of a layer of particles (erosion).

Dust carried by a flow may adhere to the inner surface of an air conduit and also to obstacles situated in the flow.

In order to prevent the settling and adhesion of particles to the bottom of an air conduit, the vertical pulsating velocity of the flow must exceed the dust-particle settling velocity. For flows no faster than about 30 m/sec the latter is only possible for particles with diameters less than 10 μ, while larger particles may stick to the bottom of the air conduit.

Relatively small dust capable of changing its motion as a result of turbulent pulsations directed normally to the surface of the wall tends to stick to the walls of air conduits.

The number of particles adhering to an obstacle is determined by the coefficient of deposition, the value of which depends both on the size distribution of the dust and the dimensions of the obstacle, and also on the flow conditions around the object and the elastic properties of the surface.

As the flow velocity increases, the proportion of large particles among those sticking to the surface diminishes since, although the large particles are capable of overcoming the inertial forces, they cannot become attached to the surface in view of the rise in F_{elas} and F_f.

The adhesion of particles, allowing for the elastic properties, is determined by condition (VI.40). The forces tending to hold particles on a surface around which an air flow is passing (or to remove them from the surface) depend on the particle size in the following way:

Forces	F_{elas}	F_f	F_{ad}
Dependence on r	$\sim r^2$	$\sim r^2$	$\sim 1/r$
Formula	(VI.39)	(VI.3)	(see §17)

Hence, with increasing particle size, the forces tending to prevent their attachment to the surface rise and the adhesion diminishes.

For a flow velocity exceeding a certain value (first critical velocity) the particles will rebound. The value of the first critical velocity depends on the elastic properties of the particle and obstacle surfaces and is inversely proportional to the particle size [see formula (V.11)]. The coefficient of deposition increases when the possibility of particle rebound is eliminated. This may be achieved by making the surface tacky or increasing the adhesive force, particularly by virtue of the triboeffect and the cooling of the molten zone of contact.

The detachment and settling of particles from or on a cylinder placed in an air flow depends on the angle of incidence (the angle between the axis of the flow and the generator of the cylinder). Detachment takes place more easily from the part of the surface placed parallel to the flow and less easily from the back. On the other hand, the minimum number of particles become attached to the part of the cylindrical surface parallel to the flow, while the maximum adhesion occurs on the frontal part of the cylinder; as the velocity of the flow increases, the proportion of large particles on this part of the cylinder becomes smaller.

Chapter VII

Adhesion of Particles in Flow of Water

§37. Determination of the Flow Velocity Ensuring the Detachment of Adhering Particles

The conditions under which a flow of water may remove particles from a solid surface are expressed by formula (VI.1) as in the case of an air flow. In order to set a particle in motion by the action of a flow of water, either the adhesive forces acting on adhering particles or the weight of lying particles must be overcome. Let us call v_p the velocity of the flow of water capable of pulling particles lying on a horizontal surface along, and v_{det} the velocity ensuring the detachment of adhering particles. In order to determine the way in which the flow acts on the particles ("pulling" or "detaching" action) we must compare the adhesive forces with the weight of the particles.

Values of the adhesive force F_{ad} in units of particle weight ($F_{rel} = F_{ad}/P$) are shown below:

d_p, μ	100	50	15	7.5
F_{rel} (ensuring 100% particle holding). .	1.3	5.6	11.5	45.5

We see from these results that the adhesive force of particles less than 100 μ in diameter is greater than the actual mass of the particles. The adhesive force of particles with diameters exceeding 100 μ will be smaller than the weight of the particles, i.e., $F_{rel} < 1$. Naturally, in order to move such particles, we must overcome their weight, i.e., the velocity of the flow of water must be v_p. In this case, we may neglect the adhesive force [330–332].

For particles less than 50 μ in diameter the adhesive force is much greater than the weight of the particles. Thus, for particles 7.5 μ in diameter, the adhesive force is 45.5 times the weight of the particle itself. In this case we may neglect the weight of the particle and in calculating the flow velocity required to detach the adhering particles (v_{det}) consider only the adhesive force.

In order to verify the effect of adhesive forces on the flow velocity at which detachment of the particles occurred, experiments were made with spherical glass particles 2-60 μ in diameter and plane steel plates 20 × 59 mm in area, the surface of which was given in Class 9 finish [224]. A monolayer of particles was deposited on the steel surfaces by free settling. The plates were fixed in the socket of a special trough and washed with a flow of water. In order to even out the flow velocities over the width of the trough, the length of the latter up to the socket was made six times the length of the plate. The depth of the flow was no greater than 5 mm. The number of particles at the beginning and end of the experiment was determined with a microscope, using the method described in § 9.

For each velocity at least 10 measurements were made and the average v_{av} was computed.

The actual velocity at which detachment of the particles occurs (v_{det}) depends on the particle size and the conditions governing the flow of liquid over the dust-laden surface. For low flow velocities and particles of a given size we may assume that the relation between v_{det} and v_{av} will be the same. The relation between the coefficient K_N and the water-flow velocity for steel surfaces placed horizontally is *

v, cm/sec	36	54	72	90	108
K_N	70	90	100	$2 \cdot 10^3$	∞

We see from the data presented that for a flow velocity of 36 cm/sec a considerable number of particles are detached (K_N = 70) only the most stubborn remaining. A water flow moving at 108 cm/sec almost entirely cleans the surface from adhering particles.

In order to determine the relation between the velocity required to remove the particles and the particle size, we must con-

*The initial number of particles averaged 4000 in 1 cm^2; the time during which the flow acted on the particles was 30-60 sec.

sider the forces exerted by the flow on a particle stuck to a horizontal surface in the same way as in §31, where we considered an air flow [333]. The particle is acted upon by pressures and also by forces tending to remove it from the surface.

The pressures in question (these include P, the weight of the particle in water and F_{ad}, the adhesive force) are experimentally determinable.

The detaching forces (these include F_f, the frontal pressure of the flow, and F_{lift}, the lifting force created by the flow) are

$$F_f = c_x \cdot d_p^2 \cdot \frac{v_{det}^2}{2g} \tag{VII.1}$$

$$F_{lift} = c_y \cdot d_p^2 \cdot \frac{v_{det}^2}{2g} \tag{VII.2}$$

where c_x and c_y are particle form factors, d_p is the diameter of the particle, and v_{det} is the velocity at which detachment takes place.

The value of v_{det} is directly proportional to v_{av}, and if this proportionality is taken into account by the coefficients c_x and c_y then $v_{det} = v_{av}$ (v_{av} is the average velocity of the flow at which detachment of the adhering particles becomes possible).

The adhering particles will be removed from the surface if the conditions expressed by formula (VI.1) are satisfied.

The relation between the adhesive forces and particle size for the case under consideration may be expressed by formula (IV.44).

Substituting the quantities given by formulas (VII.1), (VII.2), and (IV.44) into (VI.1), we obtain

$$v_{av}^2 \cdot \left(\frac{c_x + c_y}{2\mu g} \right) = \frac{k}{d_p} + \frac{P}{d_p^2} \tag{VII.3}$$

The adhesive forces of small particles greatly exceed their weight, i.e., $F_{ad} \gg P$, so that P may be neglected. Then Eq. (VII.3) takes the form*

*In view of the fact that the detachment velocity increases as particle size diminishes, Re ~ const; since $c_x = f(\text{Re})$, we may regard c_x as constant and only dependent on v_{av}.

$$v_{\text{av}}^2 \cdot \left(\frac{c_x + c_y}{2\mu g} \right) = \frac{k}{d_{\text{p}}} \tag{VII.4}$$

For large particles with $P > F_{\text{ad}}$, $P = 0.523 d_{\text{p}}^3 \rho$, and Eq. (VII.3) transforms to

$$v_{\text{av}}^2 \cdot \left(\frac{c_x + c_y}{2\mu\rho g} \right) = 0.523\, d_{\text{p}} \tag{VII.5}$$

where ρ is the specific gravity.

Denoting the expression in brackets by 1/A, we obtain from (VII.4)

$$v_{\text{av}}^2 = \frac{A \cdot k}{d_{\text{p}}} \tag{VII.6}$$

Allowing for the value of k, and assuming the complete removal of all particles for a flow velocity of 108 cm/sec, Eq. (VII.6) takes the following form:

$$v_{\text{av}}^2 = \frac{1{,}17}{d_{\text{p}}} \tag{VII.7}$$

In order to verify the validity of this relationship, the velocity corresponding to the detachment of a set of particles of constant size by a flow of water was determined experimentally and the results were compared with theory. The following represents the relation between the water-flow velocity required to detach 100% of the spherical glass particles from a steel surface and the particle diameter:

d_{p}, μ	80-100	30-40	5-10
v, cm/sec:			
experimental . .	7	17	41
calculated	11.4	19.1	39.5

We see from these results that the proposed formula (VII.7) may be used for calculating the velocity required to remove spherical glass particles from a steel surface. For other surfaces the computing relationship between v_{av} and d_{p} may be obtained from Eq. (VII.6) if we know the coefficients k (see §24) and A. The latter is determined experimentally for given v_{av} and d_{p}.

Thus, for smallish particles, it is sufficient to determine the adhesive forces between these and the surface in order to calculate the flow velocity at which all the adhering particles are removed. For larger particles, the adhesive forces of which are smaller than the weight of the particles, the "pulling" velocity of the flow v_p may be calculated from other empirical formulas [330-333].

§38. Characteristics of the Detachment of Particles by a Water Flow

Detachment and Pulling of Particles. In a liquid medium the detaching forces of adhering particles depend on the structure of the boundary layer in the same way as in the case of air (see §31). Shulyak [334] made an experimental study of the detachment of particles from a thin plate moving in a stationary medium.

The following represent the experimental values of the average flow velocities (v_{av}) required to remove particles of diameter 80-500 μ, the depth of immersion of the particles in the boundary layer (δ is the thickness of this layer), and the relative detachment velocities:

d_p, μ	80	100	250	500
v_{av}, cm/sec	4.34	6.20	8.00	10.80
$2d_p/\delta$*	0.22	0.26	0.46	<1.0
v_{det}/v_{av}	0.34	0.40	0.70	1.0

The relative velocity in the boundary layer is the ratio of the flow velocity at a level equal to the height of the particle (v_{det}) to the average velocity (v_{av}) of the flow (or in the present case, the velocity of the plate). The relative velocity shows what proportion of the average flow velocity is represented by the velocity at a level equal to the height of the particle.

It follows from these results that the particles lie in the laminar layer and their detachment is effected at a velocity lower than the mean velocity of the flow (velocity of the plate).

*The layer of adhering particles had a thickness equal to two particle diameters.

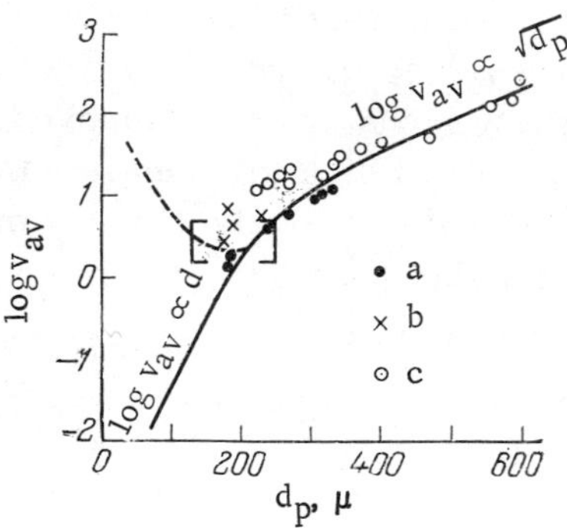

Fig. VII.1. Average flow velocity causing the motion of particles as a function of particle diameter. a,b) Shulyak's results; c) results of Tvenkhofel and Goncharov.

In view of the fact that v_{det} and the ratio v_{det}/v_{av} are as a rule unknown, it is usual to determine the dependence of v_{av} on d_p rather than that of v_{det}. According to Shulyak [334], Goncharov [331], and Tvenkhofel [335], for particles of diameter 250-500 μ the flow velocity is proportional to $\sqrt{d_p}$, which corresponds to formula (VII.5) when $F_{ad} = 0$. For particles of smaller size, these three authors consider that, in contrast to formula (VII.5), v_{av} should be proportional to d_p (Fig. VII.1). This discrepancy is apparently due to the fact that the relative velocity falls as particle size diminishes.

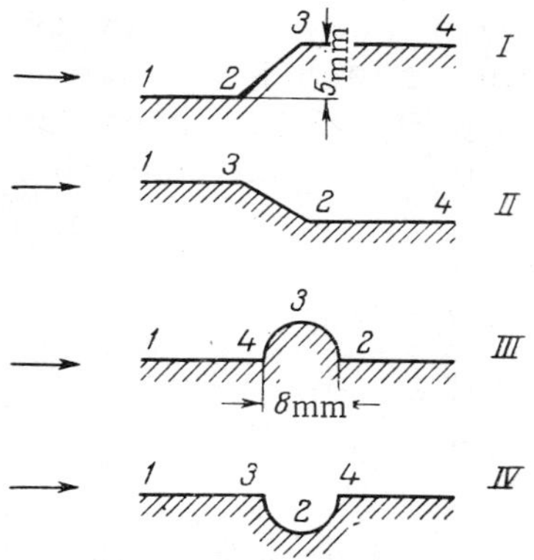

Fig. VII.2. Typical configurations (I-IV) of complex surfaces (1-4 represent sections of the surfaces).

The three authors cited took no account of the adhesion of the particles, i.e., they restricted attention to the "pulling" flow velocity. In order to calculate this velocity with respect to particles lying on a rough surface on the basis of experimental data, the following formula [334] was proposed:

$$v_p = \sqrt{2.2\,\frac{\rho_1-\rho_2}{\rho_2}\,g\,d_p + \frac{36\nu^2}{\rho_2^2\cdot d_p^2}} - \frac{6\nu}{\rho_2\cdot d_p} \tag{VII.8}$$

where ρ_1 and ρ_2 are the densities of the particle material and the liquid, while ν is the kinematic viscosity of the latter.

Formula (VII.8) is valid not only for the beginning of the pulling process, but also in the case of the uniform motion of the particles; in the latter case, v_p must be taken to mean the velocity of the particles relative to the liquid.

Table VII.1. Water-Flow Velocities at Which Spherical Glass Particles are Removed from Surfaces of Complex Configuration (γ_F = 10-20%)

Form of surface (Fig. VII.2)	Velocity of flow at various parts of the surface, cm/sec			
	1	2	3	4
I	46	88	53	49
II	46	88	49	53
III	46	91	53	49
IV	46	73	38	49

If we allow for the adhesive forces, then, in accordance with Eq. (VII.7), the flow velocity causing detachment of the adhering particles is inversely proportional to the size of the particles. Then the velocity—particle size relationship will be expressed by the curve shown by a broken line in Fig. VII.1.

Over a certain range of particle sizes, $F_{ad} + P = \text{const}$ (indicated by the brackets in Fig. VII.1) and the flow velocity v_{av} may not depend on the particle size. This is confirmed by the experimental results of [330, 336, 337].

This fact is of particular importance in studying the process of the removal of particles by a flow of water.

Up to this point we have been considering the conditions required for the removal of particles from plane surfaces. In practice, however, for example, in the washing of vehicles, surfaces of complicated shapes are often encountered (Fig. VII.2). Table VII.1 presents some data relating to the removal of glass particles from surfaces of different configurations (I-IV in Fig. VII.2).

It is hardest to remove particles from sections of type 2. The water-flow velocities at which all the particles are removed are 2-2.5 times greater than those shown in Table VII.1.

Detachment of Particles as a Function of the Position of the Surface. The detachment of particles of various sizes depends not only on the velocity of the water flow but also on the position of the surface relative to the axis of the flow.

The detachment of spherical glass particles less than 50 μ in diameter from plates 20 × 30 mm in area has been studied in two experimental troughs, one 2 × 0.2 × 0.2 m, and the other 26 × 0.6 × 0.5 m in size.*

Plates painted with perchlorvinyl enamel were used as well as unpainted glass and steel surfaces. These were fixed in a holder in such a way as to allow the position of the samples relative to the axis of the flow to be varied; the holder was placed at ⅔ of the distance from the trough entrance. The samples were deposited in air by the free-settling method to a density of about 0.3 g/m^2.

The position of the plates varying along the vertical axis we shall arbitrarily call "vertical" and that varying along the horizontal axis "horizontal."

Figure VII.3 shows the proportion of particles of different diameters remaining after being acted upon by a flow of average velocity 0.1 m/sec as a function of the position of the dust-laden plate.

For vertically arranged surfaces (see Fig. VII.3a), as the angle α varies from 0 to 90° (i.e., as the position of the plate changes from perpendicular to parallel relative to the flow axis), the proportion of particles of diameter 5 and 10 μ remaining increases, while that of particles 30, 40, and 50 μ in diameter falls. On increasing the angle α from 90 to 180°, the position of the plates changes from parallel to perpendicular; for α = 0° the flow is incident upon the dust-laden surface of the plate, while for α = 180° the dust-laden surface is situated in the opposite position to that of the surface meeting the flow. As the angle α increases from 90 to 180°, the system obeys a law opposite to that characterizing the range α = 0-90°, i.e., the proportion of particles 5 and 10 μ in diameter remaining on the surface diminishes, while that of the larger particles (30-50 μ) increases (Fig. VII.3a).

The proportion of particles 40 and 50 μ in diameter remaining for α = 0° and α = 180° is approximately the same. The small particles 5 and 10 μ in diameter are detached less easily for α = 180° than for α = 0°; this is apparently due to the characteristics

*The work in the large trough was carried out in the hydrophysical laboratory of the Physics Faculty of the Moscow State University under the direction of N. A. Mikhailova and the author wishes to express his gratitude for this.

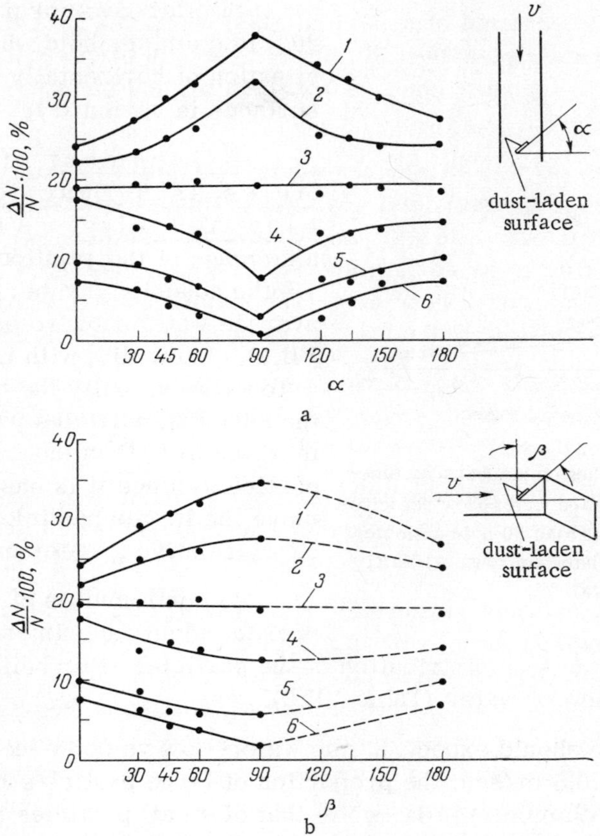

Fig. VII.3. Proportion of particles of various diameters (d_p) remaining on a dust-laden plate after subjecting to a flow of water at an average velocity of 0.1 m/sec as a function of the inclination α of the plate to the vertical position (a) and the inclination β to the horizontal position (b). 1) d_p = 5; 2) 10; 3) 20; 4) 30; 5) 40; 6) 50 μ.

of the flow around the dust-laden surface. The proportion of particles 20 μ in diameter remaining on the surface depends very little on the angle α for vertically placed plates. Particles of diameter 20 μ constitute a curious critical size, separating different regions in the graph representing the removal of particles of various sizes in relation to the position of the plate relative to the flow axis (Fig. VII.3).

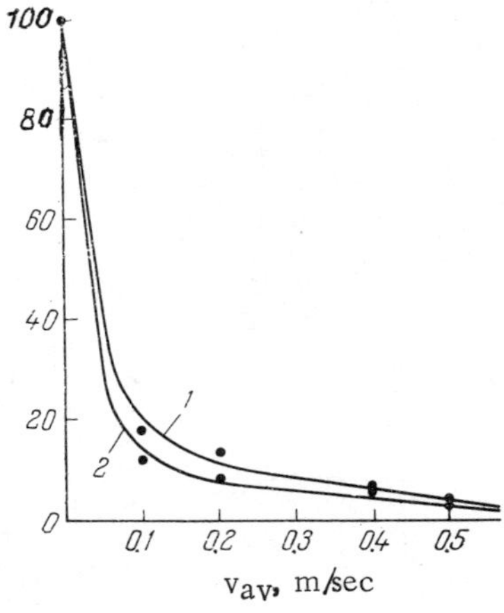

Fig. VII.4. Adhesion number as a function of the average water-flow velocity for particles less than 50 μ in diameter and different plate positions. 1) Vertical; 2) horizontal.

Similar laws for particles 20 μ in diameter hold when the inclination of horizontally placed surfaces is varied (Fig. VII.3b).

Detachment of Particles as a Function of Flow Velocity. A study was also made of the relation between γ_F (the adhesion number) and the average water-flow velocity (Fig. VII.4). Naturally, with increasing water-flow velocity the mass of the adhering particles will fall more sharply than the number of particles, since it is easier to remove the larger particles, which carry relative greater masses.

In confirmation of this we present some experimental data relating to the size distribution of the particles after being acted upon by a flow of water (Table VII.2).

As we should expect, as the water-flow velocity increases from 0.1 to 0.5 m/sec, the proportion of large particles remaining (40-50 μ in diameter) falls, while that of small particles increases. For a flow velocity of 0.85 m/sec nearly all the adhering particles are removed.

Effect of the Methods of Dust Deposition and the Properties of the Surface on the Detachment of Particles. It is well known that there are other ways in which plates become dusty as well as by free settling. Dust particles may occur in drops of rain and settle on surfaces at the same time as these. When the drops dry in the air, the adhesion of such particles becomes much stronger (see §26). The following represents some experimental data relating to the efficiency with which particles are removed by a flow of water (i.e., the variation in the coefficient K_m) as a function of the method by which the particles are deposited on the surface:

Table VII.2. Size Distribution of Particles after Being Acted upon by a Flow of Water

Average flow velocity, m/sec	$(\Delta N/N)\cdot 100$, %											
	for $\alpha = 90°$ and particles of diameter, μ						for $\beta = 90°$ and particles of diameter, μ					
	5	10	20	30	40	50	5	10	20	30	40	50
0.1	35.0	25.0	22.0	11.2	5.2	1.4	34.5	26.5	19.5	12.5	6.0	2.0
0.2	36.4/35.8	26.4/26.8	20.4/20.6	10.6/11.0	5.0/4.8	1.2/1.0	33.0/33.8	26.0/25.2	20.4/20.0	12.5/13.0	6.5/6.2	1.6/1.8
0.4	38.0/37.2	27.5/28.2	19.5/19.4	11.4/11.0	3.6/4.2	0/0	35.6/35.5	26.5/26.4	20.0/20.6	12.5/12.4	4.6/4.1	0.8/1.0
0.5	40.0/39.7	27.0/28.6	20.7/19.0	10.5/10.0	1.8/2.5	0/0	39.0/39.0	27.5/28.0	20.0/20.2	11.0/10.0	2.5/2.8	0/0

*The numerator refers to the experiments in the small trough and the denominator to those in the large trough.

v_{av}, m/sec	0.1	0.2	0.4	0.5
K_m:				
for free settling*	33/30	49/43	73/68	147/130
for settling in a drop of water	32/27	46/42	72/66	135/125

Thus the particular way in which the particles are deposited on the surface (free settling or settling in water drops with subsequent drying of the latter) has no effect on the removal of particles by a flow of water.

Below we present some data relating to the detachment of glass particles by a flow of water (i.e., the variation in the coefficient K_m) as a function of the properties of the surface:

v_{av}, m/sec	0.1	0.2	0.4	0.5
K_m for the following surfaces*:				
metallic unpainted	42/40	59/56	85/80	163/153
glass	60/57	96/98	128/118	267/258
metallic painted	33/30	49/43	73/68	147/130

The adhesive forces of glass particles to a glass surface are smaller than to a painted one. Hence it is easier to clean adhering particles from glass surfaces with a flow of water, and this is confirmed by experience.

Detachment of Particles from Oily Surfaces. The adhesion increases if there is a layer of oil on the surface. This fact inspired a study of the efficiency of removing particles as a function of the oiliness of the surface.

For a surface-oil density of 0.5 mg/cm^2 very few particles are removed by a flow of water with a velocity between 0.1 and 0.5 m/sec.

Table VII.3 presents the results of some experiments with oiliness factors of 0.03 and 0.01 mg/cm^2. Some of the plates were given an oil layer of 0.01 mg/cm^2 and then immediately coated with dust and placed in a flow of water; others of the same oiliness were left in the open air for 4-6 days and then coated with glass spheres.

*In the numerator for α = 90°, in the denominator for β = 90° (see Fig. VII.3).

Table VII.3. Coefficient K_m as a Function of the Oiliness of the Original Surface for Various Water-Flow Velocities

Average flow velocity, m/sec	K_m for the following degrees of oiliness			
	0	0.03 mg/cm²	0.01 mg/cm²	0.01* mg/cm²
0.1	33/30 †	1.3/1.2	2/2	5/4
0.2	49/43	1.6/1.5	6/5	11/10
0.3	73/68	1.9/1.8	9/8	17/17
0.4	147/130	3.0/3.0	16/15	30/28

* Surface kept in the air for 4-6 days after oiling.

† In the numerator for $\alpha = 90°$; in the denominator for $\beta = 90°$ (Fig. VII.3).

It follows from Table VII.3 that the adhesion rises sharply (K_m falls) when there is a layer of oil on the surface. For example, with a water-flow velocity of 0.1 m/sec, $K_m = 1.3$ for an oiliness of 0.03 mg/cm^2 and 2.0 for 0.01 mg/cm^2, while on a clean surface it equals 33, i.e., the efficiency of cleaning the oily surface is 15-25 times lower.

If the oily surface is kept in air first, it attracts atmospheric dust over the waiting period and this to some extent screens the oil layer. When glass particles are introduced on to this kind of surface, they come into contact with the screening layer rather than the oil. In this case, the efficiency of cleaning adhering particles from the surface in a flow of water increases, although it remains much lower than in the case of an oil-free surface.

Simulating the Detachment of Particles. The surface-cleaning factors and particle-size distributions obtained in the large and small troughs respectively after subject to a flow of water (Table VII.2) coincide. We now have to decide how far such results may be used in order to characterize other cases of the removal of attached particles in a flow of water, for example, with varying depth of flow, plate size, etc., i.e., we wish to know whether the process underlying the detachment of adhering particles by a flow of water may be modeled or simulated.

The degree of surface cleaning is determined on the one hand by the velocity and structure of the flow and on the other by the adhesion of the particles.

The adhesive force is due to the properties of the contiguous bodies and the medium and is independent of the size of the plates and the cross section and properties of the flow. Hence, in this case, the modeling or simulation of adhesion phenomena appears to be impossible.

In order to simulate the action of a flow on an adhering particle we must maintain geometric, kinematic (velocity ratio), and dynamic (mass and force) similarity. We cannot choose these conditions in detail. We can only note that for river-type flows the Froude criterion $Fr = v^2/gl$ ensures dynamic similarity and the Reynolds criterion $Re = vd/\nu$ kinematic similarity. Since the forces of gravity prevail over viscosity, the decisive factor in simulating river flows will be the Froude criterion.

In order to establish identical conditions for the detachment of adhering particles we require that the flow should have exactly the same effect on each adhering particle.

The detachment of particles under the influence of a flow of water occurs when the flow (F_{det}) is capable of overcoming the adhesion and weight of the particles, i.e.,

$$F_{det} \geqslant \mu (F_{ad} + P) \qquad \text{(VII.9)}$$

where μ is the coefficient of friction.

For $F_{ad} \gg P$, inequality (VII.9) simplifies and takes the following form:

$$F_{det} \geqslant \mu F_{ad} \qquad \text{(VII.10)}$$

The force with which the flow acts upon the particle depends on the density (ρ) and viscosity (η) of the medium, the particle diameter (d_p), the flow velocity (v), and the conditions governing the manner in which the flow passes around the adhering particles, which are taken into account by the coefficient c_x, i.e.,

$$F_{det} = f(\rho,\ \eta,\ d_p\ v,\ c_x) \qquad \text{(VII.11)}$$

Allowing for inequality (VII.10) and the fact that $c_x = \psi(d)$, we may write

$$F_{ad} = f_1(k,\ \rho,\ \eta,\ d_p,\ v) \qquad \text{(VII.12)}$$

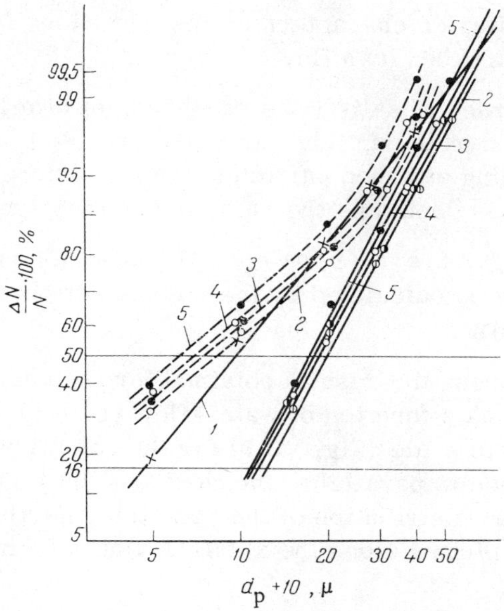

Fig. VII.5. Size distribution of spherical glass particles before and after being acted upon by a flow with the surfaces placed parallel to the latter ($\alpha = 90°$). 1) Original; 2-5) after subjection to a flow at a velocity of: 2) v = 0.1; 3) 0.2; 4) 0.4; 5) 0.5 m/sec.

The coefficient k allows for not only the flow conditions (coefficient c_x) but also other unconsidered energy losses in the flow.

In order to elucidate the laws governing the action of a flow on an adhering particle, let us use the method of dimensional analysis.

It follows from formula (VII.12) that

$$\frac{ML}{T^2} = \left[\frac{M}{L^3}\right]^x \cdot \left[\frac{M}{LT}\right]^y \cdot [L]^z \cdot \left[\frac{L}{T}\right]^q \qquad \text{(VII.13)}$$

$$\begin{array}{c|c} x + y = 1 & y = 2 - q \\ z - 3x - y + q = 1 & z = q \\ -y - q = 2 & x = q - 1 \end{array}$$

$$F_{ad} = k\rho^{q-1} \cdot \eta^{2-q} \cdot d_p^q \cdot v^q = \frac{\eta^2}{\rho} f\left(\frac{d_p v}{\nu}\right)$$

$$\mathrm{Re} = \frac{v d_p}{\nu} \qquad \frac{F_{ad}}{\rho \nu^2} = f(\mathrm{Re}) \qquad \text{(VII.14)}$$

The Re criterion characterizes the conditions for the removal of adhering particles by a flow of water.

Thus, in order to satisfy the conditions of similarity for the detachment of adhering particles we must have Re = const. The efficiency of cleaning attached particles from a surface under different flow conditions should be the same for equal Reynolds numbers.

Cleaning of a Surface. The adhesion number of coefficient K_N may be calculated by the method set out in §31 [formulas (VI.6)-(VI.29)].

Let us express the results obtained for the detachment of adhering particles as a function of water-flow velocity on the probability—logarithmic scale (Fig. VII.5) for the condition in which the dust-laden surface is parallel to the flow ($\alpha = 90°$). We see from Fig. VII.5 that the distribution of the remaining particles (broken line) when d_p is plotted along the x axis deviates from the normal logarithmic law.

In order to express the results on the normal—logarithmic scale, we use an artificial device. Instead of plotting d_p along the x axis we plot $d_p + 10$. Then the results fall neatly on the normal logarithmic distribution. In Fig. VII.5 this distribution is shown by the continuous straight lines (2-5). In this case we must understand $d_p + 10$ as the particle diameter in formula (VI.10).

The results of Fig. VII.5 enable us to find the distribution parameters (d and σ) with respect to the sizes of the particles remaining on the surface after being acted upon by a water flow at different velocities (Table VII.4).

The distribution of the particles remaining on the surface after the action of a flow of water is shown in Fig. VII.6 for plates placed perpendicularly to the axis of the flow, and the change in the distribution parameters is indicated in Table VII.5. Rectification of the resultant particle-size—distribution curves in this case demands the use of coordinates ($d_p + 5$). The distribution of the particles (after being acted upon by the flow) in coordinates of d_p is omitted from Fig. VII.6.

The velocity dependence of $\bar{d}$ and σ is given by the following formulas*:

*In (VII.15) and (VII.16) the flow velocity v is given in m/sec, and $\bar{d}$ in μ.

Table VII.4. Change in the Particle-Size Distribution Parameters under the Influence of a Flow of Water (plates parallel to the flow)

Particle distribution	$\frac{(\log \bar{d})}{\log(\bar{d}+10)}$	$\bar{d}+10$	$\bar{d}$	$\frac{(\log d_{16})}{\log(d_{16}+10)}$	$\sigma + \log\bar{d} - \log d_{16}$
Original	(1.0)	–	10.0	(0.672)	0.328
After being acted upon by a flow at a velocity (m/sec) of:					
0.1.	1.301	19.8	9.8	1.0792	0.223
0.2.	1.272	18.7	8.7	1.0607	0.211
0.4.	1.255	18.0	8.0	1.0414	0.214
0.5.	1.241	17.5	7.5	1.0294	0.211

Table VII.5. Change in the Particle-Size Distribution Parameters after Being Acted upon by a Flow of Water (plates perpendicular to the flow)

Particle distribution	$\frac{(\log \bar{d})}{\log(\bar{d}+5)}$	$\bar{d}+5$	$\bar{d}$	$\frac{(\log d_{16})}{\log(d_{16}+5)}$	$\sigma + \log\bar{d} - \log d_{16}$
Original	(1.0)	–	10.0	(0.672)	0.328
After being acted upon by a flow at a velocity (m/sec) of:					
0.1.	1.130	13.5	8.5	0.832	0.298
0.2.	1.097	12.5	7.5	0.806	0.291
0.4.	1.079	12.0	7.0	0.792	0.287
0.5.	1.041	11.0	6.0	0.763	0.278

$$\bar{d} = 10.3 - 6v \tag{VII.15}$$

for plates parallel to the flow, and

$$\bar{d} = 10 - 9v \tag{VII.16}$$

for plates perpendicular to the flow.

The quantity σ equals the tangent of the angle of inclination of the straight lines characterizing the particle-size distribution (Figs. VII.5 and VII.6). The straight lines showing the size distribution of the particles remaining are parallel (straight lines 2,

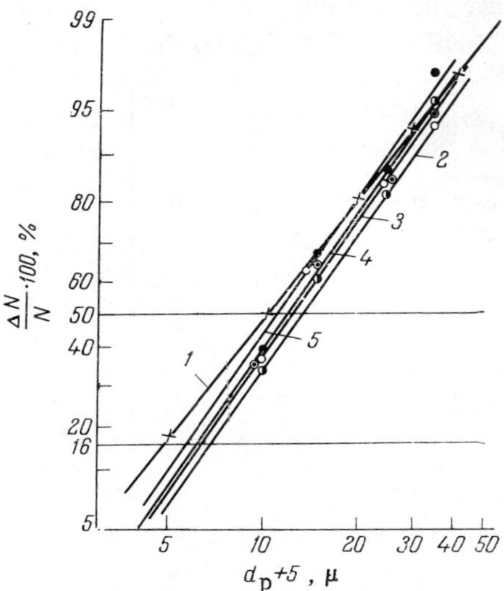

Fig. VII.6. Size distribution of spherical glass particles before and after being acted upon by a flow with the surfaces arranged perpendicularly to the flow axis (β = 90°). 1) Original; 2-5) after being acted upon by a flow at a velocity of: 2) v = 0.1; 3) 0.2; 4) 0.4; 5) 0.5 m/sec.

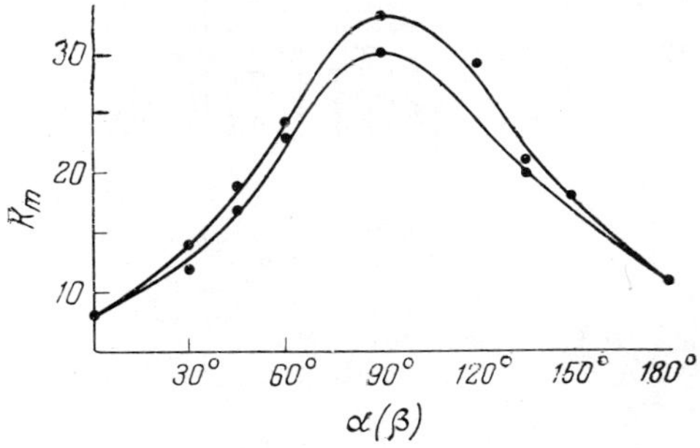

Fig. VII.7. Coefficient K_m for a flow velocity of 0.1 m/sec as a function of the inclination of the surface in the following planes. Upper curve: parallel to the flow (angle α); lower curve: perpendicular to the flow (angle β).

3, 4, and 5 in coordinates of $d_p + a$). This means that σ is constant. The deviation of σ from the mean value (Tables VII.4 and VII.5) is due to the inaccuracy of the graphical constructions. The mean value of σ may be taken as 0.220 (Table VII.4) or 0.285 (Table VII.5). In general, $\sigma = f(v)$.

The adhesion numbers may also be expressed in terms of particle diameter for various flow velocities by formula (VI.15). The following values of a and b are obtained as functions of water-flow velocity:

v, m/sec.	0.1	0.2	0.4	0.5
a.	0.80	0.45	0.90	0.5
b.	−0.16	−0.16	−0.5	0.5

Thus all the quantities in Eq. (VI.25) are known and K_N may be calculated. Below we present the experimental and calculated values of the coefficients K_N and K_m for various flow velocities:

v, m/sec	0.1	0.2	0.4	0.5
K_N:				
calculated value	4.6	8.0	12.8	19.7
experimental value	5.5	7.5	14.5	23.0
K_m:				
calculated value	15	44	69	140
experimental value	30	48	63	140

We see from these results that theory and experiment are in fair agreement, which tends to support the computing method taken.

The value of K_m is always greater than that of K_N, as it is easier to remove relatively large particles of considerable mass.

The value of K_m is shown as a function of the inclination of the plates in Fig. VII.7. According to the data presented, the minimum value of the coefficient K_m may be 3-4 times smaller than the maximum.

Thus the extent to which the adhering particles are removed from the surface may alter considerably on changing the position of the plates relative to the flow axis.

Adhesion and Autohesion in River-Bed Processes. Autohesive phenomena play a considerable part in

Table VII.6. Nonsilting Velocity of Flows for Various Soils* (particle size 0.01-0.25 mm)

Soils	Nonsilting flow velocity, m/sec, for various hydraulic radii of the river bed						
	0.25 m	0.50 m	0.75 m	1.0 m	1.5 m	2.0 m	3.0 m
Muddy	0.22	0.30	0.36	0.40	0.44	0.48	0.50
Fine-sandy	0.32	0.45	0.54	0.60	0.67	0.71	0.75
Weak loess.	0.37	0.53	0.63	0.70	0.78	0.83	0.58
Dense loess	0.49	0.68	0.81	0.90	1.01	1.07	1.13

*For brief action of the flow the velocities increase by 20-25%.

river-bed processes when, instead of interacting with a metallic or other solid surface, the particles interact with a layer composed of particles similar to those in the suspended state.*

There are two distinct characteristic velocities: the first is the transporting velocity, i.e., the lowest flow velocity at which the particles suspended in the water are not deposited from the flow and hence cannot stick to the bottom, and the second being the "pulling" velocity below which the particles are not pulled along. When discussing river and channel processes the "pulling" velocity is usually called the "scouring" velocity, and the transporting velocity is renamed "nonsilting."

In order to determine the transporting (nonsilting) velocity, a number of empirical and semi-empirical formulas have been proposed [338]. Some of these formulas are given below. Poslavskii's formula for calculating the nonsilting velocity has the form

$$v = 0.34\sqrt{NR^{1/3}} \tag{VII.17}$$

where N is the transporting capacity of the flow in kg/m^3, and R is the hydraulic radius for uniform motion in m.

Roer's formula [339] is valid for channels of rectangular cross section with a headless flow of water

$$v = A\left[\frac{m+2}{2}(\rho_p - 1)w\right]^{0.326} R^{0.473} \tag{VII.18}$$

* The part played by adhesion in the erosion of soils is considered in § 54.

where A is a coefficient (A = 39.3 for particles with a density of 2.65 g/cm^3 and roughness Δ = 0.001 m; A = 2.93 for Δ = 0.005 m); m = b/H (b is the width and H the depth of the channel in m), ρ_p is the density of the pulp (particles suspended in the water) in tons/m^3, w is the hydraulic coarseness of the particles in m/sec, and R is the hydraulic radius in m.

Table VII.6 presents Zamarin's results [338] for the nonsilting velocities of flows in various soils.

We see from the results presented in Table VII.6 that the value of the nonsilting velocity depends on the properties of the flow determined by the hydraulic radius. The larger the hydraulic radius, the higher is the nonsilting velocity. In Table VII.6 the soils are placed in order of increasing density, to which the nonsilting velocity is directly proportional.

The water-flow velocity (in the large trough) necessary for the removal of adhering particles (denominator in Table VII.2) and the nonsilting velocity (Table VII.6) for a hydraulic radius approximately equal to 0.25 m almost coincide.

The erosion (scouring) of soil takes place in the following stages: the removal of individual particles, the autohesive forces of which have been weakened, and the detachment of individual aggregates by the breaking of adhesive couplings (if the flow is taking place in an artificial trough) or autohesive couplings (in a river bed).

If the flow is moving along a trough continuously covered with a layer of adhering particles of uniform thickness, the particles may be detached as a result of the lifting force [293, 335] [see formula (VI.1)].

In actual fact, a layer or stratum of bottom deposits (sludge) constitutes a disordered distribution of soil particles of widely differing sizes. These particles work into the deposit and make the bottom rough; this causes turbulence of the boundary layer [340].

Considering the displacement of particles in loose sludge, Nikitin [340] determined the stability criterion for the stability of a layer of loose sludge with respect to erosion by a flow of water, on the basis of condition (VII.1) with $F_{ad} = 0$ and allowing for the fact that $c_x = f(\text{Re})$, as follows:

Table VII.7. Scouring Flow Velocity for Argillaceous Soil

Form of clay	Depth of flow, m	Cohesion acc. to Tsitovich (autohesive force), referred to 1 cm^2, kg	Coeff. of inhomogeneity	Size of aggregates being detached, mm	Scouring velocity, m/sec	
					exptl.	calc. from (VII.20)
Compact:						
medium-density	0.30	0.11	0.50	9.5	0.90	0.78
low-density . . .	0.42	0.13	0.71	3.5	0.84	0.73
Aggregated (low-density).	0.13	0.21	0.50	4.1	0.85	0.73

$$\psi = \frac{w}{v_d} \tag{VII.19}$$

where w is the hydraulic coarseness of the particles and v_d is the velocity of the water flow at a level equal to the height of the particles.

The stability criterion ψ characterizes the pulling (scouring) of large particles (without considering their autohesion). For sand particles 0.1-2 mm in size, this criterion is independent of the Reynolds number and equals 0.42.

The autohesion of soil particles is determined not only by the dimensions but also by the nature of the particles. Sandy soil, after drying, is transformed into loose (free-running) material. The forces appearing after the evaporation of the moisture (see §26) are insufficient to prevent the free-running state and have no effect on the change in the scouring velocity [341].

The character of the autohesion of argillaceous soils containing 3-8% clay particles and up to 40% sand particles is determined by the cementing properties of the clay particles enveloping the particles of larger size. Thus, for a soil consisting of 8% clay (particle diameter smaller than 0.001 mm), 2% sludge (0.005-0.01 mm), 50% slit (0.01-0.05 mm), and 40% sand (0.05-0.25 mm), the autohesive force between the particles equals 0.15 kg/cm^2.

On the basis of experimental data, Mirtskhulava [342] proposed a formula for calculating the scouring velocities of combined soils in a plane turbulent flow:

$$v_{\text{det}}=\log\left(\frac{8.8H}{l}\right)\sqrt{\frac{g}{1.3\rho_2}\sqrt{(\rho_1-\rho_2)\,l+1.25k\cdot\frac{m}{n}\,\text{RS}}} \qquad \text{(VII.20)}$$

where H is the depth of the flow, l is the arm of the frontal stress, i.e., the distance from the center of application of the detaching force to the plane of detachment of the aggregate, ρ_1 and ρ_2 are the densities of the aggregate and the liquid, and RS is the rupture strength of the soil for dynamic loading; the latter is determined by the coefficients m, n, and k (m is a coefficient depending on the operating conditions, n depends on the overloading, and k on the homogeneity of the soil).

The autohesion of the soils in formula (VII.20) is only indirectly taken into account by RS.

Table VII.7 shows the calculated and experimental values of the flow velocities required for the scouring of clay soils. The experimental data were obtained for the scouring of soil in a trough [342].

The values of scouring flow velocities calculated from formula (VII.20) are rather lower than the experimental values, since Mirtskhulava [342] only considered the autohesion of the adhering particles, neglecting their adhesion to the trough.

§39. Adhesion Processes in the Purification of Water

In the filtration of aqueous suspensions through grainy layers, the adhesion of the suspended particles to the grains of filtering material or to earlier-deposited particles prevails over the hydrodynamic forces in the flow, which tend to detach the adhering particles.

Figure VII.8 shows the time variation of the pressure losses (cm of water) in a layer of grainy material [343]. Curve 1 was obtained for the filtration of a suspension and curve 2 on subsequently passing pure water through the worked-out charge. The reduction in pressure losses (curve 2) may be explained by the detachment

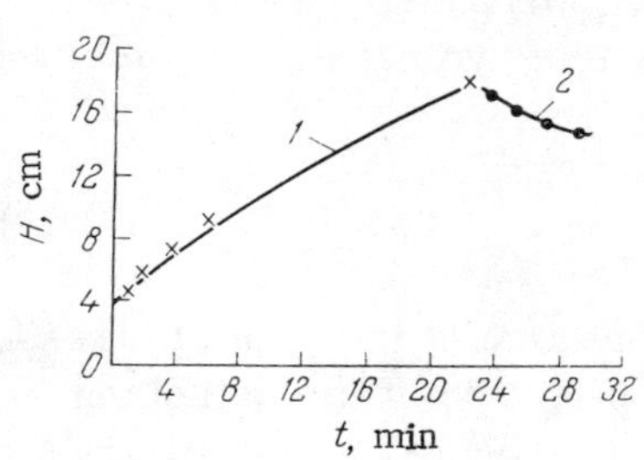

Fig. VII.8. Variation in pressure losses in a layer of grainy structure with time. 1) Filtration; 2) washing with water.

and carrying away of adhering particles, which leads to a rise in the cross section of the flow.

The results obtained by Mints [343] confirm the presence of adhering suspended particles on the surface of the sand. The process of filtration is characterized by the filtration index β, which represents the retaining power of the sand. Since the filtration process is largely determined by the adhesion of the suspended matter, the value of β characterizes the adhesion of the suspension to the grains of sand:

$$\beta = 1.13(1 - m)\frac{\lambda^2}{R} \tag{VII.21}$$

where m is the porosity of the filter, and λ is the ratio of the radius of the adhering particle (r) to the radius of a grain of sand (R).

It follows from formula (VII.21) that the retaining power depends on the porosity of the layer and the dimensions of the sand grains and adhering particles, being independent of the filtration velocity. This is true up to certain values of filtration velocity for which the flow washing the surface of the filter grains cannot erode the particles which have already become attached. The adhesion of particles in the filtering process is called "colmatation" (as in the improving of soil by silt deposition, sometimes known as "warping"; French "colmatage") and the detachment of particles already adhering is called "suffosion" (undermining) [344].

The adhesion of particles in filtration was considered in detail by the Mackrle brothers [25, 26]. These authors studied the filtration of particles in a laboratory filter of square cross section 10 × 10 cm, containing charges of various materials having a grain size of 1.0, 1.2, 1.4, and 1.7 mm. Figure VII.9 shows the variation in the ratio of the final concentration of suspended matter, $Al(OH)_3$, at a point x (coordinates represent the depth of the filtering material, i.e., the thickness of the charge layer) to the original (c_x/c_0) as a function of the constant a, which the authors called the adhesion

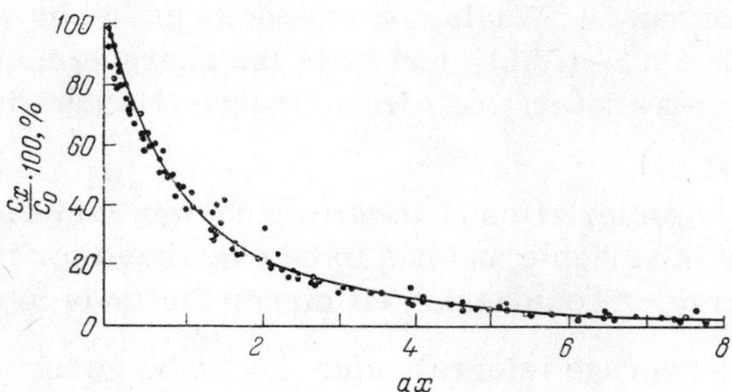

Fig. VII.9. Concentration of an $Al(OH)_3$ suspension as a function of the dimensionless parameter ax.

efficiency coefficient or factor and defined by the expression

$$a = \frac{6(1-m)\Delta H}{DmL\varphi} \qquad \text{(VII.22)}$$

where m is the porosity of the filter, ΔH is the characteristic distance determining the effects of the adhesive forces, D is the diameter of the charge grain, L is the length of the grain to which the particles adhere, and φ is the surface factor.

For the filtration of an $Fe(OH)_3$ suspension an analogous curve is obtained. The results of these investigations enable us to compare the effectiveness of the adhesion of various filtering materials. For this purpose we must calculate the Re number for a given charge and velocity of filtration:

$$\mathrm{Re} = \frac{\rho v}{\eta S}$$

where ρ is the density of the liquid in g/cm^3, v is the average velocity in the cross section of the charge in cm/sec, η is the dynamic viscosity of the liquid in g/cm · sec, and S is the specific surface of the grain referred to unit volume in cm^2/cm^3.

By using the Re number and the relation Ma = Re$^{2.16}$ (see §2) we calculate the Ma adhesion criterion:

$$\mathrm{Ma} = \frac{A}{\eta v_x \, \Delta H^2} \qquad \text{(VII.23)}$$

where A is the van der Waals constant determined by calculation [see formulas (IV.1)-(IV.6)], and v_x is the characteristic velocity. From this we may determine ΔH and thence the adhesion efficiency factor a.

If the characteristics of the filter charge vary layer by layer, the quantity a is variable and has to be calculated for the various layers of charge. The adhesion efficiency factor is in this case calculated as the average integral value $\bar{a} = \int_0^x a\,dx$, where a is a function of x. Knowing a, we may use the filtration curves to determine the change in the concentration of suspended matter as a function of the thickness of the charge layer x. Thus the filtration curve constitutes a special kind of nomogram from which c_x/c_0 is calculated in such a way that its value on the axis of abscissas should correspond to the integral $\int_0^x a\,dx$. It must be remembered that each type of suspension requires its own special nomogram.

The Mackrle paper [25] took no account of such processes as the autohesion of contaminant particles to each other and their adhesion to a layer of earlier-adhering particles, nor the detachment of adhering particles by the flow of water. These deficiencies are to some extent eliminated in the papers of Mints [345, 346], who in calculating the efficiencies of grainy filters treated the adhesion processes with due allowance for the balance of forces associated with the adhesion or detachment of the adhering particles:

$$-\frac{dc}{dx} = bc - \frac{a}{v}\rho_{sat} \tag{VII.24}$$

where $c = f(x, t)$ is the instantaneous concentration of suspended matter in the suspension, x is the distance from the first point of the charge to the section under consideration, a and b are parameters, v is the velocity of filtration, ρ_{sat} is the saturation density, i.e., the weight of suspended matter caught by unit volume of charge.

The values of c, v, and ρ_{sat} may easily be measured.

The first term on the right-hand side of the equation considers the adhesion of suspended particles (b is the adhesion param-

Table VII.8. Filtration Parameters (Al_2O_3 content 9 mg/liter)

PAD content, mg/liter	H/t	a	b	a/b	A_{sat}	$F(A_{sat})$
0	7.0	0.59	5.28	0.114	0.36	2.28
0.15	10.3	0.53	6.70	0.08	0.50	4.80
0.30	13.8	0.64	9.90	0.0645	0.58	7.95
0.40	20.7	0.72	12.20	0.0585	0.64	13.10

eter), while the second considers the detachment of these particles under the influence of hydrodynamic forces. For bc > $(a/v)\rho_{sat}$ the adhesive processes prevail over the hydrodynamic forces, i.e., filtration takes place. For $(a/v)\rho_{sat}$ > bc there is a tendency for earlier-deposited particles to be removed.

The filtration process depends on the ratio a/b, which may be determined if we know the pressure losses associated with filtration:

$$\frac{H}{t} = i_0 F(A_{sat}) \frac{a}{b} \tag{VII.25}$$

where H is the pressure loss during the flow of liquid through the filtering charge in mm of water, t is the working time of the charge in min, i_0 is the initial hydraulic gradient of the pure filter, and $F(A_{sat})$ is a function determined experimentally. The parameter A_{sat} characterizes the extent to which the threshold region is saturated with deposits, i.e., the proportion of pore volume occupied by the deposit in fully saturated layers of charge.

Table VII.8 gives the results of a determination of the parameters a, b, and A_{sat} in the course of the filtration of a flow of water containing clay particles as functions of the polyacrylamide (PAD) present [347].

With increasing PAD content the ratio a/b falls, i.e., the adhesion of the particles increases and their detachment by the flow of water diminishes, which improves the quality of the filtration.

According to Mints, the parameter b is determined at the onset of the filtration process, i.e., on the initial adhesion of contaminants to the surface of the charge (filter) grains.

Veitser and Paskutsaya [348] noted a certain agreement between the adhesion of contaminants (filtration method) and the autohesion of particles (method based on measuring the resistance of the suspension in the course of mechanical agitation) on adding polyacrylamide. This means that the parameter b should take both the adhesion and autohesion of contaminant particles into account.

The protective lifetime of the charge is an important characteristic determining the ability of the charge to ensure the adhesion of particles to the grain surface. The protective lifetime [346] is determined from the formula:

$$t_{\text{p}} = \frac{1}{k}\left(\frac{x}{v^{1.7}\,a^{0.7}} - \frac{D}{v}\right) \qquad \text{(VII.26)}$$

where k is an experimental coefficient, x is the thickness of the charge layer in cm, v is the velocity of filtration in cm/sec, and D is the charge grain diameter in cm.

With the passage of time the efficiency of filtration through a grainy filter falls and finally approaches zero, i.e., the ratio c/c_0 tends to unity (Fig. VII.10). In this case, it is essential to regenerate the filter grains, i.e., to detach the adhering particles and carry them away. Baranov [349] studied the removal of particles adhering during the filtration of flood water from the Moscow River and of artificially contaminated water containing gravel grains 2.5 to 7 mm in size. Baranov determined the amount of contaminants removed (P) on washing the filter at a specified intensity* and the total amount of contaminants P_0 held by the charge in the course of a working cycle. The relative amount of contaminants washed away by a rising flow of water depends on the rate of flow, which is directly proportional to the intensity of washing. On increasing the washing intensity from 1 to 10 liters/sec · m^2 (Fig. VII.11) the quantity of contaminants washed away increases, but on further raising the intensity of washing (from 10 to 20 liters/sec · m^2) the amount of contaminants removed changes very little. Even for this washing intensity 10–15% of the contaminants (or 0.03% of the mass of the charge) held in the working cycle remain (in essence, the curve in Fig. VII.11 is analogous to the integral curves of adhesive forces (see Figs. I.1 and I.2).

*The washing intensity (Q), i.e., the flow of material being filtered through 1 m^2 of the filter cross section in 1 sec.

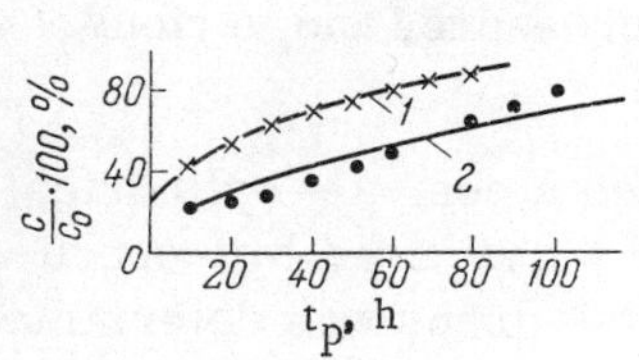

Fig. VII.10. Time variation of the concentration of a suspension at the exit from a charge layer. Layer thickness: 1) H = 4.2; 2) 11.7 cm.

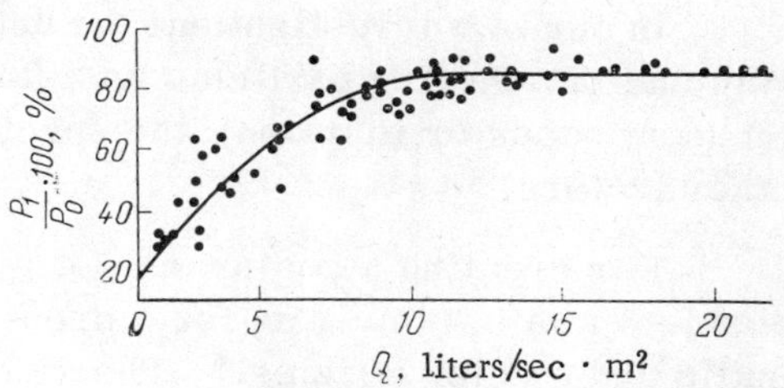

Fig. VII.11. Relative quantity of contaminants washed away as a function of the washing intensity.

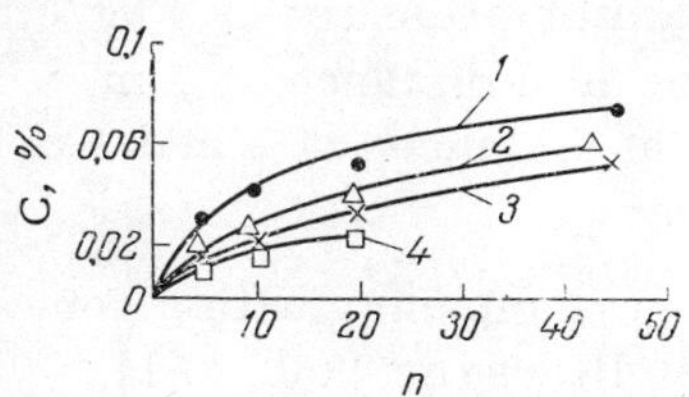

Fig. VII.12. Amount of residual contaminants C (in % of the mass of the charge) as a function of the number of cycles for various filtration intensities: 1) 9.52; 2) 11.7; 3) 13.25; 4) 15.4 liters/sec · m^2.

The amount of residual contamination (C) increases with increasing number of filtration cycles (Fig. VII.12), this rise being particularly significant for the initial stage of filtration in which $n < 10$. For $n = 45$, the residual contaminants make up roughly half the amount of adhering particles corresponding to saturation; saturation sets in when the adhering particles completely cover the surface of the original grainy material. The filtration process will then be determined by the autohesive forces between different particles of the suspension.

§40. Removal of Contamination by Washing Surfaces

Detachment of Adhering Particles by Water Jets. The washing of vehicles [350] may be mechanized by means of water jets. The effect of a water jet on a dust-laden surface is to detach the adhering particles, the first stage in the washing process. Subsequently, the detached particles are carried away from the surface so treated. The efficient use of water jets, i.e., a high degree of dust removal from the surface with a low consumption of water, may be achieved by studying the various stages in the process of washing surfaces.

In our own investigations we determined the efficiency of the removal of adhering particles as a function of the structure of the jet (continuous or in drops), the specific pressure, and various other factors.

For creating a continuous jet we used a conical—cylindrical nozzle 8 mm in diameter; for a drop-type jet we used hydropneumatic and ejector nozzles.* The size of the drops was determined by trapping them with a sticky compound. (The median diameter of the drops obtained after emerging from the hydropneumatic nozzle was 275 μ.) The specific pressure over the cross section of the jet was measured by means of a system of detectors and an MPO-2 oscillograph.

The average specific pressure of the continuous jet created by the conical—cylindrical nozzle, for a water pressure of 3 kg/cm^2 in front of the nozzle, became 0.22 kg/cm^2 at a distance of 1 m from the latter (or 0.15 and 0.03 kg/cm^2 at distances of 2 and 3 m respectively).

Under these conditions a fair degree of cleaning with a continuous jet requires a water flow of 15-20 liters/m^2 [350, 351], while K_N exceeds several thousands, i.e., an oil-free surface is cleaned almost completely. If the washing is accompanied by rubbing with brushes, the water flow may be reduced by a factor of about 10 [352].

Below we present the values of K_N and K_m for cleaning an oil-free surface with jets of drop structure generated by: a hydropneumatic nozzle (distance to painted plates, 1 m; water pressure, 2.5 kg/cm^2; air pressure, 1 kg/cm^2):

Water flow, liters/m^2	4	7	10
K_N	710	1,250	∞
K_m	9,000	14,000	∞

an ejector nozzle (distance, 30 cm; water flow, 3.5 liters/m^2):

Air pressure in front of the nozzle, kg/cm^2	3.0	4.0	4.5
K_N	20	83	$2 \cdot 10^4$
K_m	250	1000	$2.5 \cdot 10^5$

*The water was converted into drops in the hydropneumatic nozzle by means of compressed air.

We see from the results presented that a hydropneumatic nozzle producing a drop-type jet cleans the surface almost completely for a water pressure of 2.5 kg/cm^2 in front of the nozzle with an air pressure of 1 kg/cm^2; the ejector nozzle does likewise for an initial air pressure of 4.5 kg/cm^2. In this the average specific pressure of the drop-type jet was 0.002-0.006 kg/cm^2, i.e., 1-2 orders lower than in the case of the continuous jet, while the water consumption was 10 liters/m^2 of the treated surface for the hydropneumatic and 3.5 liters/m^2 for the ejector nozzle.

Since there is no need to achieve high specific pressures for the effective cleaning of oil-free surfaces, it is better to use jets of drop structure, thus reducing the water consumption.

The degree of cleaning the surface depends not only on the water consumption but also on the angle at which the jet meets the surface under treatment. Below we present some data relating to the efficiency of removing particles with a hydropneumatic nozzle as a function of the angle of incidence on the painted surface (water flow, 4 liters/m^2):

Incident angle	90°	75°	60°	30°
K_N	710	$2 \cdot 10^3$	∞	$1.4 \cdot 10^3$
K_m	9000	$2.5 \cdot 10^4$	∞	$1.6 \cdot 10^4$

In jets of drop structure obtained by means of hydropneumatic and ejector nozzles, a large number of drops of diameter no greater than 500 μ is formed; these interact with others in the jet and on striking the surface. In order to determine the efficiency of the cleaning process as a function of drop composition we therefore studied the detachment of particles secured by the free settling of single drops; this made it possible to eliminate the influence of secondary processes associated with the interaction of the drops between themselves.

Experiments were made with drops of distilled and ordinary conduit water as well as with a 0.1 N solution of NaCl and a 0.1 solution of DB; the size of the drops was no greater than 1700 μ. The drops fell from a height of 60 cm onto a glass surface placed at an angle of 60° to the horizontal. Under these conditions, the adhesion numbers of particles 40 ± 2 μ in diameter were no greater than 4%, independently of the properties of the glass surface (ordinary, hydrophilic, or hydrophobic) and the drop composition, i.e., independently of the wetting angle, which varied from 7 to 65°, i.e., under these conditions the overwhelming majority of the particles were detached. If we consider that in the free fall of drops the specific pressure at the point where these meet the surface is extremely small (less than 0.001 kg/cm^2), then the experiments in question also constitute a good confirmation of the earlier conclusions regarding the undesirability of creating high specific pressures in washing oil-free surfaces.

Removal of Particles Adhering to a Surface During the Free Settling of Drops.* In practice, similar processes of dust removal occur in periods of rain. The effectiveness of the dust removal depends on the number of drops striking the surfaces and also on their size, or on the intensity of the rain, which is expressed in millimeters of rainfall. The coefficients K_N and K_m may be determined by the earlier-developed method of calculating the degree of cleaning of a surface (see §31).

We carried out some experiments on rain imitation. We subjected a dust-laden surface† (inclined at 30°) of area S = 200 mm^2 to the impact of water drops D = 1.3 mm in diameter falling from a height of 60 cm. Under these conditions, the intensity of the "rain" equals

$$H = \frac{Wn}{S} = \frac{0.524 \cdot D^3 \cdot n}{S} = \frac{0.524 \cdot 1.3^3 \cdot n}{200} = 0.004\, n$$

where H is the height of the layer of "rainfall" in mm, W is the volume of a drop in mm^3, and n is the number of drops.

The results of experiments on the removal of adhering particles from oil-free surfaces expressed as a function of the number of drops (n) are shown in normal logarithmic coordinates in Fig. VII.13. From the straight lines characterizing the distribution we determined the parameters required for calculating the degree of cleaning of the surface [from Eqs. (VI.9) and (VI.10)]. All these quantities, as well as those required for calculating the function $\gamma_F(d)$, namely, a and b [see (VI.15)], are given in Table VII.9.

*The quality of the washing process may conveniently be monitored by using radioactive contaminants [164] (see § 9).

† In these investigations we used glass spheres up to 90 μ in diameter.

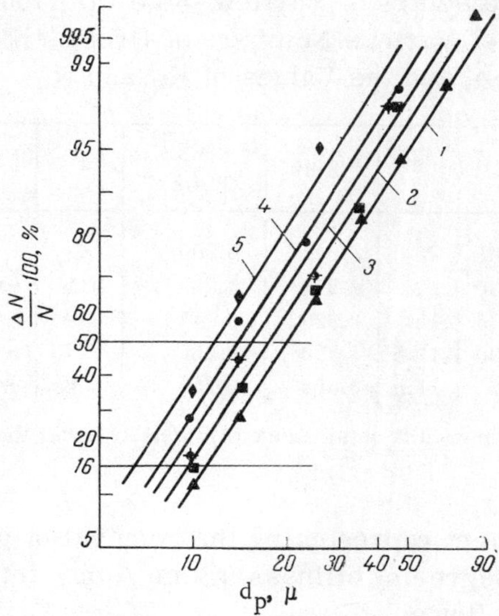

Fig. VII.13. Size distribution of spherical glass particles remaining on an oil-free surface painted with perchlorvinyl enamel after being acted upon by various numbers of drops. 1) Original; 2) 20 drops; 3) 100 drops; 4) 250 drops; 5) 500 drops.

The experiments showed that almost all the particles were removed for n = 1000; for this value of n the intensity of the shower would have to be 4 mm, which is extremely rare. There is thus no great point in calculating K_N for these conditions. The relationship between the parameters of the distribution and the number of impacts of the drops is given by the following equation:

$$\bar{d} = 20 \cdot n^{-0.09}$$

where n is the number of drops.

We see from the data presented that the calculated values of K_N are rather higher than the experimental values; however, the distance is so small that in practice it is perfectly satisfactory to use the computing formulas for determining the degree of cleaning. As we should expect, the calculated value of K_S is in all cases greater than K_N.

Table VII.9. Parameters of Particle-Size Distribution Before and After the Action of Various Numbers of Drops, the Coefficients *a* and b, and the Values of K_N and K_S

No. of drops	H, mm	$\log \bar{d}$	$\bar{d}$	$\log d_{16}$	$\sigma = \log \bar{d} - \log d_{16}$	*a*	b	K_N^*	K_S
0	0	1.301	20	1.041	0.200	–	–	–	–
20	0.08	1.230	17	0.978	0.252	1.1	−0.7	1.4/1.5	2.2
100	0.4	1.176	15	0.929	0.247	0.9	−0.7	1.8/1.7	3.3
250	1.0	1.130	13.5	0.775	0.255	0.75	−0.7	2.0/1.9	4.8
500	2.0	1.061	11.5	0.806	0.255	0.2	−0.7	8.4/8.2	23.0

*Denominator gives the results obtained experimentally under these conditions.

The coefficient representing the removal of particles from an oily surface (degree of oiliness 0.5 mg/cm^2) depends on the number of drops as follows:

n.	100	250	1000
K_N	1.6	2.2	2.7

The efficiency of removing particles from an oily surface is lower than in the case of an oil-free surface. We may expect that for n = 1000, the value of K_S will be no greater than 5.

The results just quoted were obtained under laboratory conditions in which the surface was tested for 1-2 min. Rain usually lasts much longer. We therefore also checked the relation between K_N and time under laboratory conditions. The following give the results for n = 250 drops:

Time of treatment, min . . .	1.5	5	20	40
Coefficient K_N	1.9	2.1	2.3	2.4

The greater the period of treatment for the same intensity, the higher is the value of K_N; however, this rise in K_N is no greater than 20% for a 26-fold increase in time.

Thus on the basis of all these experiments we may conclude that the efficiency of the action of rain is to a large degree deter-

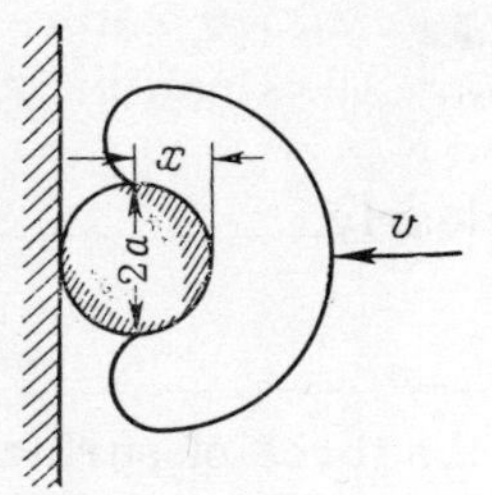

Fig. VII.14. Capture of an attached particle by a drop of water.

mined by the number of drops (or the height H) and to a smaller extent by the period of the shower.

Mechanism of the Detachment of Adhering Particles by Drops of Water. Chapka [353] considers the impact of a drop as the impact of a solid body. The drop gives out its energy to the solid surface before spreading. If this is valid, then in the detachment of adhering particles by drops of liquid the same processes occur as in the interaction between moving and stationary particles (see §32). The period of existence of a drop of water on impact is 10^{-10}-10^{-12} sec. If we start from Malyshev's results [207] [see formula (V.4)], then even for a drop falling speed of 100 m/sec (10^4 cm/sec) the period of the impact should be a few orders longer than 10^{-10} sec, i.e., the drop cannot transfer its energy to the adhering particle between the moment of impact and its deformation. The calculations of Chapka are valid to a certain extent for estimating the pressure on a solid substrate, but they cannot be extended to the detachment of particles on impact by a drop.

It may be considered that the following processes will take place when drops come into contact with a surface: the spreading of the drop after its contact with the surface, and the capture of adhering particles by the drops.

When the drops spread, detachment of the adhering particles takes place in the zone of contact between the drop and the surface and depends on the kinetic energy of the drop, i.e., on the velocity of the flow developing after the drop starts spreading (see §37).

In view of the complexity of the process underlying the removal of adhering particles when the drops strike the surface, we shall confine our attention to the simplest case: the capture of adhering particles when a drop strikes the dust-laden surface at a velocity v in a normal direction (Fig. VII.14).

The drop embraces the adhering particle, which penetrates into the drop to a distance x at a velocity v*, i.e.,

$$v^* = \frac{dx}{dt} \quad \text{or} \quad \frac{dv^*}{dt} = v^* \frac{dv^*}{dx}$$

For an acceleration of the drop equal to j, clearly,

$$-Mj = F_\sigma$$

where M is the mass of the drop, and F_σ is the force of surface tension.

Considering that $j = dv^*/dt$, we obtain

$$-M\frac{v^* \cdot dv^*}{dx} = 2\pi a\sigma \cos\theta \qquad \text{(VII.27)}$$

Using the notation $\sigma \cos\theta = \sigma'$; $a = \sqrt{x(2r-x)}$ (where r is the radius of the particle), we obtain

$$-Mv^*\,dv^* = 2\pi\,\sigma' \sqrt{x\,(2r-x)}\,dx \qquad \text{(VII.28)}$$

Integrating Eq. (VII.28) with due allowance for the boundary conditions $v_i^* = v$, $v_f^* = 0$, $x_i = 0$, $x_f = 2r$, we obtain

$$\frac{Mv^2}{2} = \pi^2 \sigma' r^2 \qquad \text{(VII.29)}$$

From formula (VII.29) we obtain the condition for the penetration of the adhering dust into the drop and its consequent capture:

$$v \geqslant 4.4\, r \sqrt{\frac{\sigma'}{M}} \qquad \text{(VII.30)}$$

We calculated the critical minimum approach velocity of a drop for the capture of adhering particles 10 μ in diameter, using experimental data relating to the free-settling velocity of drops in the air [6]; the results were as follows:

Drop diameter, cm	10^{-2}	$2 \cdot 10^{-2}$	$5 \cdot 10^{-2}$
Minimum approach velocity of drops leading to the capture of adhering particles, cm/sec	70	22	6
Free-settling velocity of the drops in air, cm/sec	72	162	403

These results show that, even under conditions of the free settling of water drops more than 10^{-2} cm in diameter, capture of the adhering particles will take place. We remember that here the angle of incidence of the drop on the dust-laden surface was taken as 90°, while the coaxiality of the impact and the probability of the drop falling on the adhering particle were not taken into account.

The fact that adhering particles are captured by water drops does not necessarily mean that the particles will be detached; it only means that the air type of adhesion may be replaced by the water type. On impact the drop is flattened and thereupon a radial component of velocity arises; this equals [354]

$$v_r = \left[\frac{v}{\pi}\cdot\frac{a^2+r^2}{r^2}\cdot\ln\frac{a+r}{a-r} - \frac{2av}{\pi r}\right] - \left[0.224\left(\frac{r}{a}\right) + 0.101\left(\frac{r}{a}\right)^3 + + 0.032\left(\frac{r}{a}\right)^5 + \cdots\right] \tag{VII.31}$$

Let us suppose that the drop is flattened into a hemisphere of radius a, i.e.,

$$W_d = W_{hs} \tag{VII.32}$$

where

$$W_d = \frac{4}{3}\pi r^3 \qquad W_{hs} = \frac{2}{3}\pi a^3$$

Taking account of (VII.32), we obtain $a = 1.145r$.

Putting the value of a into formula (VII.31), after some slight simplifications we obtain

$$v_r = 2.5\, v \tag{VII.33}$$

where v is the incident velocity of the drops on contact with the surface.

The experimental results of Enkins and Booker [355] show that the radial velocity of the spreading of a drop is 2-3 times greater than normal, i.e., they confirm the validity of formula (VII.33).

If water drops fall perpendicularly to the dust-laden surface and the radial component exceeds the velocity at which the drops come into contact by a factor of 2.5 (for surfaces not at right angles to the falling direction of the drop this factor may be more substantial), conditions are created for the detachment of the attached particles, and this is confirmed experimentally, as indicated earlier. It is therefore not surprising that the overwhelming majority of particles are removed by a hydropneumatic nozzle generating a jet of drop structure with drop velocities of the order of 1 m/sec.

In addition to the radial velocity of the spreading of the drop [formula (VII.33)], we may calculate the pressure and the force experienced by the adhering particle on spreading of the drop, and compare this force with the force of adhesion.

Concept of the Efficiency Factor (Efficiency) of Water Jets in the Removal of Particles. The capture of particles by drops is determined by the probability that a drop will fall on an adhering particle and also on the coaxiality (coaxial impact) of this incidence.

Measurements of the dimensions of the particles remaining on a surface after subjection to a water jet indicate that large particles are removed comparatively easily, while particles less than 10 μ in size tend to remain, i.e., just those particles for which the adhesive forces are greatest.

At the initial instant when the jet strikes the dust-laden surface (t_1), forces equal in magnitude to the adhesive forces arising in air ($F_{ad,a}$) are overcome. At the instant of time t_2 the dust-laden surface will lie completely in a liquid medium, and then when the particles are removed, forces corresponding to the adhesive forces in a liquid medium ($F_{ad,l}$) will be overcome. The adhesive forces in air and in liquid must be considered as two extreme possible cases of the adhesive forces overcome in the course of washing.

At the instant of time t ($t_2 > t > t_1$) there will be a liquid and an air medium between the particles and the surface, and the adhesive force will now be smaller than in air and greater than in the liquid, i.e., we shall have the inequality

$$F_{ad,a} \geqslant F_{det} \geqslant F_{ad,b} \qquad \text{(VII.34)}$$

This is confirmed by visual observations and experimental data.

When particles adhere in air there is a process of aging, i.e., adhesion increases as the period of contact of the particle becomes longer, while in the liquid no such phenomenon occurs. Thus, the effectiveness of the washing of a surface with a continuous jet created by a conical—cylindrical nozzle at a distance of 3 m from the surface, with a water flow of 20 liters/m^2, varies in the following way as a function of the period spent by irregularly shaped glass particles on the object surface:

Water pressure in front of the nozzle, kg/cm^2	2.0	2.5	3.5	4.0
K_N when the particles lie on the surface no longer than:				
30 min	21	22	100	50
24 h	12	20	14	30

Thus, the longer the particles remain in contact with the surface the more difficult it is to remove them, because of the rise in $F_{ad,a}$ [see inequality (VII.34)].

We may estimate the energy of the jet spent in overcoming the forces of adhesion, i.e., determine the efficiency of the jet (η_0) for the removal of adhering particles, in the following way:

$$\eta_0 = \frac{E_{ad}}{E_m} \qquad \text{(VII.35)}$$

where E_{ad} is the work required to remove close-packed particles from 1 m^2 [this may be calculated from (I.55)], and E_m is the energy of the jet corresponding to the mass m of the water falling on 1 m^2 of surface at a velocity v, as determined by experiment.

The following represent the calculated values of the maximum efficiency of the jet (in %), allowing for the work of the jet (obtained experimentally) required to overcome the adhesion of stubborn particles 10 μ in diameter:

Medium	Air	Water
Efficiency of a drop jet (hydropneumatic nozzle), %	16.6	$6 \cdot 10^{-4}$
Efficiency of a continuous jet (conical-cylindrical nozzle), %	0.7	$2 \cdot 10^{-5}$

The efficiency of the jet reaches a maximum value at the initial instant of contact between the jet and the dust-laden surface, when the adhesive forces corresponding to an air medium are being overcome; the energy of the jet (allowing for all losses) is sufficient for the removal of particles from oil-free surfaces. If the dust-laden surface is in water the efficiency becomes negligibly small.

Transportation of Detached Particles. The detachment of the particles is a necessary but not sufficient condition defining the efficiency of the washing. The detached particles still have to be removed from the treated surface, i.e., the transporting power of the flow formed after the impact of the jet on the surface must be sufficiently high. The main factors determining the transporting power of the flow are its velocity and structure, which in turn depend on the flow of water and the relief of the treated surface. The kinetic energy of the jet should be sufficient to carry the particles away after detaching them from the surface. We must therefore also study the efficiency of the jet in transporting the detached particles; this is given by the following expression:

$$\eta_T = \frac{E_T}{E_C} \tag{VII.36}$$

where E_T is the energy required for transporting the particles, E_C is the energy of the flow formed after the collision of the jet with the surface

$$E_T = \frac{KM_p \cdot v^2}{2g} \tag{VII.37a}$$

$$E_C = \frac{M_f \cdot v_c^2}{2g} \tag{VII.37b}$$

Here K is a coefficient taking account of the slight excess of the calculated velocity over the experimental value, v and v_C are, respectively, the velocity necessary for the transportation of the particles and velocity of the flow after the collision of the jet, and M_f and M_p are the respective masses of the flow and the particles.

From these formulas we obtain

$$\eta_T = \frac{Kv^2}{v_c^2} \cdot \frac{M_p}{M_f} \tag{VII.38}$$

The values of the velocity v necessary for the transporting of the particles were determined earlier (see §37), while the value of v_C may be measured experimentally.

Some experimental and theoretical data relating to the velocities of the flows formed after treatment of the surface with vari-

Table VII.10. Experimental and Theoretical Data for the Parameters of Jets Directed at an Angle of 75-90° to the Surface (particles 10 μ in diameter)

Type of nozzle	Distance from surface, m	Water pressure in front of nozzle, kg/cm²	Experimental data: water flow, liters per min	Experimental data: flow depth, mm	Calculated data: flow veloc.* around jet impression perimeter, cm/sec	Calculated data: effic. in particle transport, %
Conical-cylindrical.	3.0	3.0	80.0	2.0	138	1.1
Hydropneumatic . . .	1.0	3.0	10.0	1.2	72	4.1
Ejector	0.2	–	1.0	0.75	30	–

*Velocity of transporting the particles taken as equal to 32.3 cm/sec (see §37).

ous nozzles are given in Table VII.10. The calculation was carried out for wash-resistant particles 10 μ in diameter.

We see from the data presented that the jets created by the ejector nozzle, although strong enough to detach particles from oil-free surfaces, fail to provide flows of adequate velocity for removing the particles so detached. Hence, even in the course of washing there may be a secondary deposition of particles, as indeed occurs in practice.

The velocities of the jets emerging from hydropneumatic and conical—cylindrical nozzles greatly exceed those required for the transportation of the particles, and this ensures the removal of the particles from the zone of contact between the jet and the surface after being detached.

It is found by experiment that the transporting power of the flow formed after the impact of continuous jets on the surface is retained to a distance equal to two diameters of the jet cross section from the axis. This means that in the washing of vehicles it is not obligatory to overlap zones which have already been subjected to the water jet. Jets formed by a hydropneumatic nozzle have a considerably greater cross section and a fully adequate transporting velocity of the flow; they give a good washing efficiency with a low consumption of water per square meter of the treated surface.

For a jet formed by a hydropneumatic nozzle the efficiency exceeds that of a continuous jet both in the detachment of the particles and in removing them from the contact zone.

The overall efficiency of the jet may be expressed in the following way:

$$\eta = \eta_0 + \eta_T \tag{VII.39}$$

The concept of this efficiency is rather arbitrary, since a number of assumptions were made in its definition. It is considered that the particles are close packed (in removing layers of particles η_0 will increase); the stubborn particles have a diameter of 10 μ; the energy of adhesion of the particles calculated from formulas (I.53)-(I.55) is approximate, etc.

However, the efficiency factor enables us to make at least a relative estimate of the energy consumed by the jet in removing adhering particles and to choose the most economical way of washing surfaces.

The time required for the cleaning of a surface from particles is determined by the velocity of those particles reached under the action of the flow.

High-speed motion photography provides information on the relation between the velocity of the particles and the velocity of a flow 1-2 mm in depth:

d_p, μ	80-100	30-40	5-10
Velocity of liquid, cm/sec	7.0	17.0	41.0
Velocity of particles, cm/sec	5.5	12.0	34.0

Thus, as we should expect, the velocity of the particles is always smaller than the velocity of the flow formed on the surface.

It should be noted that the experimental results presented here were obtained for particles spherical in shape. In practical conditions the shape of the particles will differ from spherical; however, this does not introduce any sharp change into the efficiency of particle removal. Thus, the value of K_m in treating a painted surface with a jet from a hydropneumatic nozzle with a water flow of 4 liters/m^2 is 910 for spherical particles, and 630 for particles of irregular shape. The state of the surface and the presence of oil films has a greater effect on particle removal.

Characteristics of the Removal of Strongly Adhering Particles. The efficiency of washing depends on the nature of the contamination and the state of the surface. Black dirt (dust, soil particles, ashes, scale) is usually removed quite easily. Particles of iron and clay materials are difficult to wash even from oil-free surfaces [356]. Strongly adhering particles include silica, loam (clay contaminated with sand), loess, marl, and bituminous quartzite and other shales. Clay particles stick particularly firmly after the drying of rain drops. Particles of iron adhering in the incandescent state are also hard to remove by washing. Hence, a wet-washed surface only appears clean. On drying, particles of adhering dirt appear on the surface

It is hard to free an oily surface from contaminations, since the adhesion of the particles is much stronger (see §14). For example, in treating an oily surface with a drop-type jet (hydropneumatic nozzle) half the adhering particles are removed (γ_F = 53%), while on using the same jet on an oil-free surface nearly all the particles disappear. In order to ensure the complete cleaning of oily surfaces the velocity of the drop (100–300 μ in size) should be 60–80 m/sec; for oil-free surfaces 1 m/sec or less is sufficient.

Below we present some results on the removal of glass particles by a continuous jet in the treatment of an oily surface with a water flow of 30 liters/min as a function of the pressure in front of the nozzle:

Pressure in front of nozzle, kg/cm^2	2.5	7.5	15.0
K_N	80	140	160

As the water pressure in front of the nozzle rises six times the efficiency of removing the particles only rises by a factor of two.

Our data relating to the washing of oily surfaces agree with the experiments of Kozhemyakin [23], who determined the relation between the degree of cleaning of a surface and the form of contamination. According to these data, in order to remove weakly bound, nonoily contaminants containing up to 83% sand particles, it is sufficient to have a water pressure of 1.5–2.0 kg/cm^2 at the nozzle exit, while for removing particles containing up to 10% oily contaminants the pressure must be increased to 6 kg/cm^2. For cleaning the surface of motor vehicles contaminated with firmly bound particles of building materials (cement, alabaster, and tar), the pressure has to be 14-15 kg/cm^2. However, even these pressures cannot guarantee the removal of all the adhering particles.

Raising the efficiency of particle removal from oily surfaces by increasing the pressure in front of the nozzle for continuous jets and raising the velocity of the jet in the case of drop structure is not always convenient, since such a rise inevitably reduces η_0 and η_T [see formulas (VII.35)-(VII.39)] and leads to a greater consumption of water. Hence we must seek other ways of increasing the efficiency of particle removal from oily surfaces. It was established experimentally that the efficiency of washing oily surfaces (Avtol layer of 1.4 mg/cm^2) with a water jet of drop structure (hydropneumatic nozzle) changes very little indeed if solutions of surface-active substances (SAS) are used instead of water. For example, the K_N for water under these conditions equals 6.9, while for a 1% solution of Sulfonol and a 0.3% solution of OP-7 the K_N is equal to 3.9 and 9.1, respectively. This may clearly be explained by the fact that in the short space of time (seconds) elapsing during the removal of the contaminants the SAS cannot exert any noticeable effect, since the reduction in surface tension in the solution does not set in at once but only after a certain time measured in minutes [351, 352, 357] (see §22).

The necessity of carrying out the process in two stages arises from this. The first stage proposes the deposition of a solution of SAS on the surface and the second consists of the washing of the solution with a water jet after remaining on the surface for a certain time.

Reumuth [357] recommends placing an emulsion of SAS on the surface at a temperature of 40-65°C, holding for 3-10 min, and sometimes longer (up to 20 min), and then removing the deposited matter with a jet of water. The SAS concentrate near the contaminant particles and thus facilitate their detachment. A 0.1% Sulfonol solution may be used as SAS for this purpose [351].

It is possible to remove contamination from an oily surface with foam [358, 359] formed by SAS, but here also double treatment of the surface is required. In the present case the foam does not flow away from the surface (in contrast to solutions of SAS), but remains on the surface for a long time, thus economizing in washing materials.

Contamination may be removed from an oily surface together with the layer of oil; in this the efficiency of the washing process will be determined by the adhesion of the oil layer to the original surface. In view of the fact that surfaces become oily in the course of use, it has been proposed to carry out a preliminary treatment of the original surface with a 1% aqueous solution of Gardinol or molten glass. In order to determine the effectiveness of cleaning such surfaces with water jets after becoming oily, some special experiments have been carried out; the results are as follows.

1. A layer of Avtol was applied to a surface covered with a film of Gardinol formed after treatment with a 1% solution of the latter; the resultant K_N values were: $K_N = 90$ for cleaning with a jet of drop structure*; $K_N = 590$ for cleaning with a continuous jet.†

2. Avtol was applied to a surface treated with molten glass: $K_N = 200$ for cleaning with a jet of drop structure*; $K_N = 280$ for cleaning with a continuous jet.†

On cleaning a surface first oiled with Avtol and then treated with a 1% Gardinol solution in a continuous jet, $K_N = 830$, while for a surface not previously treated with SAS, $K_N = 80$.

We may conclude from these results that the application of a film of SAS or molten glass to the clean surface of an object before putting into service greatly increases the efficiency of subsequent washing in water jets (continuous or of the drop type) as compared with the same jets without SAS treatment. However, the advantages offered by such preliminary treatment are less marked in the case of a continuous jet than those offered by the application of a Gardinol solution to the oily surface immediately before washing with water.

*Hydropneumatic nozzle, water consumption 10 liters/m^2.

†Conical–cylindrical nozzle, water flow 30 liters/m^2, water pressure in front of the nozzle 2.5 kg/cm^2.

Chapter VIII

Adhesion in Gas-Purifying Apparatus

§ 41. Adhesion of Dust in Electric Filters

Rise in Adhesive Forces. The process of removing dust from air or any other gas in electric filters includes the following stages: the charging of the particles, their transportation to a settling electrode, their adhesion to the electrode, and the regeneration of the original surface.

The first two stages will not concern us at present; they are fully covered in the relevant literature [143, 144, 360-362].

As a rule, the adhesion of particles to a surface in an electric field is greater than the adhesion associated with free settling [363]. The rise in adhesion is due to the image forces, the magnitude of which (see § 12) is determined by the charge given to the dust particles in the field of the corona discharge of the electric filter. The maximum charge on particles in the air depends on the field strength and the size and nature of the particles themselves. The following table indicates the maximum charges and maximum adhesive forces calculated by Brandt [364] for particles of different sizes with a density of 1 g/cm³ and a field strength of 4.75 kV/cm:

d_p, μ	100	10	1	0.1
q_{max}:				
in units of elementary charge	$2.5 \cdot 10^6$	$2.5 \cdot 10^4$	$2.5 \cdot 10^2$	2.5
in C	$4 \cdot 10^{-13}$	$4 \cdot 10^{-15}$	$4 \cdot 10^{-17}$	$4 \cdot 10^{-19}$
F_{max},* units of g:				
F_i	3	920	3000	100,000
F_{se}	0	0.07-3	70-135	70,000

*F_i is the component due to image forces; F_{se} is that due to supplementary electric forces.

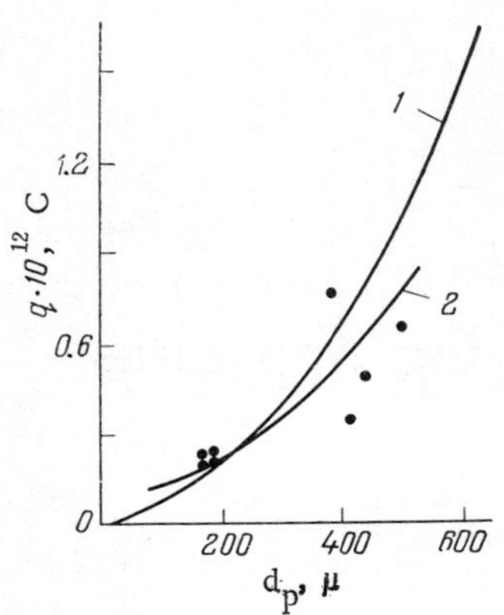

Fig. VIII.1. Charge received in a corona discharge as a function of the diameter of the metal particles. 1) Calculated; 2) experimental.

Brandt's calculations are approximate; in practice [143], there is a deviation from the theoretically possible charges. We see from Fig. VIII.1 that for conducting particles the charge received in the field of a corona discharge is smaller than the theoretical value. For insulators the reverse is the case [144].

Dependence of Adhesion on the Specific Resistance of a Dust Layer. A particle reaching the surface of the electrode may either give up its charge or take charge from the electrode; in some cases it may even leave the electrode surface again. Similar processes take place in the autohesion of particles to a layer of earlier-deposited dust; they are determined by a supplementary electric force. This force depends on the specific electrical resistance of the dust layer and may be expressed by the equation

$$F_{se} = \frac{1}{2} \varepsilon_0 \left[E^2 - \left(\frac{I \Omega \varepsilon_1}{\varepsilon_0} \right)^2 \right] \qquad \text{(VIII.1)}$$

where F_{se} is the supplementary electric force referred to 1 cm^2 of the dust-laden surface, ε_0 and ε_1 are the dielectric constants of the medium and the dust, respectively, E is the electric field potential gradient in the surrounding medium in kV/cm, I is the current density in $\mu A/cm^2$, and Ω is the specific resistance of the dust layer in Ω · cm.

For

$$E = \frac{I \Omega \varepsilon_1}{\varepsilon_0}, \quad F_{se} = 0$$

$$\text{if } E < \frac{I \Omega \varepsilon_1}{\varepsilon_0}, \text{ then } F_{se} < 0 \qquad \text{(VIII.2)}$$

and

$$\text{if } E > \frac{I\Omega\varepsilon_1}{\varepsilon_0}, \text{ then } F_{se} > 0 \tag{VIII.3}$$

For a negative value of F_{se} the particles adhere and for positive values they rebound. The sign and magnitude of F_{se} are determined, other conditions being equal, by the specific resistance of the dust layer. In view of this we may distinguish three groups of particles by reference to their specific resistance.

Dust particles having a specific resistance of 10^3-10^4 $\Omega \cdot$ cm (group 1) easily acquire a charge similar to that of the settling electrode and adhere poorly to its surface.

Particles with a specific resistance of over 10^{10} $\Omega \cdot$ cm (group 3); these are unable to discharge themselves on the settling electrode and adhere firmly to the latter as a result of F_{se}.

Particles with a specific resistance of 10^4-10^{10} $\Omega \cdot$ cm (group 2); these occupy an intermediate position.

When particles belonging to the third group come into contact with a surface, an additional contact potential difference is created. Penney [362, 365] calculated the magnitude of the supplementary electric force from the equation:

$$F_{se} = \frac{1}{2}\varepsilon_0 E^2 \tag{VIII.4}$$

for parallel electrode surfaces in vacuum (or air) when a potential difference of 1 V is applied between them; he obtained the following values for different distances between the electrodes:

Distance between electrodes, mm	0.01	0.1	1.0	10.0
F_{se}, g/cm	450	4.5	$4.5 \cdot 10^{-2}$	$4.5 \cdot 10^{-4}$

These results are valid when the surface of the layer of adhering particles is parallel to the electrode surface. Actually, the adhesion of particles is accompanied by the formation of local contacts, the total area of which is considerably smaller than the area occupied by the adhering layer. Hence the value of F_{se} referred to 1 cm^2 is higher for the actual contact area than for the area of the dust layer adhering to the surface.

Table VIII.1. Contact Potential Difference (referred to 1 cm^2 contact area)

Material of the particles	Contact potential difference V for air humidity:	
	50%	90%
Talc.	0.870	0.420
Ash	0.558	0.238
Mica	0.210	0.120
Nickel	0.450	0.280
Silicon	0.800	0.500

Table VIII.2. Specific Resistance of Dust

Material of the particles	Specific resistance of dust, Ω · cm				
	at 70°C	at 150°C	at 250°C	at 350°C	at 450°C
Sulfonated coal.	$8 \cdot 10^5$	$8 \cdot 10^5$	$8 \cdot 10^5$	$1 \cdot 10^6$	$1 \cdot 10^6$
Cement.	$8 \cdot 10^7$	$7 \cdot 10^8$	$7 \cdot 10^{10}$	$3 \cdot 10^{11}$	$9 \cdot 10^9$
Chromium oxide	$2 \cdot 10^8$	$4 \cdot 10^8$	$2 \cdot 10^{10}$	$9 \cdot 10^{10}$	$3 \cdot 10^{10}$
Coal.	$3 \cdot 10^8$	$5 \cdot 10^9$	$1 \cdot 10^{11}$	$4 \cdot 10^{11}$	$1 \cdot 10^{11}$
Lime	$1 \cdot 10^8$	$1 \cdot 10^9$	$1 \cdot 10^{11}$	$3 \cdot 10^{11}$	$1 \cdot 10^{11}$
Aluminum oxide	$3 \cdot 10^8$	$3 \cdot 10^{11}$	$2 \cdot 10^{12}$	$5 \cdot 10^{10}$	$8 \cdot 10^8$
Sublimation from nickel furnaces.	$3 \cdot 10^{10}$	$8 \cdot 10^9$	$6 \cdot 10^9$	$5 \cdot 10^8$	$1 \cdot 10^8$

For particles of group 3 the value of F_{se} may be calculated not only from formula (VIII.4), but also from the known value of the potential difference (see §11) arising in the particle–surface contact zone; for this purpose we may use the data presented in Table VIII.1.

We see from Table VIII.1 that the contact potential differences attain such values that the attractive forces to which they give rise will have a considerable effect on the adhesion of dust (§11).

The supplementary electric force may be repulsive for fairly low values of the specific resistance of the dust. Then the adhesion is reduced and the efficiency of trapping dust in the electrical filters diminishes accordingly. The harmful effect of the adhering layer on the operation of the electrical filter vanishes when the resistance of the layer exceeds 10^8-10^9 Ω · cm. On the other hand, if the specific resistance of the dust is large, a considerable local potential gradient is created and reverse corona may take place;

this leads to a local break in the adhering layer and also harms the efficiency of the electrical filter. The mechanism of the reverse corona is based on the discharge of negatively charged particles by positive ions [366]. This effect vanishes on reducing the specific resistance of the layer to 10^{10} or 10^{11} $\Omega \cdot$ cm. Thus, electrical filters operate best for a specific dust resistance of 10^{8}–10^{11} $\Omega \cdot$ cm.

Table VIII.2 presents some results [367] relating to the specific resistance of dust layers of various materials at various temperatures.

We see from the results presented that the specific resistance of the dust is not determined by the material of the particles only, but also by the temperature of the surrounding medium. Winkel and Schutz [368] observed the same* for iron oxides, e.g., for Fe_3O_4:

Temperature, °C	20	125	1000	1320
Specific resistance, $\Omega \cdot$ cm	$2.5 \cdot 10^2$	$4.7 \cdot 10^2$	2.2	$7.7 \cdot 10^{-1}$

By varying the temperature the specific resistance of the dust, and hence the adhesion, may be varied.

The specific resistance of a layer of dust deposited in an electric filter is determined not only by the characteristic specific resistance of the material at the specified temperature, but also by the particle size, the density of the dust layer, and the surface conductivity of the particles, which varies as a result of the adsorption of gases and vapors.

Duffield and Grotenhuis [363] and later Eishold [369] showed that the specific resistance of copper oxide powder fell with increasing particle size and increasing density of the layer; the density of the layer played the principal part in this.

Usually the specific resistance of a dust layer is 2–3 orders above the specific resistance of the material itself measured under the same ambient conditions.

The effect of gases and vapors in the surrounding medium is particularly strong at low temperatures, i.e., under conditions favoring the condensation of vapor, and also for a high original spe-

*It should be noted that Winkel and Schutz failed to mention the conditions under which the specific resistance was measured or the material from which the particles were prepared.

cific resistance of the dust. However, gases present in the surrounding medium may not be indifferent to the deposited dust, and then the surface conductivity of the dust layer may increase with increasing temperature [370, 371].

Thus in copper-melting and lead—zinc works, ZnO, PbO, and other oxide dusts are easily trapped by electric filters.

An increase in the adhesion may here take place for two reasons. First, as a result of capillary forces at the temperature corresponding to the condensation of SO_3, i.e., at the temperature at the entrance into the electric filter, which equals 145-170°C. On raising the temperature the SO_3 evaporates and the adhesion increases still further (see §26). Secondly, the increased adhesion may be due to the fact that the SO_3 contained in the waste gases interacts with the metal oxides at higher temperatures, so that the dust particles become larger, the electrical conductivity of the layer increases, and the adhesion does likewise.

On roasting molybdenum sulfide concentrates there is a reduction in the trapping of dust by the electric filter as temperature rises. This may be explained by the fact that there are no substances capable of interacting with SO_3 in the dust formed.

The efficiency of the trapping of dust by an electric filter may be increased by artificially humidifying the gas to be cleaned or introducing a mist or spray of sulfuric acid at ordinary temperatures [162, 366]. It is noteworthy that the deposition of gold-bearing dust in electric filters increases as the content of sulfuric acid vapor rises from 4 to 12% [372]. In this case the adhesion of the dust to the settling electrode increases as a result of capillary forces (see §13). One must remember, of course, that this makes subsequent purification more difficult. The presence of SO_3 in the nickel dust is also clearly the main reason for the poor cleaning of the electrodes [373].

There are some well-known methods of raising the efficiency of electric filters by increasing the autohesive forces between the particles. Ferromagnetic iron oxide particles merge and become larger under the influence of an electric field; they then stick to the electrode surface in the shape of acicular formations [374]. The enlargement of the particles promotes a rise in the specific resistance and hence in the adhesion of the layer.

The operation of an electric filter depends not only on the properties of the particles being trapped, but also on those of the settling electrode surface. Thus the forces of adhesion to the surface of the settling electrode may be increased by using plastics for the electrodes [375] and subjecting the plates to preliminary treatment in a special solution [376].

§ 42. Cleaning the Settling Electrodes of Electric Filters from Attached Dust

The dust which has already settled on the electrode screens its surface and makes it less efficient in purifying a gas flow [377]. First of all the attached dust may be removed mechanically: by blasting the dust-laden surface, scrubbing it, subjecting it to vibrations or blows, and also by washing it with water.

The removal of the adhering layer by air-blasting the surface is not very convenient in practice, since it requires an air blast of a very high velocity, and even then the dust may settle again on the surface.

The scrubbing method of cleaning electrodes has not been very widely used, since it necessitates putting the filter out of action.

The vibration method is used in practice, but it cannot always produce the required degree of cleansing. Thus the use of vibrators (frequency, 50 cps; amplitude, 1.2 mm) creates a detaching force [378] of 1200 g. However, in shaft furnaces containing 50-60% lead and 10-15% zinc particles 0.8-1.5 μ in size, a force of this kind only breaks the autohesive bonds in the dust layer, while a monolayer remains on the surface of the electrode. In order to increase the detaching force, and hence the cleaning efficiency, we may use the vibration method together with an arrangement for varying the amplitude of the vibrations [379] and the direction of the shock associated with the vibration [380]. Brandt [380] and later Dieter [381] showed that, for the same actual force of the blow, the acceleration received by the adhering layer was greater for a normal blow (normal to the electrode) than for a tangential blow (along the electrode). The following represents some experimental data obtained for single metal plates 7500 × 770 × 1.5 mm in size on striking with a hammer at a frequency of 67 cps:

Direction of blow	Tangential	Normal underneath	Normal in the middle
Acceleration, g units:			
maximum	550	9300	65,000
minimum	64	540	500
taken as average	215	2950	4000

For a tangential blow only up to 7% of the energy is transformed into acceleration, while the rest is expended in deforming the electrode material. A normal blow communicates an acceleration 8-18 times as much as this to the adhering layer. Hence, detachment of the adhesive as well as the autohesive type takes place, as indicated by the high degree of cleaning of electric filters treated in this way: $K_m = 10^4$ (see §31).

The force of the blow is in general determined empirically when cleaning by the vibration method.

Lowe and Lucas [382] obtained a semi-empirical formula for determining the thickness of a layer of dust H_l detached by shaking

$$H_l = \frac{4 \cdot 10^{-8} \cdot F_{aut}}{\rho_{av} \cdot jD} \qquad \text{(VIII.5)}$$

where F_{aut} are the autohesive forces of the dust later, ρ_{av} is the mean density of the layer, j is the acceleration on shaking, and D is the diameter of the dust aggregates.

If the diameter of the detached dust aggregate equals H_l or is close to the latter, then, in accordance with Eq. (VIII.5), it is sufficient to communicate a slight acceleration to the adhering layer on detachment in order to remove it. This is valid for an adhering layer consisting of aggregates of large dimensions.

Owing to the indeterminacy and large spread of the value of H_l we cannot calculate the value of the shaking blow precisely.

In addition to periodic impact shaking, one may subject the settling electrodes to continuous vibration in order to clean them. If the settling electrodes are set into oscillatory motion with a frequency of 8-9 cps (mean amplitude, 3 mm; acceleration of the electrode, 5-6 m/sec^2) during the operation of the electric filter, the settling dust will "flow off" from the electrode surface continuously, i.e., the electrode will remain practically clean, independent of the original dustiness of the gas undergoing purification [383].

The vibrational method, like all mechanical methods of cleaning settling electrodes, has a passive character in the sense that the value of the adhesive forces remains comparatively high and suffers no change in the course of filtration. A reduction in the adhesive forces may be achieved on replacing the air adhesion by the liquid type, as for example in wet electric filters. At the same time this method offers the possibility of avoiding electrical breakdown, i.e., the electrodes may be under voltage [384] while being washed, and the cleaning may be carried out continuously. In addition to this, as a result of the attraction of charged dust particles to the drops of water the particles will become larger, and this will improve the purification of the gases [385].

The washing systems may incorporate the deposition of a SAS solution and its subsequent removal as in the washing of motor vehicles [386].

The efficiency of the process of detaching adhering dust from the surface of the settling electrode is raised by increasing the velocity of the washing liquid; for this purpose the liquid may be introduced tangentially to the surface of a tubular electrode [387] and flow over its sides in a helical manner.

Wet electric filters are sometimes used for trapping radioactive dust [388], since the use of dry electric filters is not very efficient in this case owing to the effect of the radioactive isotopes on the electric field.

§43. Adhesion During the Purification of Gases in Filtering Apparatus

Filtration and Adhesion. In purifying dust-laden industrial gases or air, the particles suspended in the gas pass through porous filtering material and stick to the latter. The dust deposit so gathered in turn becomes a filtering medium for subsequent particles. As the deposit builds up, the porosity of the medium diminishes, preventing the free flow of gas. At some point this deposit has to be removed.

Thus, adhesion appears at two stages of the filtration process: in holding the particles when these touch the filtering element, and preventing them from being carried away on subsequent passage of the gas, and also in connection with the regeneration of a spent filter.

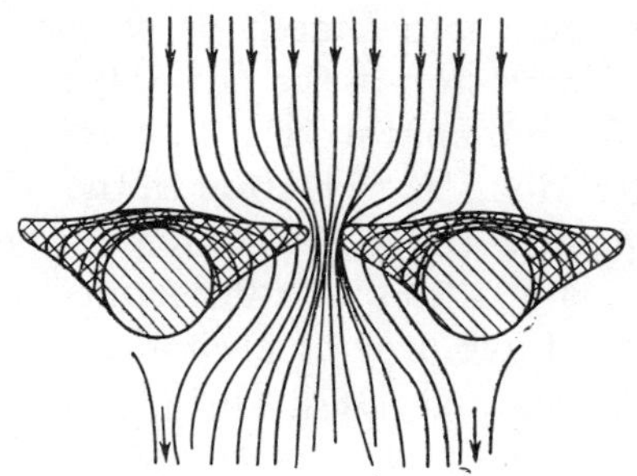

Fig. VIII.2. Formation of a layer of adhering dust on individual filaments (schematic).

Frequently it is precisely the relation between the adhesive properties of the dust and the filtering material which determines the desirability of using one particular kind of cloth, etc., or another as well as methods of regeneration and the optimum conditions for the working of the filters.

In order to discover the part played by adhesion in the filtration process, let us consider the deposition of particles on a single cylindrical filament placed in a flow of aerosol. The way in which the dust deposit is formed on individual cylindrical filaments of a filtering barrier for a rate of flow of 1 m/sec is shown in Fig. VIII.2. The clearly visible local side growths of lead and zinc oxide dust particles about 1 μ in size are directed at an angle of ±110-120° to the axis of the flow. When more aerosol passes the growths may come together forming a continuous layer which plays the part of a "secondary" filtering medium [389].

The velocity of the air flow affects the adhesive forces of finely dispersed dust. Yankovskii [390] photographed the deposition of litharge spheres about 1 μ in diameter in air on polystyrene filaments and also on micron Plexiglas filaments and copper wires. For air speeds of 1.5 m/sec or over, adhesion occurred mainly on the frontal part of the wires and filaments. The sticking of dust to the sides started after the formation of the frontal layer. It should be noted that the buildup of dust on the side surface took place in a nonuniform manner. For considerable flow velocities (30-40 m/sec), particles with sizes greater than the diameter of the filament were torn off by the air flow. Particles smaller than the filament diameter stuck to the filaments and remained attached to these.

Small particles with diameters an order lower than the diameter of the filament adhered in a dense layer to the whole filament surface. For low flow speeds (up to 0.15 m/sec) the distribution of particles around the perimeter of the filament cross section was practically uniform, even large particles sticking to fine fibers.

We must emphasize the difference in the character of the adhesion to a dust-free (fresh) and a dusty filter, particularly for highly dispersed dusts. The dust-holding capacity of filters improves sharply as they gather dust; the secondary porous barrier formed by the particles of dust deposited in the pores between the fibers or grains of the filter are more efficient in trapping particles as a result of diffusion and contact.

Gillespie [208, 209] carried out analogous experiments on wires with aerosols of paraffin, stearic acid, and lycopodium powder. For low flow velocities the actual and theoretical* amounts of trapped dust agree closely. However, as velocity increases the actual amount of trapped dust becomes considerably smaller than the theoretical. For wires treated with viscous silicone liquids the number of trapped particles is close to the theoretical, while on untreated wires it is only half this. Under some conditions some of the particles are charged and this leads to trapping and better adhesion.

The deposition of particles on single filaments and wires is to be distinguished from the filling of a filter with dust particles. Whereas, in the first case, deposition depends on the flow conditions around the obstacle and the elastic properties of the surface as well as adhesion (see §34), in the second (the filtration process) the whole volume of the pores in the filter is filled with particles and clogging takes place.

In order to increase the adhesion of particles, it is sometimes necessary to prepare the filter material (wetting with sticky substances, charging the fibers, etc.) or dust in a special way.

The quality of the filtration is conditioned not only by the adhesion but also by processes preceding it. It is thus hardest to catch the finest dust particles, particularly at high filtration rates. In order to settle these particles better, they must first be made larger, i.e., they must be artificially coagulated by increasing the forces of autohesion. Enlargement of particles may be carried out in air conduits and dust extractors by kinetic coagulation, under the influence of an ultrasonic or electric field, and by condensation

*The theoretical amounts were calculated from the conditions governing the behavior of an air flow around an obstacle (see §34).

of water drops on the particles. For example, in the lead production process a dust with particle diameters of about 0.3 μ is produced. On treating such particles with electrically charged drops of water the particles are captured by the latter; the water then evaporates and the particles stick firmly to each other [391, 392].

Choice of Filtering Material. The main requirement for filtering cloths is that they should effect purification at a maximum filtering velocity and with a minimum hydraulic resistance. In addition to this, the cloth must be resistant to high temperature and aggressive media, it must be mechanically strong and have a fair dust capacity, i.e., hold on its surface a certain amount of adhering dust in the form of a layer which may easily be removed by mechanical action.

These requirements may be satisfied by suitably choosing the structure and form of the fiber used for the filter cloth and also by treating its surface. Until recently, industrial dust filters made extensive use of cloth (cotton fabric and wool) and felt materials. At the present time synthetic fibers are coming into use. Thus, a good filtering material at temperatures of 150 to 315°C is glass cloth [395] treated with silicones. At lower temperatures (under 150°C) cloth materials such as Lavsan, Nitron, Capron, and chlorinated polyvinyl chloride fiber are being used. These have high strength, heat resistance, and resistance to aggressive media, and the adhering dust may easily be removed from their surfaces. However, synthetic-fiber cloths are less efficient in purifying gases than natural cloth, since the surface of the synthetic material contains no microscopic asperities such as usually occur on natural fibers.

Thus the adhesion of highly dispersed lead dust to a cloth made from "Nitron" fiber is smaller, i.e., the degree of purification is lower, than in the case of TsM cloth (mixture of 70% wool and 30% Capron), since the former has a less developed surface [396]. Yet the Nitron cloth has a high heat resistance, hydrophoby, and moisture resistance, which ensures good regeneration. The adhesive capacity of Nitron and other synthetic materials may be increased by giving them a special structure [397].

Dust is easily removed from filter cloths based on polyfluorethylene fibers.

In order to ensure high air-purifying efficiency, cloths made of synthetic fibers should not be cleaned too intensively; if they are cleaned too intensively, a large amount of the dust layer is removed and filtration of the dust-laden air flow is less effective.

In the classical theory of filtration, the category of factors causing dust to adhere to the surface of the filter material includes not only contact, gravitational settling, inertial and diffusion trapping, thermo- and diffusiophoresis, etc. [398], but also the electrical interaction of oppositely charged bodies. Electrical forces, however, are only able to appear if either the filtering material or the dust particles are charged. Thus, neutral or weakly charged dust is attracted to charged surfaces and sticks firmly to these. An analogous effect occurs for charged dust and an uncharged surface.

Charging of the particles and filtering cloth may be effected as a result of friction (triboelectricity) during filtration [399]. For example, when an air flow passes through a filter made of synthetic or glass fibers comparatively large particles are held purely mechanically, while small particles are electrified and, depending on their charge, are attracted either to the positively or to the negatively charged fibers. Such filters make it possible to achieve efficient dust elimination at fairly high flow velocities and low hydraulic resistances [400].

Certain synthetic fibers acquire electric properties in the course of formation and drawing. In addition to this, a charge may be given to the fibers or particles by, for example, applying an electric field. Thus, in order to effect polarization, fiber layers based on glass fiber or synthetic materials are placed in a high-voltage field. Cloths used as filter materials may be arranged in a triboelectric series, according to their capacity to become electrified on friction [401]. The triboelectric series must only be regarded as an arbitrary specification of filter cloths, since the technological characteristics of the production process or a slight modification of the fiber surface may change the position of any particular type of cloth in the triboelectric classification.

Methods of determining the position of a cloth in the triboelectric series are based on determining the charge which it acquires by friction with another fiber, taken as standard.

The classification of dusts by reference to their electric activity and conditions of adhesion to charged surfaces was indicated earlier (see §28).

Removal of Adhering Dust. The regeneration of filters, i.e., freeing them from adhering dust, may be carried out in several ways. The adhering layer may be removed by blowing back, although this method is not very efficient. Thus, for an air-flow speed of 2 m/sec through a layer formed on 1 m^2 of the surface of a filter cloth from 500 g of quartz dust smaller than 10 μ in size, no more than 0.2% of the particles are blown out [402]. Only the weakly attached particles are removed in this way. The majority of the adhering particles are detached at higher air-flow velocities, up to 50 m/sec (see §31).

More efficient methods of removing adhering dust include intermittent blowing back at considerable air speeds (pulsating, impulse, etc.) with the simultaneous breaking up of the dust layer, and mechanical shaking (shock, vibrational, impulse, etc.).

A reverse blast at a velocity of 10-25 m/sec, accompanied by deformation of the cloth, beats the attached dust into easily removed aggregates. The pressure created by the flow of air ensures the removal of dust which has penetrated into the layer of cloth [403]. This should ensure close contact of the scavenging rings with the material of the filter (sleeve or bag) and also the breaking of the dust layer as the ring moves.

The dust layer may be removed extremely efficiently from the cloth surface by means of pulsating air blasts, the direction of these changing each time. In this way a filter may be freed from the most strongly adhering dust [404], namely, volatile ash with a mean particle diameter of 0.6 μ, fumes of silicon oxide ($d_{av} = 0.3 \mu$), and iron oxide ($d_{av} < 0.1 \mu$) for a dust concentration of over 2 g/m^3 in the gas.

It is possible to clean bag filters by means of an air flow directed parallel to the adhering layer at a velocity of 20 m/sec.

The efficiency of the cleaning of bag filters by shaking is determined by the amplitude and frequency of the vibrations. Increasing the detaching force does not always lead to the removal of residual particles; this is because of the indeterminacy of the adhesive forces and the difficulty in detaching a small number of residual particles (see § 16).

The filter bags may also be cleaned by ultrasonic vibrations. The level of sonic pressure should not be lower than 125 dB for a frequency of 200–4000 cps; its precise value is determined by the construction of the filter and the properties of the adhering dust. The acceleration received by the dust layer sticking to the cloth (and hence the degree of cleaning) depends on the distance between the source of the vibrations and the filter bag. As this distance increases to 20 cm or more, it becomes less easy to remove the attached layer. The weakening of the vibrational forces may be compensated by twisting the filter bag through 30–40°, which helps break up the adhering layer [405].

Sometimes filters may be cleaned with hot water ($t = 60$–$70°C$); if the dust contains oil, wetting agents are added to the water.

In order to improve their cleaning from adhering dust, glass-cloth filters are treated with silicones. Glass cloths impregnated with silicones are used for trapping soot with a particle size of under 0.1 μ and also for cleaning the fumes from electric-arc furnaces in the melting of lead, zinc, and aluminum [390].

§ 44. Autohesion Properties of Some Industrial Dusts and Characteristics of Adhesion in Dust-Trapping Processes

The operating efficiency of many gas-purifying systems depends on the nature and particle size of the dust trapped. Very often poor operation of dust traps is due to the fact that no proper allowance has been made for the characteristics of the dusts, especially their particle size and their capacity to stick to each other and to other surfaces, i.e., their autohesion and adhesion. The autohesive and adhesive properties of various dusts have not yet been studied and applied sufficiently.

Industrial dusts are separated into four classes: nonautohesive (alumina and slag dust), weakly autohesive (coal ash, slate dust, dry dust of the magnesite, blast-furnace, and apatite types), moderately autohesive [ash, peat dust, cement dust, dust from the concentrates of nonferrous metallurgy, soot, and also dust of the

Table VIII.3. Wettability of Various Dusts (half the particles have a diameter smaller than 10 μ)

Dust material	Density of dust material, g/cm³	ρ, g/cm³	F_{aut} (referred to 1 cm²), dyn	φ,°
Room	2.0	0.69	40	105
Random fine rock	2.38	0.92	78	101
Shale	2.38	1.07	159	79
Limestone	2.74	1.17	167	95
Gypsum.	2.38	0.97	294	70

previous (second) class after wetting], and strongly autohesive (wet cement, gypsum, and alabaster dust, flour, and dust from fibrous materials) [406]. The division of dusts into classes is quite arbitrary and subjective. Thus ashes may give particles and dust layers with different autohesive properties, depending on the properties of the fuel and the conditions of combustion.

It must also be remembered that real dusts have a very complicated composition. Thus the dust in cotton-cleaning works contains 17-40% of organic substances (fibers, leaves) and 60-83% of soil and loess particles adhering to the organic particles when the cotton is gathered [407]; this dust is inhomogeneous and constitutes an organic—mineral complex (fungi, bacteria, mineral inclusions). The dust of molybdenum concentrates contains MoS_2, 94.5%; and SiO_2, 5.5%; titanium concentrate contains 54.7% TiO_2, 38.7% FeO, 2.16% SiO_2, 1.09% MgO, the rest being impurities; beryllium concentrate contains 90% $Be_3Al_2(Si_6O_{18})$ and 10% SiO_2 [408]. In the dust of Bessemer slags, vanadium, chromium, and manganese may be found

Davies [62] carried out a detailed investigation into the autohesive properties of dusts, measuring the forces associated with the breakup of dust layers; these results constitute an objective characteristic of the autohesion of dust particles (Table VIII.3).

We see from Table VIII.3 that, in the case of the types of dust considered, the autohesion varies from 40 to 294 dyn.

The autohesive forces associated with any particular type of dust may also vary with the conditions of formation and the nature of the dust. For example, according to Thouzeau and Taylor [24], the autohesive force of limestone dust is 520 dyn. If we consider

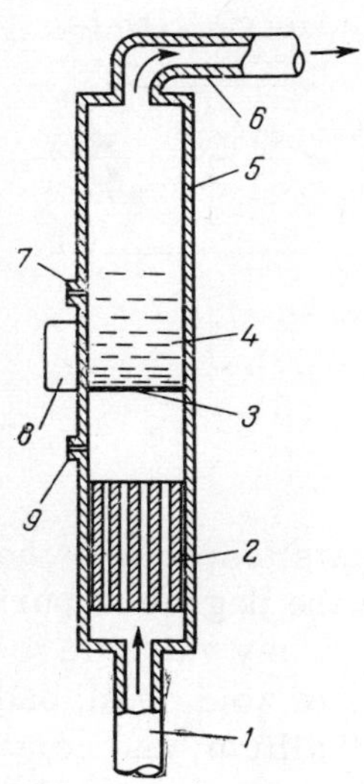

Fig. VIII.3. Arrangement of apparatus for the electrical filtration of air. 1) Tube for introducing the gas; 2) rectifier; 3) grid; 4) filtering layer composed of spheres; 5) cylinder; 6) tube for removing the gas; 7, 9) tubes for manometers; 8) vibrator.

that the methods of determining the autohesive forces used by Davies and by Thouzeau and Taylor were exactly the same, then we can only explain this discrepancy in the autohesive force by reference to the physicochemical characteristics of the dust, or by a difference in their particle size.

Owing to the differences in the shapes and sizes of particles it is difficult to secure a correlation between such parameters as autohesion, the bulk density of the adhering layer, the angle of rest (φ) and the air permeability, although in principle such a correlation should be possible. Thus the relation between the autohesive force and the apparent specific gravity (ρ) of limestone dust may be expressed by the empirical formula [24]:

$$F_{\text{aut}} = 2.8 \cdot 10^3 \rho \sin \alpha \qquad \text{(VIII.6)}$$

where α is the slope at which autohesion-type detachment of the dust layer takes place (see § 6).

When aeration takes place, i.e., when powder particles are transferred into a suspended state, the air pressure is expended in overcoming both external and internal forces, which include the autohesive force between the powder particles [409]:

$$p_a = k(p + \rho H + F_{\text{aut}}) \qquad \text{(VIII.7)}$$

where p_a is the air pressure required for the aeration of a dust layer with an apparent specific gravity ρ and thickness H; k is a coefficient allowing for the expenditure of air energy on the relative motion between the particles of the material; p is the external pressure; and F_{aut} is the autohesive force.

Table VIII.4. Characteristics of Some Industrial Dusts

Solution	Wettability of the dust, %		
	quartz-cericite schists	porphyrites	pyrites
Water	38	89	67
0.05% OP-10	100	100	52
0.5% DB	100	100	100

By using the autohesive properties of dusts, especially those of the highly dispersed type, we may increase the degree of purification of gases in dust catchers and recapture many valuable products from the purified gas, such as compounds of zinc, lead, cadmium, selenium, tellurium, indium, gallium, thallium, and germanium [404]. The efficiency of dust trapping in dust catchers and multiple dust catchers may be increased by raising the forces of autohesion tending to increase the particle size and reducing their adhesion to the inner walls of the dust catcher.

The first problem is solved by introducing particles of a type which will readily become charged and autohere with the other particles in the gas flow being purified. The adhesion of the aggregates to the inner surface of the dust catcher is reduced by covering the latter with a layer of plastic. Static electricity is generated when the dust particles come into contact with this layer, and the charge is removed when the dust catcher is emptied [410].

The efficiency of purifying aerosols by passing through a layer of spherical particles may be increased by raising the adhesive force between the dust and the filter particles [411]. For this purpose a triboelectric charge is imparted to the layer particles by contact with dielectric spheres circulating in the layer. The spheres of the filtering layer are made of plastic (for example, polystyrene), which easily acquires and holds a triboelectric charge.

In the filtering zone the body of the cylinder (Fig. VIII.3) is made of polystyrene (with a high electrical resistance). The part of conductor is played by a conducting copper grid or mesh. On switching the filter off, the triboelectric charge flows off along the copper grid, reducing the adhesive forces between the dust particles and the filtering spheres, and facilitating the cleaning of the filter.

Mixtures of marble with quartz, coal with sand, and clay with coal may be separated in this way. The separation process takes from a few seconds to several minutes. The quality of the separation depends on the material and the shape of the charged surfaces, as well as on the temperature and humidity, i.e., on just those factors which determine the adhesion of the powder to the surface [412]. In choosing the material of the particles used as insulators we may make use of the triboelectric series.

In order to increase the purification efficiency we may also use a combination of a cyclone dust catcher (for trapping the large particles) and a bag filter for catching the small ones [413].

In order to prevent or reduce the adhesion of dust, the inner surface of the dust catcher may be wetted. Under the influence of the centrifugal force the wet solid particles and liquid in the form of slurry are thrown toward the inner walls of the apparatus and fall in a thin film into the bunker [414]. In order to clean the inner surface of a cyclone-type dust catcher, the inner tube is sometimes made movable; on moving downward it comes into contact with the spiral, which cleans off the adhering dust [415]. The dust catcher may also be cleaned with compressed air after shaking [416].

In order to detach the adhering layer and empty the bunkers of dust catchers more completely, the latter are sometimes made of elastic material. On reducing the air (or liquid) pressure, the volume of the bunker is also reduced, so that the dust layer adhering to its surface is distorted and flies off [417].

When particles fall into a liquid medium (in the washing of vehicles or in rain showers), the wettability of the dust is very important. The greater the wettability of the dust, the more completely does the liquid penetrate into the zone of contact, i.e., adhesion in air is replaced by adhesion in the liquid, and this makes removal of the attached particles easier (see §24). The wettability of the dust depends on the particle size, the properties of the surface, and the properties of the solution [418]. For example, in the case of ash, coal, and soot particles of the same size, the wettability is approximately the same [419].

Table VIII.4 presents some data relating to the wettability of 1 g of dust of various materials in 150 ml of solution over a period of 1 min [420].

In conclusion, let us once again mention the factors producing the adhesion of dust in the course of gas purification.

In the operation of electric filters the main property of the dust determining its trapping characteristics is the specific resistance of the adhering layer, which depends not only on the material of the particles but also on the density of the layer and the temperature of the medium.

When purifying gases in cloth filters the adhesion of the particles is determined by their size and physicochemical properties as well as the properties of the cloth itself.

An objective indication of the autohesive (cohesive) properties of dust is provided by the autohesive force, which also governs the efficiency of gas purification.

Chapter IX

Some Characteristics of Adhesion Processes Under Industrial Production Conditions

§ 45. Adhesion in Beneficiation Processes

Triboelectric-Adhesion Method of Beneficiation. The differences in the adhesive forces of powders are used in a recently developed method of beneficiating (enriching) free-running material: the so-called triboelectric-adhesion or triboadhesion method [421, 422, 423]. The arrangement of the apparatus for the triboadhesive enrichment of minerals is shown in Fig. IX.1. The original material is fed from the bunker 1 via a trough 5 onto the drum 7. As the particles move they rub against the bottom of the trough and against each other and acquire a charge. Those powder particles which increase their adhesive force as a result of the charge so acquired, so that the force becomes greater than the centrifugal force in the system (the gravitational forces may be neglected), remain on the drum until they are removed from the latter by the brush 8 and collected in the receiver 9. The rest of the particles fall into receiver 10.

The triboadhesive method of enrichment may be successfully used in two cases: for separating powdered materials into different grades and for separating mineral particles from impurities. In the first case this relates to the separation of quartz, barytes, magnetite, hematite, pyrites, fluorspar, coal, asbestos, graphite, periclase (crystalline magnesium oxide), pegmatite, and iron ore. In the second case the triboadhesive method may be used, for example, in separating asbestos fiber (the falling-out product) from mineral dust adhering to the surface of the drum (captured or held product) [423].

By varying the adhesive forces of the various powders, we may widen the possibilities of the triboadhesive method and extend it to a large number of minerals. This may be achieved by chang-

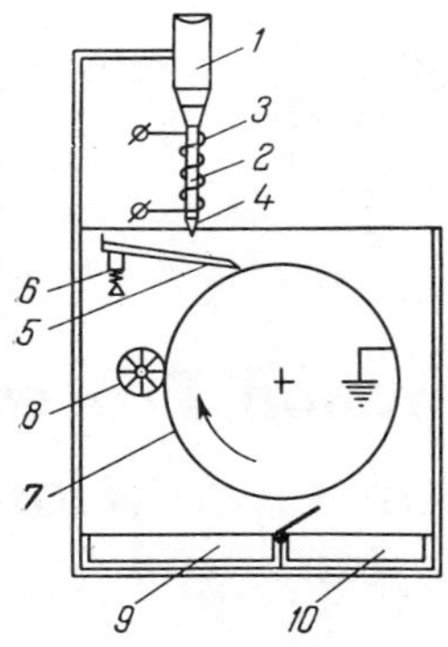

Fig. IX. 1 Scheme of apparatus for triboadhesion of the beneficiation of minerals: 1) hopper; 2) pipe; 3) heating coil; 4) nozzle; 5) chute; 6) vibrator; 7) drum; 8) brush; 9, 10) dust bins.

ing the molecular component of the adhesive forces (see §4) or changing the electrical charge on the particles, and hence the electrical component of the adhesive force (see §§11, 12).

The molecular component of the adhesive forces can only be changed by selecting the material of the drum 7 (Fig. IX.1) or modifying the drum surface. For example, fluorspar sticks more easily to an aluminum surface [422] than to tin plate. However, changing the adhesion by virtue of the molecular component has so far proved of only limited used in the tribohesive method. The practical use of this method is at present based on varying the adhesive forces by virtue of their electrical components.

Let us consider in more detail how changes in electrical forces due to the charging of the particles are embodied in the triboadhesive method and how they depend on the surface properties.

Plaksin and Olofinskii [423] found that the separating capacity of minerals depended on the properties of the materials composing the trough or launder (see Fig. IX.1) and explained this relationship as being due to changes in the charge on the minerals as these passed through the launder.

Balabanov [143] measured the charge on refined material falling through a glass funnel and then into a metal receiver mounted on insulators (the motion of the particles corresponds to the motion of the falling-out product in Fig. IX.1). The average charge referred to one particle (in unit charges) has the following value for various minerals at a grain temperatures of 18-20°C and relative air humidity 70-75%:

Apatite	$4.8 \cdot 10^4$	Hornblende	$1.2 \cdot 10^5$
Bauxite	$4.8 \cdot 10^4$	Sulfur	$1.5 \cdot 10^5$
Magnetite	$5.6 \cdot 10^4$	Garnet	$1.9 \cdot 10^5$
Chalcopyrite . . .	$9.1 \cdot 10^4$	Quartz sand	$1.9 \cdot 10^5$
Ilmenite	$9.2 \cdot 10^4$	Zircon	$2.5 \cdot 10^5$
Marble	$1.2 \cdot 10^5$		

Although Balabanov did not give the size distribution of the particles, these results indicate a considerable charge on the particles, much greater than that on similar particles in the ordinary state. The image forces become very appreciable in this case, and by varying these the process of adhesion may be controlled (see § 12).

Glazanov [71] also determined the sign and magnitude of the charge acquired by minerals passing through a launder under various conditions. Glazanov determined the sign of the charge on particles of various minerals on passing these through launders made from various materials (dielectrics and metals). The results of these measurements are set out in Table IX.1. It follows from these data that the particles of many minerals may change the sign of their charge according to the material composing the launder. For copper, cadmium, mica, cellulose, and cardboard more than half the particles studied acquire a negative charge; for paraffin the ebonite they become charged positively.

In some cases the body with the larger dielectric constant is charged positively (Cane's law). Thus sulfur particles ($\varepsilon = 3.6$) are charged positively after friction with paraffin and negatively after friction with minerals having a dielectric constant larger than their own. An analogous relationship occurs in the case of quartz. The larger the dielectric constant of the launder material (Table IX.1), the smaller number of minerals receives a positive charge. The change in charge due to friction may have a great influence on adhesion (see §§ 11, 12).

Unfortunately, Glazanov [71] failed to give the values of the charges and only indicated that the materials in Table IX.1 were arranged in order of the positive charges acquired and that the charge on particles 200 μ in diameter was 10^{-3}-10^{-9} C/g.

The electrical component of the adhesive forces may be changed by heating the particles. The values of the charges on the

Table IX.1. Signs of the Charges on Various Minerals on Electrification by Friction*

Trough (launder) material / Particle material	Paraffin (2.1-2.2)	Ebonite (2-3.5)	Cardboard (2.5-4.0)	Mica (4-7.5)	Celluloid (4-10)	Glass (5.5-10)	Copper ∞	Cadmium ∞
Sulfur (3.6)	+	–	–	–	–	–	–	–
Calcite (6.3)	+	+	–	–	–	+	+	+
Quartz (6.5)	+	+	–	–	–	–	–	–
Gypsum (6.8)	+	+	–	–	–	–	–	–
Microcline (6.9)	+	+	–	–	–	–	–	–
Magnesite (7.0).	+	+	–	–	–	+	–	–
Fluorite (7.1)	+	+	+	+	+	+	+	+
Rock salt (7.3)	+	+	+	–	+	+	–	–
Zircon (7.8)	+	+	+	+	+	+	–	–
Barytes (7.8).	+	+	–	–	–	+	+	+
Biotite (9.3)	+	+	–	–	–	–	–	–
Cuprous oxide (10.0) .	+	+	+	+	+	+	+	+
Zinc oxide (11.0) . . .	+	+	–	–	–	+	–	–
Selenium (11.7)	–	–	–	–	–	–	–	–
Barium oxide (12.9) . .	+	+	+	+	+	+	+	+
Nickel oxide (22.0) . .	+	+	–	+	–	–	–	–
Aluminum oxide (23.0)	+	+	–	–	–	–	–	–
Cupric oxide (36.0) . .	+	+	+	–	+	–	–	–
Hematite (81.0)	+	+	–	–	–	–	–	–

*Dielectric constants of the minerals in parentheses.

particles of various minerals heated to 400°C and passed through a glass tube at a temperature of 300°C are given in Table IX.2 [71]; the measurements were made in a Faraday cylinder connected to an electrometer and referred to 1 g.

We see from Table IX.2 that the passage of heated particles through a heated launder leads to a considerable change in the magnitude and even the sign of the charge, and this cannot fail to have an effect on the adhesive forces associated with the dust. As temperature rises, so does the magnitude of the charge on quartz, barium oxide, and to some extent calcite. In this case we should expect the adhesive force to rise with increasing temperature. In a number of other cases the electric charges become smaller with rising temperature. The experimental results obtained by Glazanov are

Table IX.2. Magnitude and Sign of the Triboelectric Charge on Various Minerals as Functions of Temperature

Mineral	$q \cdot 10^8$, C/g	
	at 20°C	at 300-400°C
Calcite	+2.32	+3.60
Quartz	+0.31	−3.95
Barium oxide . . .	+0.23	+3.85
Zinc oxide.	+2.73	−0.83
Cupric oxide . . .	+0.75	−0.63
Lead sulfide	+0.82	−0.66
Fluorite.	+3.00	−1.16

Table IX.3. Charge* on Particles Smaller than 60 μ in Diameter on Rolling Along a Trough and Being Blown Through a Tube

Mineral	$q \cdot 10^9$, C/g	
	rolling	blowing
Calcite	2.2	93
Quartz	1.2	240
Microcline. . . .	0.6	16
Aluminum oxide	0.5	140
Barium oxide . .	2.0	230
Cupric oxide . .	1.1	80
Zinc oxide. . . .	1.0	140

* The author gave no indication of the sign of the charge.

confirmed by those of Balabanov [143], who also measured the triboelectric charge on mineral particles as a function of their temperature.

Thus a change in the temperature of contiguous bodies may either increase or reduce the adhesion.

The charge on the particles depends considerably on the means employed for their transportation. Glazanov [71] measured the charges associated with the rolling of particles along a copper trough and the blowing of particles through a copper tube at a velocity of 5-10 m/sec (Table IX.3) and came to the conclusion that on blowing through the tube (the diameter was not stated) the particles acquired a considerably greater charge than on simply rolling over the trough, since the probability of contact between the particles and the tube and between different particles was greater in the tube.

The charge on the particles may also be changed by applying a potential to the trough or launder. Figure IX.2 shows the way in which the charges on quartz and fluorite particles 50-60 μ in diameter varies with the potential applied to copper and cadmium surfaces. The charges were measured by letting the particles fall onto a screened metal cylinder connected to an electrometer. On grounding the cadmium surface (surface potential zero), fluorite and quartz particles nevertheless still acquired a charge.

For a certain specific surface potential, an isoelectric point occurred, such that the charges formed on account of the double layer became neutralized and the electric component of the adhesive force vanished. In a number of cases the particles became saturated with charges (Fig. IX.2b). For a negative potential the value of the charge increased much more rapidly than for a positive. The charges on particles deposited on a substrate at a negative potential were tens of times greater than those found on a grounded surface (Fig. IX.2); this indicates a change (in the present case a rise) in the electrical component of the adhesive forces (see §§11, 12).

Hence, in the triboelectric method of beneficiation the charges on the dust particles, and hence the adhesive forces, may be changed by selecting a different material for the launder, changing the method of transporting the particles (launder or tube), changing the temperature of the original minerals, or applying positive or negative potentials to the launder surface.

Other Methods of Beneficiation. Analogous to the triboadhesive method of separation is the thermoadhesive method [424, 425], based on the fact that different particles of the original mineral are heated differently on irradiating with infrared light, nontransparent particles being heated more than transparent ones. If, therefore, the heated mineral is fed onto a belt covered with polystyrene resin, the particles at the temperature corresponding to the plastification of the resin will stick to the surface as a result of the tackiness of the layer, while the colder particles will fail to stick. This method is used in industrial establishments for purifying rocksalt from dolomite and anhydrite (2-5% impurities); the output is 1 ton/h with a consumption of 50 g of polystyrene per ton of product.

In one variant of the magnetic method of enriching and desulfurizing coal it is proposed to mix coal siftings and dust in a corrugated rotating drum with a small quantity (1 wt.%) of fine ferromagnetic additives, such as magnetite, iron filings, pyrite cinders, etc. The coal particles stick to the magnetic additives and are extracted from the rock fragments together with these in a magnetic field [426].

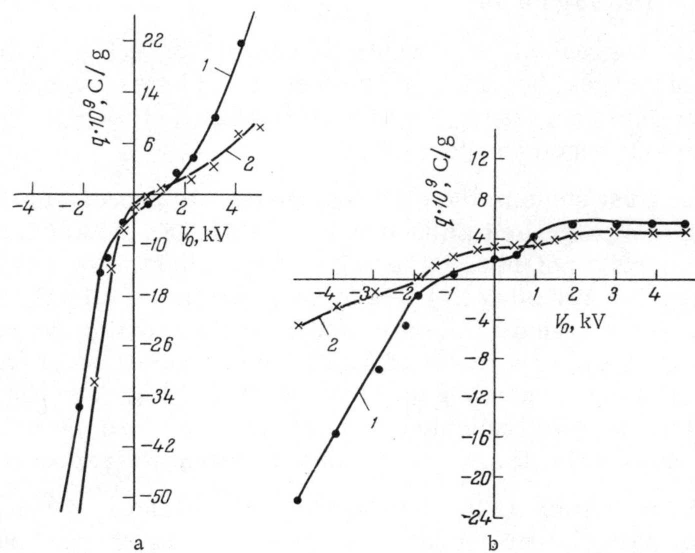

Fig. IX.2. Charge received by quartz (a) and fluorite (b) particles on rolling along a cadmium (1) or copper (2) trough as a function of the potential on the latter.

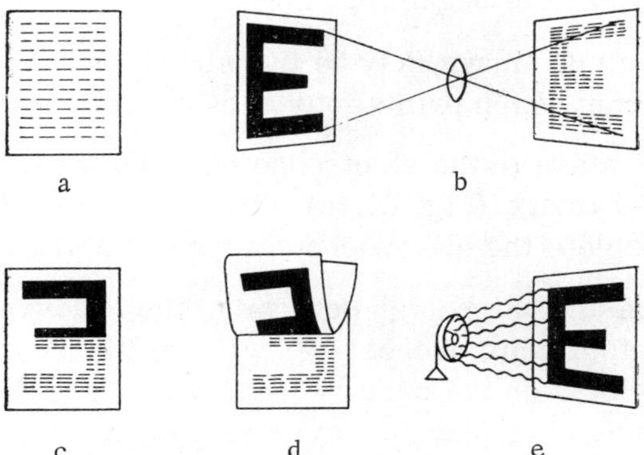

Fig. IX.3. Stages in the electrophotographic process.

§ 46. Adhesion Processes in Electrophotography

Electrography is an essentially new principle for the reproduction of images, being based on electrical, electromagnetic, or other physical processes. An important branch of electrography in wide use is electrophotography.

The first stage in the electrophotographic process is the charging of a photosemiconducting layer (Fig. IX.3a), which acquires a certain potential. The value of the potential is determined by the specific resistance of this layer and falls off slightly with time; however, a sharp fall only occurs on illuminating the semiconducting layer, as a result of which the charges flow off from this layer onto a conducting substrate 2 (Fig. IX.4). The charges retained on the nonilluminated parts of the layer form an invisible (latent) image (Fig. IX.3b). This stage is called the exposure.

The next stage is the development of the image. When the developer particles are applied they adhere to the charged parts of the layer (i.e., to those earlier darkened) (Fig. IX.3c). In contrast to ordinary photography, a positive image of the object photographed appears immediately. If, however, the particles have charges opposite to the charges of the semiconducting layer, they stick to the uncharged parts, i.e., those not earlier darkened; in this case the image is negative.

The resultant image may be transferred to paper (Fig. IX.4d) or to a base from which prints may subsequently be obtained.

The last stage in the electrophotographic process is fixing the transferred image (Fig. IX.4e). The developing and fixing of the image complete the electrophotographic picture.

Adhesion and autohesion determine the following stages of the electrophotographic process (Fig. IX.4): the fixing of the developer particles 3 on the carrier particles 4, and the formation of the developing complex; the development, i.e., the detachment of developer particles and their retention on the charged parts of the semiconducting layer 1; the detachment of the particles from the semiconducting layer, and the transfer of the resultant image to the paper 5; the fixing of the image 7.

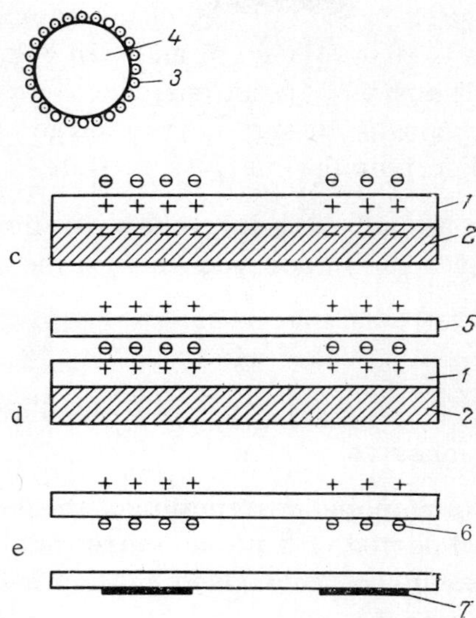

Fig. IX.4. Adhesion in electrophotography. 1) Semiconducting layer; 2) conducting substrate; 3) developer particles; 4) carrier particles; 5) paper or base; 6) print; 7) developed image (c, d, and e correspond to the stages shown in Fig. IX.3).

Electrography and electrophotography are still being studied; the results of such investigations have been generalized, in particular, in certain monographs [427, 428]. Very careful attention has been devoted to the physical processes associated with the charge on the semiconducting layer, the charge distribution over the thickness of this layer, and the change in the surface density of the charges at different stages of the electrophotographic process, etc. Unfortunately, published papers contain no data relating to the direct measurement of the adhesive forces acting on the developer particles in the course of electrophotography (Fig. IX.4).

We shall try to consider the adhesive processes associated with the main processes of electrophotography on the basis of our earlier analysis of the causes of adhesion [429].

Adhesion of the Developer to the Surface of the Carrier Particles. In the course of mixing the developer, particles stick to the surface of the carrier particles. A developing complex is thus formed (3 and 4 in Fig. IX.4). The carrier particles fulfill auxiliary functions: they give the developer particles a charge opposite in sign to the charge on the semiconducting layer 1 and deliver the charged particles to this layer.

Let us denote the adhesive force between the developer and carrier particles by F''_{ad}. In accordance with the earlier argument (see Chapter III):

$$F''_{ad} = F''_{M} + F''_{e} \quad \text{(IX.1)}$$

where F''_M and F''_e are, respectively, the molecular and electrical components of the adhesive forces.

Since the formation and destruction of the complex take place over a limited period of time (no more than 1-2 min), the capillary forces may be neglected (see §13). The Coulomb forces are also negligible (see §12).

The electrical forces which arise as a result of the charges in the double layer are given (see §11) by the formula:

$$F''_{e} = 4\pi S_2 \cdot \sigma_2^2 \quad \text{(IX.2)}$$

where S_2 is the area of contact between the particle and the surface and σ_2 is the surface charge density.

If we neglect the molecular component in the adhesion of developer particles to the surface of the carrier particle,* the developer particles will be fixed on condition that

$$4\pi S_2 \cdot \sigma_2^2 \geqslant P_{dev} \quad \text{(IX.3)}$$

where P_{dev} is the weight of the developer particle.

Zhilevich [430] expressed the surface charge density in terms of the electrical capacity of the carrier particle—contact—developer particle system. In order to calculate this capacity we must know the extent of the gap between the contiguous particles, which has not yet been experimentally determined.

In view of the fact that the developer particles have been adjacent to the carrier particles before contact, the charge arises not only in the contact zone, but also on the surface of the developer.

*Friction of the developer particles leads to their charging and a rise in adhesion on account of the electrical forces, which may exceed the molecular component.

Dividing the left- and right-hand sides of inequality (IX.3) by the weight of a carrier particles (P_c), we obtain [429]

$$\frac{P_{dev}}{P_c} \leqslant \frac{24\, S_2 \cdot \sigma_2^2}{D^3 \cdot \rho_c} \qquad \text{(IX.4)}$$

where D and ρ_c are the diameter and specific gravity of the carrier.

If we confine attention to the equality sign, it follows from Eq. (IX.4) that with increasing contact area and surface charge density and with diminishing carrier-particle diameter, the proportion by weight of the developer particles per unit weight of carrier particles will increase. The ratio P_{dev}/P_c depends on the diameter of the carrier particles to a greater degree than on the other quantities determining it. In order to reduce the amount of carrier used, it is desirable that the diameter of the carrier particles should be as small as possible. In the last stage of the development process the carrier particles are removed from the developed layer. In order to facilitate this process it is desirable to have carrier particles as large as possible (see §16). Hence a suitable compromise is required in selecting the optimum carrier-particle size.

A suitable carrier is provided by a pearl copolymer of styrene with particles 200-600 μ in diameter, and a suitable developer comprises artificial and synthetic resin particles such as polystyrene, phenol formaldehyde, etc., 5-15 μ in diameter.

The adhesion of irregularly shaped developer particles to the surface of the carrier is greater than that of regular particles. However, in both cases (particles or both regular and irregular shape) the area of true contact will be much smaller than the surface of the carrier particle.

The fixing of the developer particles to the surface of the carrier even takes place under the influence of molecular forces alone (§§4, 10). The force F_e'' still further increases the adhesion. However, the charging of the developer particles is required not simply to increase their adhesion to the carrier particles but to intensify the sticking at the next stage of the process, the development. The magnitude of the charges on the carrier particles is determined by the surface charge density σ_2, which is in turn due to the choice of the material for the developer and carrier particles.

The further apart the carrier and developer materials stand in the triboelectric series, the greater is the charge which they acquire, and hence the greater is the charge density σ_2.

Dravin studied the distribution of powders with respect to their triboadhesive properties [429]. In order to classify the adhesive capacity of powders used as developer, the powders were placed in an electrical-contact series limited on one side by polychlortrifluoroethylene, which is electrified negatively for all substances, and on the other by rosamine C, which is electrified positively under the same conditions.

The electrical-contact properties also depend on the dimensions of the particles. As particle size diminishes, for example, as styrene particles pass from 40 to 4 μ, the electrical-contact properties move to the negative side of the series [429, 431]. In view of this the developing powder must be of a single particle size; otherwise its electrical-contact properties will vary and hence so will the adhesion to the substrate and the quality of the image.

Electrical-contact properties may also be varied by modifying the surface of the contiguous bodies (see §11).

Dravin and Kirillova [432] developed a sorption method of tinting, enabling the electrical-contact properties of the polymers to be varied. In order to displace the properties toward the negative part of the series, alcohol and fat-soluble dyes may be used, while for the positive part basic dyes are required.

In the formation of the developing complex, it is essential to eliminate the formation of aggregates of developer particles and the adhesion of these to the carrier surface, since in the development such aggregates may easily break up and adhere to the blank parts of the image, i.e., to the surfaces to which the developer is not intended to stick.

The formation of aggregates may be prevented by reducing or eliminating the effects of capillary forces (this may be done by reducing the air humidity, hydrophobizing the surface of the particles, or by the other methods considered in §§4, 10, 13, and 14) and using single-size fractions with particle sizes no smaller than 10 μ (see §16).

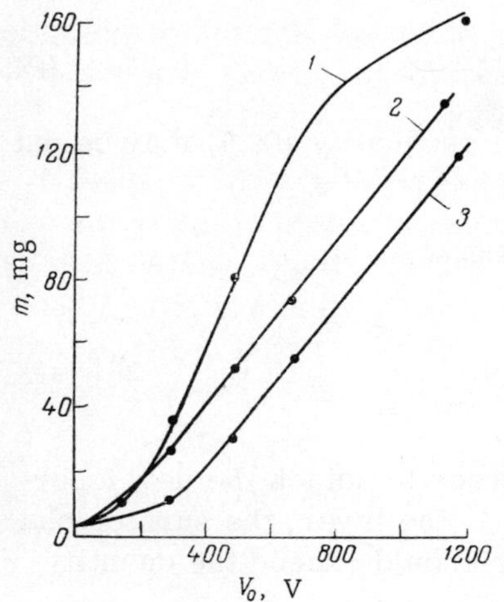

Fig. IX.5. Amount of adhering developer powder as a function of the potential of the latent image (carrier pearl copolymer SNP-15). 1) Developer polystyrene emulsion with orange dye PS-5; 2) with dark red PS-4; 3) with violet PS-6.

The electrophotographic process may also be effected without any carrier; here the developer particles may be charged by spraying [433] or superimposing an electric field [434]. In the first case the electrophotographic plate is placed in a jet of gas containing suspended developer particles [435].

Adhesion Processes in Development. Adhesive forces arise between the developing complex and the charged semiconducting surface; these may be expressed in the form

$$F'_{ad} = F'_{M} + F'_{e} \qquad \text{(IX.5)}$$

where F'_M is the molecular component of the adhesive forces and F'_e is the force of interaction between the charges of the latent image and the triboelectric charge on the developer particles, which equals [429]

$$F'_e = \sigma_1 \cdot S_1 \cdot E \qquad \text{(IX.6)}$$

where S_1 is the area of the developer—particle surface and E is the field strength of the latent image.

In addition to the forces considered, there is another force F_e in the contact zone [see formula (III.7)], directly proportional to the area of the actual contact between the particles and the surface, which is considerably smaller than S_1. Then $F_e \ll F'_e$ and may be neglected.

The developing complex moves under the influence of a force F_d. The developer particles should thus stick to the layer and be detached from the carrier particles. The conditions for development are as follows:

$$F'_{ad} > F''_{ad}$$

or

$$F'_{M} + F'_{e} > F''_{M} + F''_{e} \quad \text{(IX.7)}$$

If we consider $F'_{M} = F''_{M}$, then inequality (IX.7) may be put in the form

$$\sigma_1 \cdot S_1 \cdot E > 4\pi S_2 \cdot \sigma_2^2$$

where

$$\sigma_1 > \frac{4\pi S_2}{S_1 \cdot E} \cdot \sigma_2^2 \quad \text{(IX.8)}$$

We see from (IX.8) that, in order to detach the developer from the carrier and make it stick to the layer, the surface charge density of the semiconducting layer should exceed the quantity $(4\pi S_2/S_1E) \cdot \sigma_2^2$.

In order to remove the developer particles the force F_d must exceed F''_{ad}. If F_d is directed tangentially to the surface, the frictional force must also be taken into account.

In estimating the values of the forces F'_{ad} and F''_{ad}, we have made the assumption that no discharging of the double layer in the zone of contact between the developer particles and the surface of the layer or the carrier particles takes place. This is not altogether true (see §12) when in calculating the values of F'_{e} and F''_{e} we have to allow for the resistance in the zone of contact and the possibility of a discharge of the double layer.

The residual charge on the attached layer of developer powder was considered by Zhilevich with due allowance for the charge on the substrate (Fig. IX.3) and the electrical capacity between the adhering layer and the surface.

The surface charge density [formula (IX.8)] and the adhesion of the developer powder, the amount of which determines the quality of the image, are determined by the potential of the developing layer. The potential of the blank layer should prevent sticking of the powder. Hence the development effect depends on the potential difference between the shaded (rastered) and blank elements. Dravin [429] experimentally determined the relation between the amount of adhering developer powder and the potential of the latent image (Fig. IX.5) for three developing compositions, the carrier of which was a pearl copolymer of styrene with the nitrile of acrylic

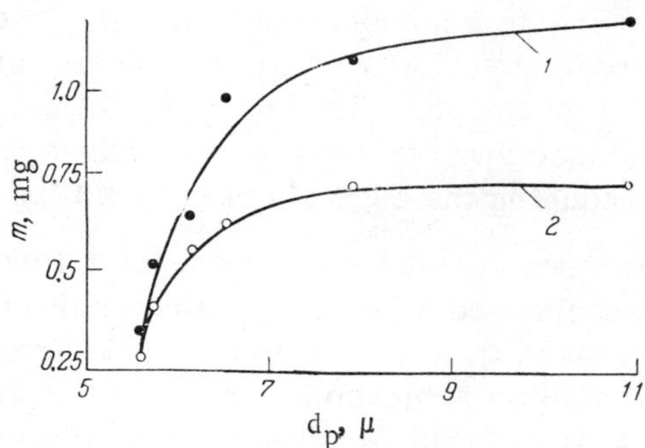

Fig. IX.6. Amount of developer powder sticking to the layer at a potential of 630 V as a function of the mean particle size. 1) Ordinary polystyrene; 2) colored with a fatty dye.

acid and butadiene styrene rubber (SNP-15). The developer particles are prepared from a crushed and then tinted emulsion of polystyrene (PS).

For a zero layer potential the amount of developer powder sticking to the layer is still finite. Adhesion in this case is due to molecular forces, which also affect the process involved in the development of the image [see formula (IX.5)]. Hence, it is not at all correct to neglect the molecular component of the adhesive forces, as was done in [429], when considering image-development processes. However, the amount of developer powder sticking to a charged semiconducting surface as a result of molecular forces is much smaller than that due to electric forces, i.e., $F'_M \ll F'_e$.

The amount of adhering developer depends not only on the potential of the layer but also on the size of the powder particles [429]. We see from Fig. IX.6 that as the size of the polystyrene particles rises from 5.5 to 8.0 μ the amount of powder sticking increases, since the mass of the particle is proportional to the cube of the diameter. As the size of the particles increases further, however, the amount sticking remains fairly constant.

The value of the potential V_0 is not constant over the whole area of the developed image but reaches a maximum value at the edges. An "edge" or "end" effect increasing the adhesion at the edges of the image takes place.

In order to create a homogeneous field one uses a movable developing electrode carrying a potential opposite in sign to the potential of the latent image. The electrode moves parallel to the layer a short distance away from it and increases the field strength over the layer, reducing the edge effect [436, 437].

In the processes considered, the adhesive forces of the developing powder increased principally as a result of the charging of the contiguous bodies, i.e., as a result of electric forces. The same effect may also be achieved by changing the wettability of the surface (see §§4, 10). In this case panchromatically sensitized zinc oxide deposited on an aluminum foil is used as conducting layer. Since the layer itself is hydrophobic, hydrophilic particles, for examples, metal oxides, stick better to the illuminated, conducting parts. In order to obtain negative images, the photoconducting layer is made hydrophilic and hydrophobic substances are deposited upon it [433].

In development of the magnetic-brush method, iron powder used as carrier particles is mixed with a readily fusible pigmented crystalline powder, which becomes charged positively on friction with the iron. When a magnet acts on this mixture, the iron particles stick to it. Since the forces of interaction between the developer particles and the charged surface exceeds the forces of adhesion to the carrier particles, the developer is held strongly on the original surface, ensuring due development of the image [438].

When using liquid developers, the motion of the developer particles is a result of electrophoresis. The liquid must have a high specific resistance (over $10^{10}\ \Omega \cdot \text{cm}$) and a low dielectric constant (under 2.5). Suitable developers include alkyd resins, talc, magnesium oxide, and soot, which are capable of adhering strongly. Liquid developing media include gasoline, kerosene, turpentine, benzene, or carbon tetrachloride [439]. A liquid electrophotographic developer is prepared by dispersing the developer in the liquid.

Adhesion on Transferring and Fixing the Image. On transferring the image from the electrophotolayer to the paper or molding base (see Fig. IX.4) one has to overcome the adhesion of the powder to the layer, and ensure the sticking of the developer particles to the paper. The latter process is called fixing.

In order to transfer the image one uses a steady electric field creating a potential difference between the electrophotolayer and the electrode placed over the paper [429, 440]. There are also other methods of communicating electric charges to a molding base having a fair specific resistance. One such method is based on a corona discharge, under the influence of which the sheet of paper acquires a charge opposite in sign to the charge of the adhering particles.

The molding base may be a metal foil on which charges capable of transferring an image are induced.

In addition to the use of an electric field, other methods may also be used for transferring images. Thus the molding base may be covered with a thin layer of hydrophilic adhesive later subjected to moistening. This base is placed on the adhering layer of the developed image. By applying pressure the particles are "embedded," and their adhesion to the tacky layer then exceeds their adhesion to the developed plate. The powder image may be transferred to a paper strip with an adhesive coating (adhesion by virtue of tackiness) and fixed with a transparent plastic introduced onto the tacky surface of the paper [441]. Transfer of the image is effected by rolling the mold onto the layer with a rubber roller [442]. If the roller rubber is conducting and is connected to a dc source [440], transfer of the image to a nonconducting base is effected by the combined action of the electric field and the applied pressure.

In using various transfer methods the proportion of powder transferred varies widely, e.g., between 40 and 77%, i.e., by a factor of 1.5.

After the development process is finished, the excess of developer powder has to be removed (the powder held on the adhering layer or on the undeveloped surface). In this, the adhesive properties of the powders have to be considered. The simplest methods are blowing [433] and vacuum suction [443].

The plates may be cleaned from excess developing powder, in particular, by depositing granulated salt (sodium chloride, aluminum chloride), the particles of which have a charge opposite in sign to the charge on the developing powder. The particles thus merge, become larger, and are easily removed [438].

After depositing the charged powder on the paper (see Figs. IX.3 and IX.4) the charges may leak off and the adhesion of the powder may therefore become weaker. It is thus important to fix the deposited powder, and this may be done by heating, applying pressure, and using a solvent.

On heating, the adhering powder particles melt and are fixed on the paper. Vapor from a solvent (acetone, carbon tetrachloride, ethyl ether, etc.) partly dissolves and also fixes the developer particles on the paper.

Electrophotography has come into use comparatively recently [433]. The main directions for the development of this technique include improving the quality of the image and simplifying the process by eliminating or merging several stages of the process. The solution of these problems must make due allowance for the adhesion characteristics of the powders used in electrophotography.

§47. Adhesion in Powder Metallurgy

In powder metallurgy autohesive and adhesive processes are involved in extrusion and pressing techniques and in the production of finished articles from the original powdered material; they also affect the free-running properties of powders.

The free-running properties (the "running" of a powder are measured by its rate of flow, i.e., by the amount of powder passing through a calibrated aperture, and also by its pulverization or dissemination quality. This property is largely determined by the particle size. Thus, the running of reduced iron and copper powders and also aluminum powder (the volumetric flow of which is determined in air by passing through funnels of radius 100 and 200 mm) increases with diminishing particle size [444, 445] and reaches a maximum at 100 μ, after which it falls rapidly. This is because autohesive forces appear in particles smaller than 100 μ.

Evidently the small particles merge with each other, forming strong aggregates, and this also reduces the running of the powders. Since autohesive forces may be smaller in vacuum than in air (§16), the best running properties occur for smaller particles under vacuum conditions.

The volumetric flow of powder is also affected by the roughness of the surface, which in turn affects the adhesion. Thus the volumetric flow of powder being transported over a rough surface is 10-12% smaller than that obtained on transporting over a smooth surface [444, 445].

The running also depends on the autohesion of the powders, particularly on the electrical and molecular components.

In the treatment of tin powder (particle diameter 1.8-5.9 μ) with Aeroflok 552 (a surface-active substance), the particles merge, their charge increases, and the completeness of their pulverization is reduced. Pulverization worsens with increasing concentration of SAS (on passing from 0.005 to 0.05 wt.%). For quartz powder the reverse applies [446]. Hence, SAS solutions may increase or reduce the charges on the particles, producing either coagulation or dispersion.

The forces of adhesion and autohesion may be reduced by hydrophobization of the powder surface (§10), thus increasing the free-running properties. Hydrophobization of the surface of sand may be effected by treatment with suspensions or solutions of wood pitch [447].

In order to create a hydrophobic layer on the surface of such powder particles as the silicates, sulfides, and sulfates of various metals as well as asbestos, bentonite, kaolin, talc, mica, and stone flour, these are impregnated with a small quantity of a monomer which turns into a polymer after heating. Suitable monomers include styrene, divinyl benzene, methyl acrylate, etc. [448].

The bond strength of the particles in the original mass, i.e., the autohesion, also determines the packing density of the powder mass and its sintering properties. The packing density is given by the shrinkage of the powder.

In the tungsten—copper and tungsten—silver systems, shrinkage and hence autohesion take place as a result of capillary forces as the pores are filled with a molten mass of the more easily melting component [445] and reach a maximum when the degree of filling equals 50-60%.

The vibrational method is used in order to condense powders formed from hard and elastic metals and alloys (titanium, molybdenum, silicon carbide, etc.). Here not only are the fine particles distributed between the large ones, but the autohesion is increased at the same time. The packing density of the particles depends on the frequency and amplitude developed by the vibrator. Thus, for the vibrational condensation of powders a vibrator with a frequency of up to 30 kc/sec is employed; the powder is wetted with a 6% solution of glycerin in methyl alcohol or some similar liquid [449] so as to ensure that the adhering agglomerations of powder should not prevent subsequent shrinkage.

The packing density may be sharply raised by adding zinc stearate as lubricant (0.3-1% of the mass of the powder) [450].

The shape, roughness, and size of the particles determine their contact area and hence affect the packing density and autohesive interaction. The greatest autohesion occurs for powder prepared from spongy material, for example metallic sponge, and the least for powders consisting of spherical particles. The shape of the majority of metallic powders differs appreciably from spherical.

The shape of powder particles largely depends on the methods of production [451], which may be divided into mechanical, physicochemical, and chemical. Mechanical methods include milling, granulation, and spraying (atomization); physicochemical methods include condensation and the electrolysis of aqueous solutions and molten salts; chemical methods include the production of a very fine material on separating from the reaction mixture (melt or solution). For example, in the milling of Sb, Bi, Mn, Cr, Co, Fe, Ti, Ni, and Cu in ball or vibrational mills particles are obtained in the form of irregular polyhedra; on milling such alloys as Fe—Al, Al—Ni—Co, and Al—Si—Fe in vortical mills, the particles have a plate-like shape. On the condensation of nickel and iron carbonyls spherical particles are formed; in the reduction of metals (W, Mo, Fe, Ni, Co, Cu, Pt, Sn, Ag, Au, etc.) from their salts or oxides, and also on electrolysis of salt melts or solutions, the metal particles produced have the shape of dendrites.

The forces of adhesion and autohesion may be changed by varying the micro- and macroroughness of the particles (see §14). Thus, while a powdered substance (kaolin, talc, etc.) is being mixed with a crystalline substance, the latter is firmly fixed on the surface of the particles, reducing the area of contact and preventing agglomeration [452].

A new method of creating protective metallic (silver, gold) coatings proposed by Besidovskii [453] may be included within the framework of powder metallurgy, although rather arbitrarily; this method is as follows. Highly dispersed silver powder is deposited on a metal surface and then rubbed or ground into the latter mechanically. This forms a coating held firmly on the surface by adhesive forces, the adhesion of the powder to the substrate increasing if the particles are plane in shape. The adhesion may also be strengthened by previously removing the oxide film from the contiguous surfaces.

§48. Adhesion of Ash and Slag

The adhesion of ash and slag to the heating surfaces of boiler plants greatly reduces their efficiency. Deposits on the boiling pipes may interrupt the circulation of the heat carrier, corrode or rupture the tubes, interfere with the passage of the gases, and lower the efficiency of the whole system. Deposits may be distinguished by three characteristics: their location, their chemical composition, and their structure. Depending on the point of adhesion, ash and slag may be arbitrarily divided into the following forms: pipe incrustation adhering to the heating surfaces, wall incrustation forming on the walls of the furnace, flue incrustation clinging to the sides of the flues, and hearth incrustation settling on the hearth of the furnace.

The conditions for the formation and removal of wall and flue incrustation were considered earlier (see §§32, 33). In this section we shall therefore devote principal attention to pipe incrustation.

As regards chemical composition we may distinguish the following types of deposit: aluminum silicates (compounds of aluminum and silicon), alkali-combined (alkali compounds form an easily melting base cementing refractory particles together), sulfate ($CaSO_4$ and in part SiS, Al_2S_3, FeS, the cementing effect of which is similar to that of alkali compounds), and phosphate with a high content of phosphorus compounds such as P_2O_5.

As regards structure, the adhering layer may be classified as free-running (ash and soot), tacky, dense (sintering and cementing itself), and liquid. The properties of the adhering layer are determined by the stage at which the formation process ends, this process consisting of the deposition of the volatile ash, the adhesion of the ash particles to the heating surface, the sintering of the ash particles, and the melting of the outer layer of the deposit. We shall be considering three forms of deposits: free-running, tacky, and dense.

A free-running or loose deposit is an adhering layer of solid particles. A tacky deposit is caused by the presence of tacky (liquid, oily) constituents. Dense deposits are formed, for example, in the combustion of Estonian shales at 500-1000°C and also of the poor coal under Moscow. Dense deposits are also formed in the combustion of certain kinds of fuel oil. In practical conditions, all forms of deposit may be present at the same time, and it is sometimes difficult to observe the boundary between them.

Free-Running Deposits. The composition (as regards particle size) and quantity of free-running deposits precipitated from a gas flow depends on the size distribution and concentration of the particles in the flow, the operating time of the plant, the velocity and direction of the flow, and also the diameter and mutual arrangement of the pipes.

The particle-size distribution of deposits usually differs from that of the dust in the gas flow, since small particles tend to stick to the pipes most, so that the proportion of these particles in the deposits is greater than in the flow. This is confirmed by particle-size analysis [454] (carried out by the method of air separation) of the ash from Moscow coal, both in the flow and in the deposits on pipes 38 mm in diameter:

d_p, μ	<10	10-20	20-30	30-50	> 50
Content, %:					
in the flow	26.3	7.7	13.5	11.5	41.0
in the deposits	52.0	14.2	9.9	1.6	22.3

The number of particles falling out of the flow depends both on their concentration in the flow and on the time of operation of the plant. The longer the plant has been working, the more deposits are formed; however, this increase is not without limit, and for any particular conditions a state of saturation is reached after a certain time, i.e., the quantity of deposits formed reaches a maximum. It has been found [454] that this saturation usually sets in after 7 h operation, the deposition taking place, not uniformly, but chiefly in the first 1-3 h. Thus, half the maximum amount of deposits are formed in 2.5 h if their concentration in the flow is 7 g/m^3 and in 1.5 h if the concentration equals 21 g/m^3. Hence, the main bulk of the deposit is formed most rapidly in the case of the most contaminated flow.

The relation between the amount of dust settling and the velocity of the flow may be expressed by the equation [454]:

$$\varepsilon = c_D \cdot c_d \cdot 10^{-k_s \cdot v} \qquad \text{(IX.9)}$$

where ε is the coefficient of contamination, i.e., the difference in the thermal resistances of the contaminated and clean surfaces, in $m^2 \cdot h \cdot deg/kcal$, c_D and c_d are coefficients depending on the diameter of the pipes and the fractional composition of the deposits, k_s is a coefficient allowing for the mutual positioning of the pipes (the step), and v is the rate of flow.

We cannot calculate the true quantity of adhering dust by using this formula; however, we may use the coefficient ε to characterize the adhesion process, since the amount of contamination settling from the flow is directly proportional to the thermal resistance of the wall, i.e., the value of ε.

According to Eq. (IX.9) the value of the coefficient ε falls with increasing velocity of the air flow; this is in fact observed experimentally (Figs. IX.7 and IX.8).

The formation of free-running deposits is possible when the process of dust sticking predominates over the detachment of adhering particles under the influence of the air flow.

If we consider that the amount of dust settling is directly proportional to the flow velocity while that removed by the flow varies as the cube of the velocity, we may express the amount of dust settling (in grams per m^2 per sec); thus [455]:

$$m' = (r_0 - r_i)\, c_V \cdot v - k_1 \cdot r_i \cdot c_V \cdot v^3 \qquad \text{(IX.10)}$$

where r_0 is the relative number of particles with a radius below which no rebounding takes place (see §26), r_i is the relative number of particles with a radius below which inertial settling is improbable, v is the velocity of the flow in m/sec, c_V is the concentration of volatile ash in g/m^3, k_1 is a coefficient characterizing the detachment and removal of deposits, inversely proportional to the autohesion of the deposited matter, expressed in sec^2/m^2.

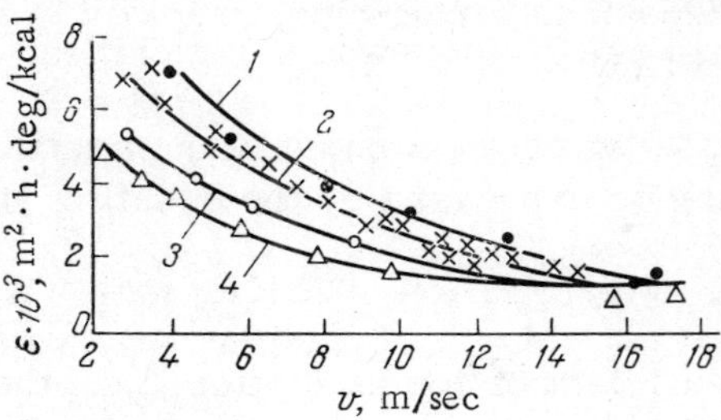

Fig. IX.7. Coefficient of contamination of pipes 38 mm in diameter as a function of flow velocity for fractions containing different amounts of particles with a radius of under 30 μ: 1) 75%; 2) 66 %; 3) 37%; 4) 24%.

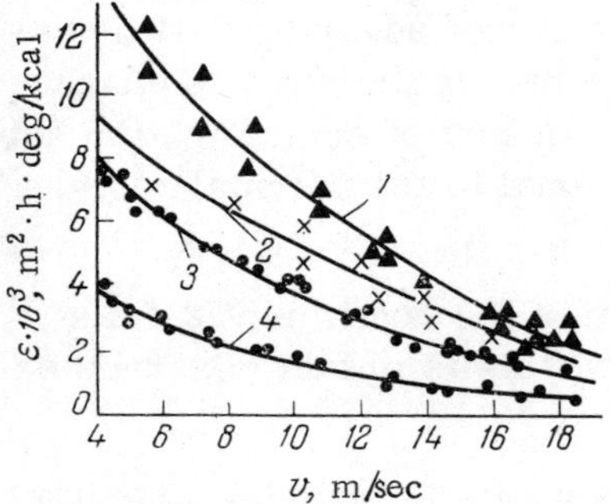

Fig. IX.8. Coefficient of contamination (residue of screens r_{30} was 33.7%) for a checkered arrangement of the pipes as a function of the flow velocity for various pipe diameters: 1) D = 76; 2) 51; 3) 38; 4) 25 mm.

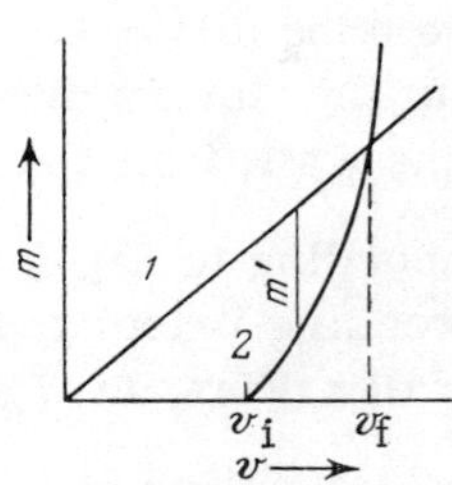

Fig. IX.9. Amount of free-running deposits as a function of the flow velocity calculated from Eq. (IX.10). 1) First term of the equation; 2) second term of the equation.

Fig. IX.9 shows the relation between the first and second terms of Eq. (IX.10) (curves 1 and 2, respectively) and the flow velocity.

The difference between the ordinates characterizes the quantity of deposit (m'). We see from the figure that this amount diminishes with increasing flow velocity, which agrees with the earlier data (see Figs. IX.7 and IX.8).

It has been shown experimentally that there are two special velocities: the first is v_i, below which detachment of the deposits is impossible; and the second is v_f, above which no sticking of particles occurs. Hence the frontal surface of the pipes is contaminated if $v_i < v < v_f$.

By using the contamination coefficient ε we may estimate the quantity of deposits not only as a function of the velocity of the air flow but also as a function of the particle dimensions, the diameter of the pipes, and the pipe-laying pattern.

As the proportion of small ash particles (under 30 μ in radius) increases the coefficient of contamination does likewise, particularly for low velocities of gas flow (see Fig. IX.7), which corresponds to the earlier views on the adhesion of particles arising from an air flow (see §34).

As the diameter of the pipes falls from 76 to 25 mm (see Fig. IX.8), the coefficient of contamination falls by a factor of 4.

The quantity of deposits is expressed in relation to the mutual disposition of the pipes (in practice the corridor and checkered arrangements are most frequently used) by the coefficient c_D in Eq. (IX.9). This coefficient may be calculated from the formulas

$$c_D = 1+3\log\frac{D}{D_0} \qquad \text{(IX.11)}$$

for checkered arrays of pipes, and

$$c_D = 1+3.3\log\frac{D}{D_0} \qquad \text{(IX.12)}$$

for corridor-type arrays, where D and D_0 are the actual and gauge diameters of the pipes ($D > D_0$). For $D = D_0$ we have $c_D = 1$; as the gauge diameter we take [454]: $D_0 = 38$ mm.

Generalizing the experimental data, Kuznetsov [454] gave computing formulas for determining the coefficient of contamination in the case of checkered (ε'_{ch}) and corridor (ε'_c) pipe arrangements (laboratory determination):

$$\varepsilon'_{ch} = 0.0126\left(1 - 1.18\log\frac{r_{30}}{33.7}\right)\cdot\left(1+3\log\frac{D}{38}\right)\cdot 10^{-k_s\cdot v} \qquad \text{(IX.13)}$$

$$\varepsilon'_c = 0{,}039\left(1 - 1.7\log\frac{r_{30}}{33.7}\right)\cdot\left(1+3.3\log\frac{D}{38}\right)\cdot 10^{-k_s\cdot v} \qquad \text{(IX.14)}$$

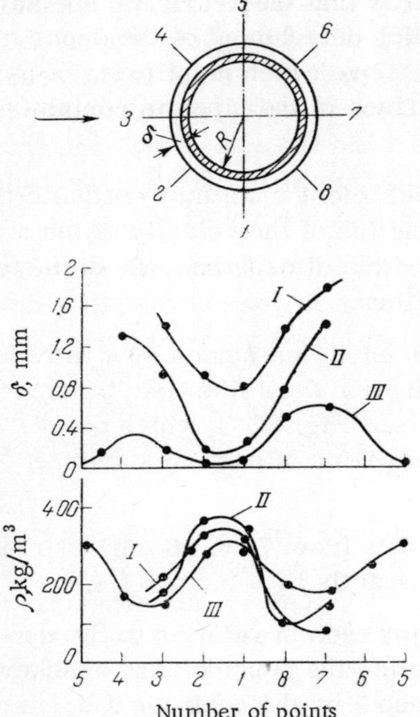

Fig. IX.10. Distribution of the density and thickness of the deposits around the circumference of the pipe. I, II) On a screen pipe after 30-h and 12-h operation at 180°C; III) on a sample selector after 4-h operation with a water-cooling temperature of 40°C. 1-8) Points around the pipe.

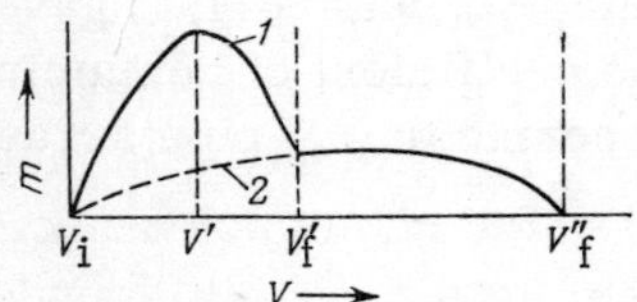

Fig. IX.11. Amount of deposits as a function of the flow velocity. 1) Total, i.e., free-running plus dense; 2) dense.

where D is the diameter of the pipes in cm, r_{30} is the dust residue on a mesh with an aperture of 30 μ or the number of particles* with diameters larger than 30 μ, in %.

The quantity k_s depends not only on the mutual disposition of the pipes but also on the distance between them, the step or pitch of the tubes (L). For a checkered arrangement of the pipes, when

$$\frac{D}{L} = \frac{1}{2} - \frac{1}{3}$$

$$k_s = 0.052 + 0{,}094\left(\frac{D}{L}\right)^4$$

We note that in the present case L is understood to mean the transverse and longitudinal steps of the pipes.

For the corridor type of arrangement, $k_s = 0.08$.

For industrial conditions a correction must be introduced into the value of the coefficient of contamination in order to allow for the difference in dimensions between the laboratory apparatus and the industrial plant:

$$\varepsilon = \varepsilon' + \Delta$$

*There may be inaccuracies in the characteristics of the particle-size distribution of the dust, since the mesh analysis of particles of such sizes is extremely rough owing to aggregation.

It has been found that the particle-size distribution, number, and shape of the contaminations falling out of the flow depend on the direction of gas flow. Thus, small particles stick in the rear zone. The shape of edge contaminations depends on the manner in which the dust-laden stream flows around the obstacles. The frontal part of the pipes is subjected to the impact of the large particles; hence, sticking to this part only occurs at low velocities.

On increasing the ratio D/L from 0.5 to 1, a component of velocity directed tangentially to the outer surface of the pipes appears in the rear part of the latter; this promotes the blowing away of the attached ash from the rear parts of the pipes, i.e., from those places at which the layer of deposits is greatest. It is noteworthy that in a descending vertical flow 10-20% less particles stick than in an ascending flow.

The distribution of the thickness and density of the adhering layer over the circumference of the pipe is not uniform (Fig. IX.10). On the rear parts of the pipes (points 6-8) the deposits have their maximum thickness and minimum density (170-250 kg/m^3), owing to the eddying of the furnace gases. On the side surfaces (points 1, 2, and 5) the density of the deposit increases to 400 kg/m^3, i.e., it becomes a maximum, while the thickness of the adhering layer falls to a minimum.

Tacky Deposits. If the deposits contain particles with tacky surfaces, the adhesion of such deposits increases sharply, and the deposits themselves become tacky rather than free-running. Tacky deposits may be distinguished from the free-running type by eye.

Thus deposits on the surface of boiler plants working with Vorkutinsk coal [456] consist of three clearly distinguished layers characterized by different stickiness. The first (inner) layer, up to 0.5 mm thick, is a core of dirty brown color which may be removed with a sharp scraper and consists of particles having a mean diameter of 1 μ (minimum 0.1 μ). This first layer is tacky, since the easily melted and easily evaporated ash components (50-60% SiO_2, 10-15% $NaHCO_3$, 15-20% Fe_2O_3, and 5-10% fuel) condense on the heating surface; it has a very low thermal conductivity.

The second (intermediate) layer 0.1-0.25 mm thick is loose, breaks up easily, and contains up to 45% of unburned carbon; the mean particle diameter is 1.5 μ and particles of diameter 0.1-0.2 μ are never found.

The third (outer) layer, 1-3 mm thick, consists of easily coagulating ash particles of dark brown color, detachable in the form of scales which are easily ground into powder.

Crossley [457] carried out a microscopic study of the properties of deposits associated with coal heating. The first (inner) layer closest to the surface consists of sulfates (and sometimes thiosulfates) of the alkali metals with small proportions of iron and aluminum sulfates. As well as the sulfates and thiosulfates, this layer may contain 10-45% sulfur, computed in terms of SO_3, and also 10-30% potassium and sodium in terms of $(K, Na)_2O$. If the deposit contains pyrosulfates of the alkali metals $(K, Na)_2S_2O_7$, the layer grows more rapidly than a layer of sulfates $(K, Na)_2SO_4$, since the melting point of the pyrosulfates is lower than that of the sulfates and the tackiness is greater. Red iron oxide Fe_2O_3 and SO_3 from the flue gas react with the layer of sulfates, yielding complex sulfates such as $Na_3Fe(SO_4)_3$ and $KAl(SO_4)_2$; these make the surface tacky and may be observed by x-ray diffraction. If the coals contain a considerable amount of chlorine-containing substances (up to 0.9% on conversion to Cl), the alkali-combining substance may be $CaCl_2$, which has a melting point of about 770°C and easily evaporates and condenses on the pipes.

Dense Deposits. A reason for the formation of dense deposits is the chemical reaction between the solid particles of the ash and certain components of the furnace gases. The process of forming dense deposits is quite complicated.

If dense deposits are accompanied by free-running deposits, the sum of these may be calculated, with due allowance for (IX.10), from the following formula [455]:

$$m = m_1 + m_2 = [(1 - a)(r_o - r_i) c_V \cdot v - k_1 \cdot r_i \cdot v^3] + b [a (r_o - r_i) c_V \cdot v - k_2 \cdot r_i \cdot c_V \cdot v^3] \qquad \text{(IX.15)}$$

where m_1 and m_2 are, respectively, the quantities of free-running and dense deposits, a is the relative proportion by weight* of the

*The symbol a denotes only the proportion of dense deposits formed as a result of the reactivity of the particles; the influence of tackiness on adhesion is not taken into account.

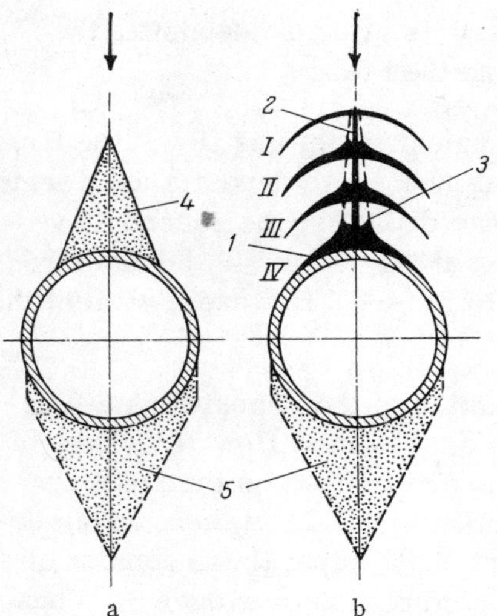

Fig. IX.12. Form of deposits on a pipe 32 mm in diameter for a flow velocity of 8-11 m/sec (a) and more than 11 m/sec (b). 1) Base; 2) central ridge; 3) side ridges of dense deposits; 4,5) loose deposits on the front and back of the pipe, respectively. I-IV) Measurement of the form of the dense deposits as a function of the period of contact between the pipe and the flow.

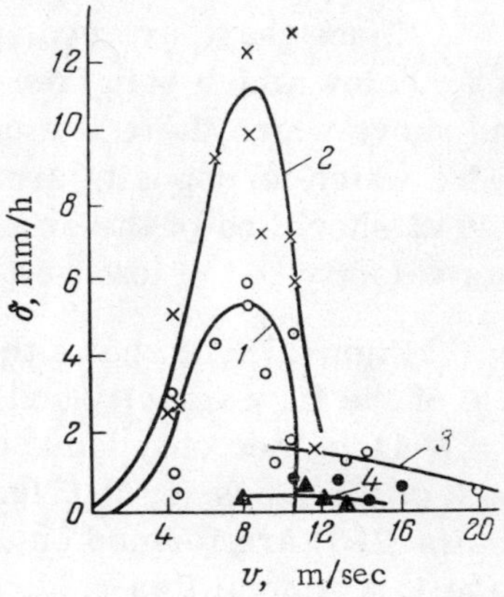

Fig. IX.13. Rate of growth of the thickness of the deposits (δ) on the front of a pipe in a transverse flow as a function of the flow velocity. 1) Average over the length of the pipe; 2) growth of the ridge of free-running deposit; 3) growth of the ridge of dense deposit at a temperature of 640-664°C; 4) the same at 565-588°C.

dense deposits, and b is a coefficient accounting for the growth of the deposits resulting from reaction with furnace gases (for the formation of dense sulfate deposits, b = 2.43).

The amount of dense deposits (m_2) varies with the velocity of the flow in the same way as the amount of free-running deposits (see Fig. IX.9). However, the laws governing the variation in the total amount (free-running plus dense) of the deposits, which is characterized by the quantity m in formula (IX.15), has its own special characteristics. As the velocity of the air flow rises from v_i to v' the amount of the total deposit increases principally on account of the free-running components (Fig. IX.11). Further increasing the velocity of the flow, subject to the condition $v' < v < v_f'$, increases the proportion of the dense deposits. For $v \geq v_f'$, the main bulk of the adhering layer consists of dense deposits, the total amount of which first rises and then falls.

Hence there are two special velocities of gas flow: the first is v_i', below which both free-running and dense deposits are formed and above which there are only dense deposits; the second is v_f'', above which no deposits are formed at all. Épik [455] considered that v_f'' should be of the order of 220 m/sec. However, we feel that this velocity is too low (see §31).

Figure IX.12 shows the formation of the deposits as a function of the flow velocity and the time. Thus for flow velocities of v = 8-11 m/sec only loose deposits are formed on the front and back of the pipes (4, 5) (Fig. IX.12a); for v > 11 m/sec, dense deposits (1-3) are formed on the front of the pipe; these consist of a base 1, a central thin ridge 2, and side combs or ridges 3. Thus a velocity of 11 m/sec may be regarded as v_f'.

Figure IX.13 also shows the experimental relation between the rate of growth of the deposits and the rate of gas flow. In all cases the rate of growth of the free-running deposits was many times greater than that of the dense deposit. For a velocity of 8-11 m/sec there was a sharp transformation from free-running to dense in the form of the deposits.

On pipes subjected to a flow in the longitudinal direction, dense deposits 2-4 mm high and 10-20 mm long developed in the form of islands. The rupture strength of the deposits not removed by blowing in shale furnaces was 100-1000 kg/cm^2. However, in view of the tackiness of the ash, it was difficult to remove deposits formed in the combustion of fuel oil.

We have considered examples in which the formation of a dense deposit was accompanied by the formation of a free-running one. It is possible to have other cases in which tacky and dense deposits are formed at the same time as, for example, in the case of the combustion of fuel oil. The ash of fuel oil consists of metal-corrosion products (iron salts and oxides), the remains of substances used in the acid and alkali treatment of petroleum, salts of bore-hole water, contaminant particles, and particles of unburned carbon (soot and carbides). When the individual ash components interact with each other and with the gas medium, both dense and tacky deposits are formed. Despite the fact that in the combustion of fuel oil 50 times less ash is formed than in the combustion of anthracite, 100 times less than in the combustion of peat, and 200 times less than in the combustion of shale, it is harder to remove the deposits so formed because of their tackiness.

Removal of an Adhering Layer of Deposits. The deposits so formed may be removed by both dry and wet methods. Dry methods include air blowing, cleaning with sand, chains, and shot, and also by cutting the incrustation off; wet methods include washing with a water jet and aqueous washing solutions, and blowing with a mixture of steam and ammonia.

Since the form of the deposits produced and their adhesive properties are determined by different factors (ash composition, particle size, temperature, state of the surface being cleaned, etc.), the best means of cleaning must be considered separately in each specific case. The ash from different fuels may in fact differ considerably. Whereas the maximum size of the ash from coal-dust furnaces is no greater than 120 μ, that from shale furnaces reaches 300 μ, while the mean diameter of the ash particles from Pechorsk coal [459] fluctuates between 3.5 and 125 μ. On the other hand, the light-brown deposits of aluminosilicates are fine and easily blown away, while the dark Fe_3O_4 and Fe_2O_3 are very hard and the boiler has to be put out of service in order to remove them [457]; the deposits formed in the combustion of fuel oil are also extremely hard to remove.

We may combat incrustation not only by optimizing furnace construction, but also by reducing the tackiness and insulating the original surface. Thus adhesion may be reduced by graphiting the surface or treating it with a lime solution.

One of the most effective methods of combating the formation of tacky and dense deposits lies in mixing certain additives with the fuel so as to change the subsequent composition and properties of the ash. Thus, the introduction [460] of additives changing the amount of $SiO_2 + Al_2O_3$ in the ash of Reinsk brown coal also changes the corresponding melting point.

By introducing slaked lime (50–60% CaO) into a fuel-oil boiler to the extent of 1.1% of the mass of the fuel burned, loose deposits are formed instead of dense ones, since the ash is enriched with magnesium oxide [461]; this makes subsequent removal of the deposits easier.

An adhering deposit may be made more freely running and removed more easily if ammonia is introduced into a fuel-oil boiler [462]. Particularly serious complications arise in the burning of fuel oil in gas turbines, when the ash sticks to the vanes of the turbine. By adding substances containing silicon, aluminum, magnesium, and zinc compounds to the heavy fuel oils used for this purpose, the melting point of the ash is considerably raised and its tackiness is thereby reduced. It is also recommended in [463] that up to 0.15% kaolin powder should be added to the fuel oil; this ensures the formation of loose, easily removed deposits and reduces the adhesion of particles to the turbine vanes by a factor of several times.

Chapter X

Detachment of Sticking Particles on Application of an Electric Field

§ 49. Detachment of Particles Under the Influence of a dc Electric Field

The detachment of particles under the influence of a steady electric field has already been considered in § 46, which was devoted to electrophotography. The same method of detachment may also be used for cleaning dust contamination from various surfaces.

Conditions for the Detachment of Particles. For studying the effect of the conditions governing the detachment of particles in a high-voltage electric field on the degree of cleansing of a dust-laden surface, we set up the apparatus shown schematically in Fig. X.1. Low voltage was converted into high (in the present case 50 kV at a current of 10^{-7} A) in a converter based on a semiconductor triode of the P4B type [440].

For studying the detachment of particles in vacuum or in an atmosphere of some particular gas, for example, nitrogen, we used a special arrangement (Fig. X.2) in which the electrode and dust-laden surface were placed inside a bell connected to a vacuum pump for removing the air and also to a source of gas.

It was found experimentally that the degree of cleansing of the surface depended on a large number of factors: the shape of the electrode, the intensity of the field created (E), i.e., the potential V_0 applied to the electrode and the distance H between the electrode and the dust-laden surface, the properties of the material and the state of the surface to be cleaned, the type of dust in question, etc.

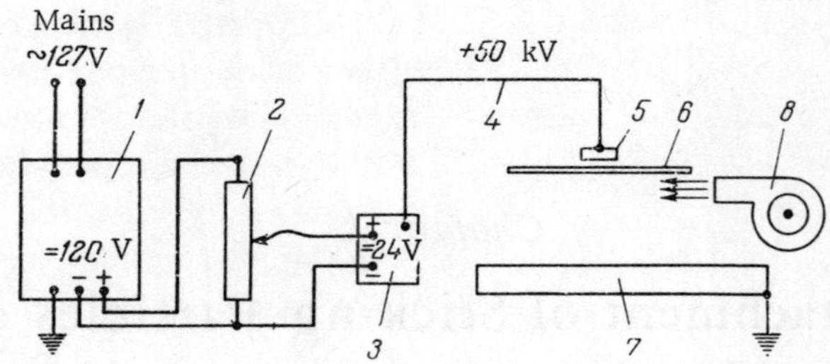

Fig. X.1. Schematic circuit for detaching particles in a high-voltage field. 1) Rectifier; 2) potentiometer; 3) high-voltage converter; 4) high-voltage conduit; 5) electrode; 6) intermediate plate; 7) dust-laden surface; 8) fan (blower).

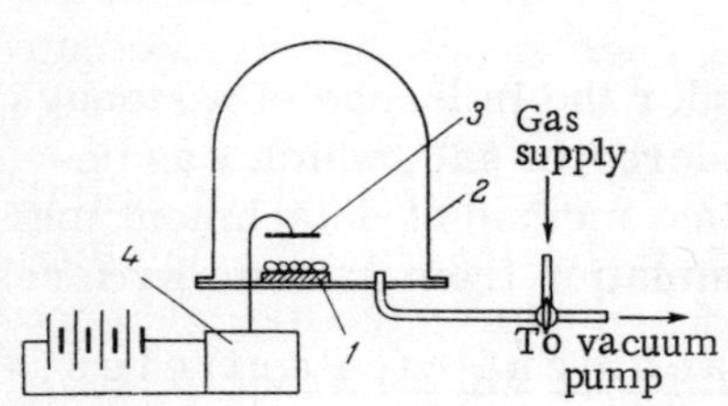

Fig. X.2. Apparatus for studying the detachment of adhering particles in a nitrogen atmosphere under the influence of a high-voltage electric field: 1) Dust-laden surface; 2) bell-jar; 3) electrode; 4) converter.

In cleaning surfaces from dust by means of an electric field one usually employs needlelike, continuous (solid rectangular plate) or rod (cylindrical rods joined together) electrodes.

The degree of cleaning of dust-laden surfaces in a high-voltage field (under identical conditions) using a rod electrode is a little higher than on using a continuous electrode, mainly in relation to the removal of the larger particles. Thus the maximum on the differential particle-size distribution curve lies at 50μ for the original dust; after cleaning in a high-voltage field created by a continuous electrode it moves to $30\ \mu$, and after similar treatment with a rod electrode it moves to $10\ \mu$ (see Fig. X.3).

When using a needlelike electrode electrical breakdown may occur. In order to avoid this, an intermediate plate is placed between the electrode and the dust-laden surface; this naturally affects the degree of dust removal from the latter. The effect of the needlelike electrode thus becomes analogous to that of the continuous plate. When using continuous or rod electrodes, the danger of breakdown is absent, and no intermediate plate need be placed in the apparatus.

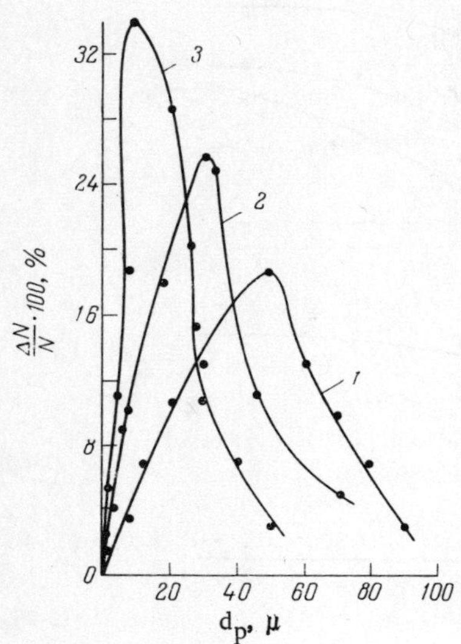

Fig. X.3. Differential particle-size distribution curves. 1) Original dust; 2,3) dust remaining after the imposition of an electric field by means of a continuous and rod-type electrode, respectively.

In order to determine the effect of the properties of the intermediate-plate material, namely the dielectric constant (ε) on the degree of cleaning of the dust-laden surface, experiments were made with plates made from organic and aluminosilicate glass (diameter of glass particles, 40–60 μ; V_0 = 20 kV; H = 1 cm; E = 20 kV/cm) (see below).

We see from the data presented that values of K_N are directly proportional to the dielectric constant of the intermediate-plate material.

It was found experimentally that the degree of cleaning increased with increasing potential applied to the electrode (Fig. X.4). We may consider that the optimum field strength is approximately 20 kV/cm. For a lower field strength it is difficult to detach the particles, while for a higher field strength electrical breakdown may take place. However, in no case (see Fig. X.4) did the value of K_N exceed 10, even for E = 20 kV/cm.

	ε	K_N
Organic glass	3.5	3.9
Other glass of various composition. .	6.7	5.0
	8.0	5.9
	14.0	6.7

In order to determine the dependence of the degree of cleaning of the dust-laden surface on the electrode material, special experiments were made with a field strength of 18 kV/cm and continuous electrodes of aluminum, copper, brass, and tin plate; the following results were obtained for different dust-laden surfaces:

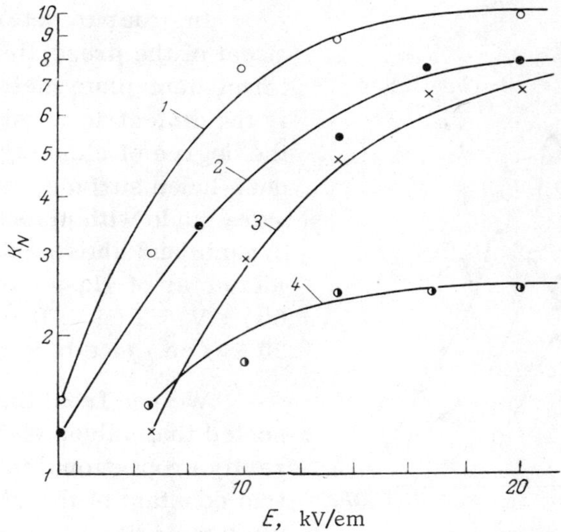

Fig. X.4. Value of K_N as a function of the electric-field strength for dust-laden surfaces (wide spread of particle sizes) made from different materials. 1) Tin plate; 2) brass; 3) aluminum; 4) painted surface.

Electrode material	Aluminum	Copper	Brass	Tin plate
K_N for a surface of:				
aluminum	5.7	5.9	5.9	6.0
brass	6.1	6.5	6.7	6.9
tin plate	12.3	11.3	11.1	11.1

Thus we see that the electrode material has no effect on the degree of cleaning. However, the same experiments show that the same electrode gives different values of K_N if the dust-laden plates are made of different materials. We may conclude from these results and those presented in Fig. X.4 that the properties of the material of the surface being cleaned have a distinct effect on its degree of cleaning. The effect of the state of the surface on the cleaning factor was also studied in order to confirm this.

The following results were obtained for cleaning a steel surface from spherical glass particles 20-30 μ in diameter (field strength, 18 kV/cm):

	K_N		K_N
Worked to a Class-9 finish	5.6	Moistened	2.1
Painted with perchlorvinyl enamel .	4.3	Oily	1.7

Thus the state and properties of the surface greatly affect its degree of cleaning. However, we see from the results presented that even a hard steel surface cannot be cleaned from dust by imposing an electric field.

All these results confirm once again that, under otherwise equal conditions, the degree of cleaning is determined by the adhesion of the particles to the surface being cleaned. The greater the adhesion, the harder it is to clean the surface. In view of this, the efficiency of cleaning may be increased by reducing the adhesion, for example, by modification of the original surface. In fact, whereas the cleaning of a surface painted with perchlorvinyl enamel from spherical glass particles 20-30 μ in diameter with an electric field of 20 kV/cm is characterized by the coefficient $K_N = 3.2$, the value of K_N for the same surface after modifying with stearic acid is 8.3.

It was also established that, as the particle size became smaller, the number of particles removed on imposing a steady electric field also diminished:

d_p, μ	20-30	30-50	40-60	60-100	100-120
K_N	2.9	3.2	4.0	4.2	6.5

In all these investigations, even under the most favorable conditions, the value of K_N never exceeded 10, i.e., no surface could be cleaned completely. This may be explained by the fact that only the particles lying in the top layer were fairly easily detached under the influence of an electric field, whereas the particles in closer contact with the surface were more stubborn.

The particle material also affects the efficiency of the cleaning process. Thus, the value of K_N corresponding to the cleaning of a painted surface in a steady electric field of 18 kV/cm from dust having a wide spread of particle size (particle diameters under 100 μ) is as follows: for particles prepared from glass No. 23, sand, alumina, and glass of the phosphor type, K_N respectively equals 334, 143, 82, and 72. In these experiments the dust-laden surface was wiped with a hair brush in order to facilitate the re-

Table X.1. Charge on Dust Particles Detached by the Vibrational Method (positive potential on the electrode)

Surface material	d_p, μ	q (in C) after detachment of the dust particles: without imposing an electric field	q (in C) after detachment of the dust particles: after being in an electric field for 3 min
Glass	20-30	$+1.5 \cdot 10^{-17}$	$-1.5 \cdot 10^{-17}$
	50 ± 2	$+1.4 \cdot 10^{-16}$	$-1.2 \cdot 10^{-16}$
	80-100	$+6.0 \cdot 10^{-16}$	$-9.0 \cdot 10^{-16}$
Brass	50 ± 2	$+2.5 \cdot 10^{-16}$	$-5.0 \cdot 10^{-16}$

moval of the particles during the application of the electric field. The particles were not actually swept off the surface, their removal being solely attributable to the action of the electric field.

Mechanism of Particle Detachment. The charging of the adhering particles under the influence of a steady electric field may occur as a result of the polarization of the particle material and the sorption of ions on the particle surface [143, 144]. In order to understand the mechanism of particle detachment more clearly, some special experiments were made; in these, the charges on the particles were measured and the effect of cleaning the surface in different gas media was determined.

Apparatus of the type described in [440] was used to measure the charges remaining on the surface after the detachment of adhering particles by the vibrational method, i.e., the charges in the contact zone between the particles and surface. Under these conditions the charge on the particles was equal in magnitude but opposite in sign to the charge remaining on the surface. The charges were measured by detaching spherical glass particles from the same surfaces under different conditions: either without the application of any field or after removing a field which had acted for a period of 3 min. On supplying a positive potential to the detaching electrode, the sign of the charge on the particles subjected to the action of the electric field in the experiments under consideration reversed and became negative. On detaching glass particles 80-100 μ in diameter from an aluminum surface under the influence of a steady electric field with a positive potential applied to the electrode, the charge on these particles was $-6.3 \cdot 10^{-16}$ C; for a negative electrode potential it was $+8.0 \cdot 10^{-16}$ C.

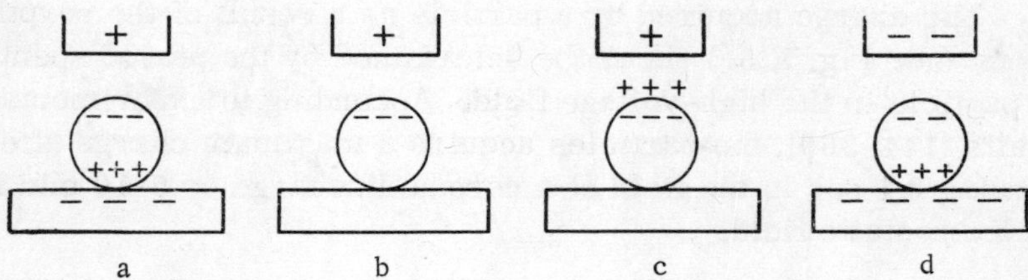

Fig. X.5. Charging and discharging of the particles. a,b,c) On application of a steady electric field; d) on application of an alternating field. a) On an insulator; b) on a conductor; c) on a conductor with the sorption of ions.

The absolute value of the charge measured after removing the electric field (Table X.1) was very small and differed little from the charge observed on detaching particles from a surface not subjected to the action of an electric field. From this we may conclude that at the time of measurement there was no polarization charge on the particles. Apparently elastic rather than volume polarization takes place during the period of action of the electric field (3 min); the lifetime of this effect after the removal of the voltage is no longer than 10^{-2} sec, i.e., the polarization charge on the surface of the particle only exists in the electric field and vanishes shortly after removing the voltage [464]. In all probability charges of different signs are created in the contact zone at the initial moment of application of the electric field (Fig. X.5).

If the surfaces of the particles and substrate are capable of conducting charges (see §12), the opposing charges on the particles and substrate are discharged, and the particle surface retains charges opposite in sign to the sign of the electrode (Fig. X.5b). In this case, a force of electrical interaction tending to detach the particles from the surface develops between the particles and the electrode.

If the particles have become stuck to the surface of an insulator (Fig. X.5a), the discharge in the contact zone is impeded or takes place very slowly. The charges on a polarized particle interact with each other; the particles are not discharged, and no force of electrical interaction develops. Under these conditions the cleaning of surfaces subjected to a high-voltage field is very difficult or even impossible.

The charge acquired by a particle as a result of the sorption of ions (see Fig. X.5c) should be determined by the period spent by the particle in the high-voltage field. According to experimental results [144, 360], the particles acquire a maximum charge after spending 0.1 sec in the field of a corona discharge or 5-10 min in an electrostatic field.

The charge acquired by a particle as a result of the sorption of ions should also depend (other conditions being equal) on the mobility of the ions formed during the ionization of the medium, i.e., it is determined by the properties of the medium in question. Since the mobility of ions formed in an atmosphere of pure nitrogen is two orders greater than in air, we should expect that the degree of cleaning a surface achieved by imposing an electric field in nitrogen would differ from that obtained in air; this is confirmed experimentally* for E = 20 kV/cm:

Medium	Air	Nitrogen
γ_F, %:		
electrode positive	20.5	100.0
electrode negative.	13.0	100.0

We see from the data presented that the particles are not detached in a nitrogen atmosphere ($\gamma_F = 100\%$). This is because the ions are adsorbed on the polarized particles more fully owing to their greater mobility; this screens the charge (Fig. X.5c) and hence reduces the detaching force. Another reason for the poorer cleaning of the surface in the nitrogen atmosphere is the fact that the latter contains hardly any moisture, so that the surface conductivity in the contact zone is reduced. Thus, by choosing a medium with a low ion mobility we may improve the cleaning of conducting surfaces.

The foregoing arguments may also explain the increase in the degree of cleaning by means of a needle-shaped electrode as the dielectric constant of the material composing the intermediate plates is increased (see earlier). This is probably associated with an increase in the electron work function, which reduces the number of free positive ions in the air gap between the intermediate

*In our experiments the dust-laden surface was held in an electric field for more than 10 min, so that the sorption of ions on the surface may be expected to have taken place.

plate and the dust-laden surface for the same voltage applied to the electrode. This reduces the neutralization probability of the negative polarization charges on the particle and increases the detaching force.

Estimation of the Forces Causing the Detachment of Particles. A dust particle is acted upon by a detaching force due to an electrical interaction defined by the equation

$$F_{ei} = QE \tag{X.1}$$

where Q is the charge on the particle in the high-voltage field.

The charge on the particles will depend both on the field strength and the size of the particles. Allowing for this dependence, the force F_{ei} for a conducting sphere of radius r placed in the field of a plane condenser may be put [458] in the form

$$F_{ei} = 1.37\, E^2 \cdot r^2 \tag{X.2}$$

According to this formula, for conducting particles 10 μ in diameter and E = 20 kV/cm, $F_{ei} = 1.5 \cdot 10^{-3}$ dyn, and for particles 100 μ in diameter it equals 0.15 dyn. In this case, the forces of electrical interaction exceed the adhesive forces of the particles (see §16) measured without the intervention of the high-voltage field. However, the detachment of the particles is opposed not only by the adhesive forces (F_{ad}),* but also by the Coulomb forces (or image forces F_i), the forces due to the double electric layer (F_e) allowing for the charge on the particles in the electric field, and the ponderomotive force (F_{pon}).

The value of F_{ad} is determined experimentally and that of F_e from formula (III.8).

The image force depends on the charge of the particles, which may be calculated as follows for a conducting particle in a high-voltage field [458]:

$$Q = 1.64\, Er^2 \tag{X.3}$$

*In the present case, the quantity F_{ad} is understood to mean simply the molecular and capillary components.

Then the image force [see (III.10)] will be equal to

$$F_i = \frac{(1.64\,Er^2)^2}{l^2} \approx 0.67\,E^2 \cdot r^2 \quad \text{(X.4)}$$

By comparing formulas (X.2) and (X.4) we see that the image force is commensurable with the force of electrical interaction.

As noted earlier, the image force only acts at the initial instant of contact between the particles and the surface (§ 12). Later there is a considerable reduction in the value of this force; the value of F_i is particularly minute on metallic or other conducting surfaces. This also explains the fact that in practice metallic or other conducting surfaces are cleaned much better under the influence of a high-voltage electric field.

The greater the distance (l) between the centers of the equipotential charges [see formula (X.4)], the smaller the value of F_i, and hence the more effective the detachment of particles under a high-voltage field. Hence the upper layer of adhering particles is easier to remove than a monolayer of particles; this is also confirmed experimentally.

The ponderomotive force due to the inhomogeneity of the electric field in a direction perpendicular to the dust-laden surface may be determined from the following formula for spherical particles:

$$F_{pon} = \varepsilon_1 \cdot r^3 \frac{\varepsilon_2 - \varepsilon_1}{2\varepsilon_1 + \varepsilon_2} E \operatorname{grad} E \quad \text{(X.5)}$$

where ε_1 and ε_2 are the dielectric constants of the medium and the particle respectively, and r is the particle radius.

This force is very small (see below).

In addition to the forces mentioned, on using a rod electrode (which may be regarded as a cylinder), a force due to the inhomogeneity of the electric field in a direction tangential to the dust-laden surface acts on the particle. A ponderomotive force may also arise as a result of this inhomogeneity. The following results represent the approximate values of the ponderomotive forces as functions of the particle size and electrode construction or the direction of the ponderomotive force (V_0 = 20 kV, H = 1 cm):

d_p, μ	5	25	50	75	100
F_{pon}, dyn:					
plane electrode (normal to the surface)	$2 \cdot 10^{-9}$	$4 \cdot 10^{-6}$	$4 \cdot 10^{-5}$	$1 \cdot 10^{-4}$	$3 \cdot 10^{-4}$
rod electrode (tangential to the surface)	$8 \cdot 10^{-13}$	$3 \cdot 10^{-8}$	$6 \cdot 10^{-8}$	$5 \cdot 10^{-7}$	$2 \cdot 10^{-6}$

We must remember the following facts in this connection. First, we have taken a rather abstract case. We are considering one cylinder (rod) and are not allowing for the total effect of neighboring parts of the rod electrode on the particle; if we consider this, the resulting F_{pon} may rise by one or two orders. Secondly, the particles may touch one another and form aggregates; for 20-30 particles (diameter 100 μ) forming an aggregate, F_{pon} becomes commensurable with F_{ei}. Thirdly, the detachment of particles by a tangentially directed force is three or four times more effective than by a force directed normally to the dust-laden surface (§3).

Thus, the inhomogeneity or nonuniformity of the electric field, and the consequent ponderomotive force directed tangentially to the surface being cleaned, offer the best possibility of removing dust particles completely; this is in fact confirmed by experiment in that the detachment of dust particles by rod electrodes is more effective than by continuous plates.

Naturally dust particles may also be detached when the resultant of the detaching forces is greater than that of the forces resisting detachment.

Figure X.6 shows the dependence of the detaching forces and the forces resisting detachment on the particle diameter for a distance of 1 cm between the electrode and the dust-laden surface. The charge on the particle was calculated from formula (X.3) and the adhesive force was determined by reference to the retention of 50% of all the dust particles. The forces resisting the detachment of the particles are plotted above the axis of abscissas in Fig. X.6 and the detaching forces are plotted below. The resultant force is shown as ΣF.

For large dust particles (more than 70 μ) the detaching force $\sim 10^{-1}$ dyn, i.e., an order higher than the forces retaining the dust particles. As the particle size diminishes, the relation between the detaching and retaining forces changes. For particles smaller than 15 μ, the forces resisting detachment exceed those promoting

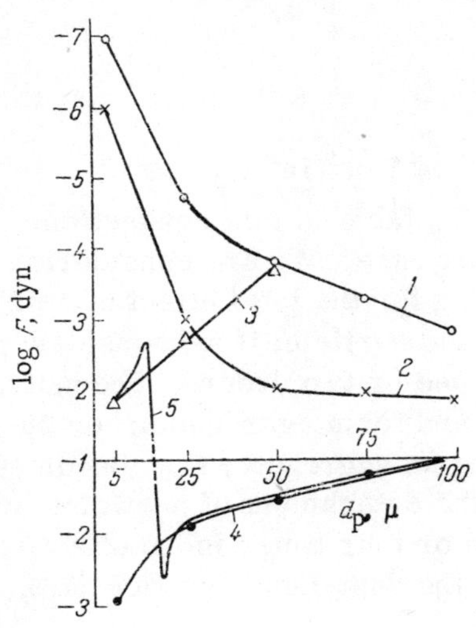

Fig. X.6. Detaching and retaining forces as functions of the dimensions of adhering spherical glass particles. 1) F_e; 2) F_i; 3) F_{ad}; 4) F_{ei}; 5) ΣF.

it, and this is confirmed by experimental results.

Thus the efficiency of the process of detaching particles by means of a constant electric field increases with particle size.

§ 50. Detachment of Particles Under the Influence of an ac Electric Field

Conditions for the Detachment of Particles.

An alternating high-frequency field is sometimes used for removing contaminations from a surface [465], and in particular for cleaning the rails in front of a moving locomotive [466].

In order to discover how the efficiency of an ac electric field in this respect depended on the conditions of particle detachment, the properties of the electrode, the properties of the surface being cleaned, and other factors, some special experiments were carried out with an apparatus (Fig. X.7) delivering a current with a frequency between 50 and 2000 cps and a voltage of 3-5 kV and enabling either one or several electrodes of the needle type to be used at the same time, and also allowing the distance between the electrode and the treated surface to be varied from 1 to 5 mm.

In all the experiments the surface (first made dusty to the extent of 3-5 g/m^2 by free settling) was held for 20-30 min before imposing the electric field. The dimensions of the surface to be treated were no greater than 40 × 60 mm.

The efficiency of the operation of the ac electric field was estimated by reference to the area, or rather the diameter, of the sharply defined dust-free spot formed on the grounded surface opposite to the detaching electrode.

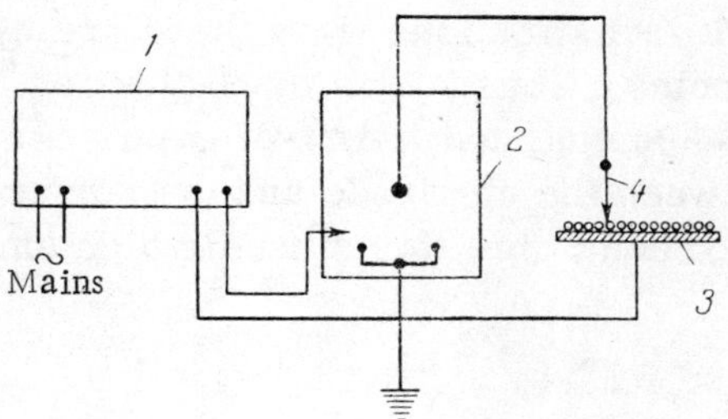

Fig. X.7. Block diagram of the apparatus for cleaning a surface by superimposing an ac electric field. 1) Sound generator; 2) transformer; 3) surface under treatment; 4) electrode.

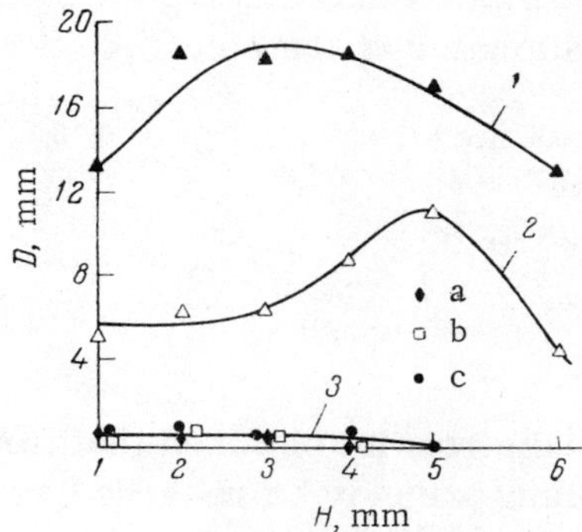

Fig. X.8. Diameter of the spot freed from dust as a function of the distance from the surface being treated and the nature of the surface material (electrode diameter, 0.35 mm; frequency, 400 cps; voltage, 3.0 kV; period of action of the electric field, 4 sec). 1) Plexiglas; 2) painted metal; 3) metal (a, copper; b, tin plate; c, aluminum).

By varying the field frequency it was established that the optimum frequency, for a voltage of 2.2 kV and a treatment time of 10 sec, was 400-600 cps:

Frequency, cps	100	200	400	600	800	1000
Spot diameter, mm:						
for H 1 mm.	5.3	4.2	4.8	4.2	4.7	2.2
for H 2 mm.	5.0	5.2	4.8	4.5	4.0	2.1

Subsequent investigations were therefore mainly carried out at these frequencies. Thus it was found that for $V_0 = 2.2$ kV, frequencies of 400-800 cps, and a time of treatment equal to 10 sec, the distance between the electrode and the surface under treatment (H) should not exceed 5 mm (for a needle diameter of 0.35 mm):

H, mm	1	2	3	4	5
Spot diameter, mm:					
for 400 cps.	4.8	4.8	5.2	8.8	8.6
for 600 cps.	4.2	4.5	6.4	7.8	–
for 800 cps.	4.7	4.0	6.2	4.8	–

Under the same conditions (frequency, 400 cps; H = 3 mm) we established the relation between the spot diameter, the electrode potential, and the diameter of the electrode needle:

Diameter of electrode needle, mm	0.3	0.35	0.5	1.0	1.5
Spot diameter, mm:					
for 3 kV	–	5.0	5.0	0	0
for 4 kV	4.0	6.0	4.0	5.0	–
for 5 kV	5.5	6.0	10.0	7.0	6.0

We see from the results obtained that for potentials of 3-5 kV, the optimum diameter may be regarded as 0.35 and 0.5 mm. However, in view of the fact that the most stable spot size on varying the potential was obtained for a diameter of 0.35 mm, all subsequent investigations were carried out with a needle of this diameter.

The following represents the relation between the spot diameter and the electrode material for a voltage of 2.5 kV, a frequency of 400 cps, a needle diameter of 0.35 mm, and a treatment time of 10 sec:

Electrode material	Copper	Nichrome	Nickel	Steel
Spot diameter, mm	7.1	6.9	7.2	6.8

We also studied the relation between the area of the spot freed from dust by means of two electrodes 0.35 mm in diameter, the distance separating the electrodes, and the electrode potential:

l, mm	1	2	3	4	5	6
Spot diameter referred to one electrode, mm:						
for 3 kV	7.0	7.6	7.6	7.6	5.1	5.1
for 5 kV	8.2	10.0	10.0	9.9	9.9	4.9

Hence under these conditions the optimum distance between the electrodes is 2-5 mm.

The degree of cleaning a surface depends considerably on the properties of the surface to which the dust has adhered. For metallic surfaces (Fig. X.8) cleaning is difficult (curve 3); painted metal surfaces (curve 2) and dielectrics (curve 1) are cleaned fairly easily, i.e., the law is exactly opposite to that associated with the action of a steady (dc) electric field on adhering particles (see §49).

Mechanism of Particle Detachment. Whereas on subjection to a steady dc electric field the particles are removed instantaneously from the surface and fly to the electrode, on subjection to an ac field the adhering particles move over the surface from the center to the periphery of the cleaned spot, and the dust is blown off very much as in the case of an air jet. This is because the gas is ionized at the point of the needle electrode and an ionic wind develops [467].

The cleaning process takes place over a certain interval of time (Fig. X.9) and practically finishes after 4 sec, although even over the next 26 sec the cleaned area continues to increase, so that the surface is almost completely freed from adhering particles.

In order to determine the change in the force of the ionic wind on a plane surface as a function of the voltage applied to the electrode at a frequency of 400 cps, we used a torsion balance with metal and glass pans. The force of the ionic wind was proportional to the potential applied to the electrode (Fig. X.10).

If the only cause of dust-particle detachment were the ionic wind, the efficiency of the removal of attached particles should rise with increasing electrode potential. However, this is not in fact the case (Fig. X.11). For relatively short distances from the surface under treatment (1-3 mm), the diameter of the cleaned spot increases with electrode potential until the latter reaches 3.2 kV, after which the tendency is reversed. For a distance of 4 mm the reverse is the case. The maximum spot diameter is obtained for

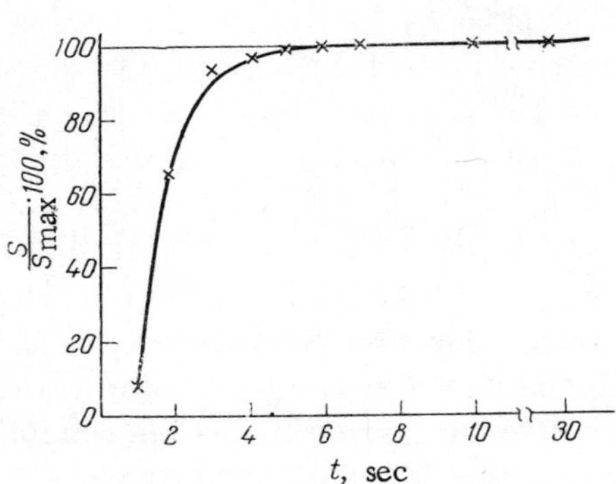

Fig. X.9. Relative cleaned area (in % of the maximum possible) as a function of the period of action of the ac electric field.

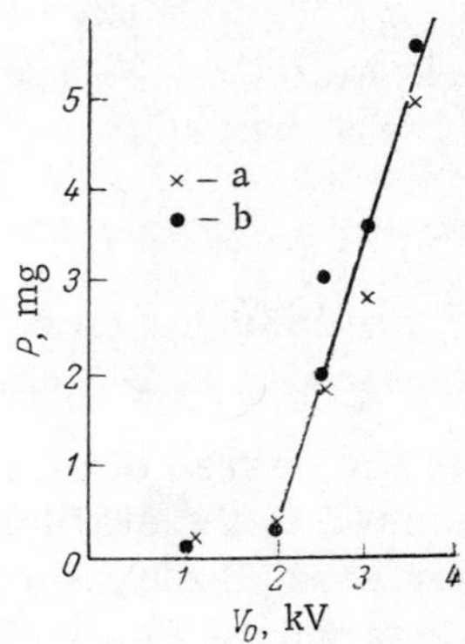

Fig. X.10. Force of ionic wind as a function of the potential applied to the electrode during the cleaning of glass (a) and metal surfaces (b).

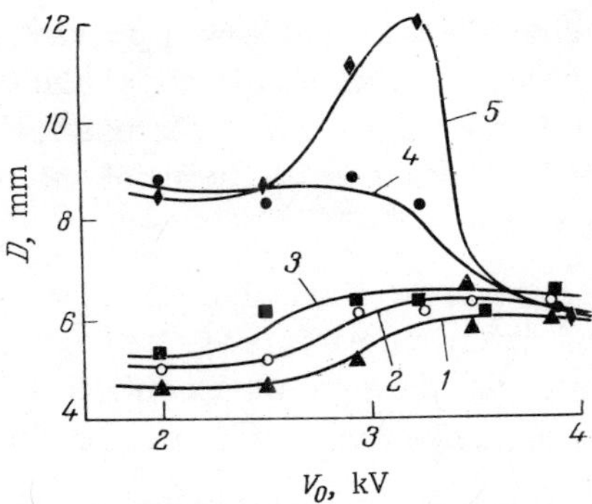

Fig. X.11. Diameter of the cleaned spot as a function of the electrode potential for various distances (number on the curves indicate the distance in mm) from the surface being treated (a painted metal surface) for a frequency of 400 cps and an electrode diameter of 0.35 mm.

H = 5 mm and V_0 = 3.2 kV. This is why a potential of the order of 3-3.2 kV must be regarded as optimum, i.e., the value for which the greatest possible area is freed from dust. On further raising the potential the size of the cleaned area diminishes.

The experimental results strongly suggest that the attached particles are acted upon by some other force (arising as a result of the polarization of the particles) in addition to the ionic wind. Actually, the charge on the particle (see Fig. X.5) remains constant for no longer than 10^{-2} sec after removing the electric field. On varying the sign of the potential applied to the electrode (when the frequency of the current is 400 cps or over), the sign of the charge on the particle is unable to change (Fig. X.5). In this case, the force of electrical interaction arising between the similarly charged particle and electrode will be repulsive (in contrast to the force arising in a dc field), and the particles will as it were fly apart (a dust-free spot will be obtained), i.e., the force of the ionic wind and the repulsive force developing between the electrode and particle will add, and the cleaning efficiency will increase. In view of this, the necessary condition for the detachment of particles under the influence of an ac field is that the period of action of the polarization charge on the particle after removing the field should exceed the time required for the recharging of the particles (i.e., a quantity inverse to the frequency of the ac field).

In an ac field a metallic surface may also become recharged, so that charges of different sign will be formed on the particle and the metallic surface; these will interact and may increase the adhesive force. Hence, it is more difficult to clean metal surfaces than insulators or semiconductors in an ac field.

§51. Transportation of Detached Dust Particles

In cleaning a surface it is very important not only to detach the adhering particles, but also to remove them from the area, i.e., to prevent the possibility of their reattachment. A process of this kind occurs when particles are detached in the dc electric field created by a continuous electrode; the detached particles execute a shuttle motion between the substrate and the detaching electrode.

Let us consider the forces acting on a dust particle being detached from the surface. Let us first make an assumption. We shall not consider forces acting tangentially to the dust-laden surface; this is equivalent to the assumption that the electric field is uniform in a direction perpendicular to the surface. Then only electric and gravitational forces act on the particle after detachment. The height z to which the particles are raised may be determined from the equation of motion of the particle:

$$F_z = m\frac{d^2z}{dt^2} = QE - mg \tag{X.6}$$

Integrating Eq. (X.6), we obtain

$$z = \left(\frac{QE}{m} - g\right)\frac{t^2}{2} \tag{X.7}$$

where Q is the charge on the particle, and t is the time during which it is being raised.

Remembering that the value of Q is directly proportional to r^2 [see formula (X.3)] and that the mass varies as r^3, we may consider that the height to which the particle is raised is inversely proportional to its radius. This means that under otherwise equal conditions small particles will be lifted higher than large ones.

The following represent the "lift" of the particles as a function of their size when using a rod electrode (rod diameter, 2 mm; distance between rods, 6 mm) with an electrode potential of 20 kV and a distance of 10 mm between the electrode and the dust-laden surface:

d_p, μ	80-100	75 ± 2	70 ± 2	50 ± 2	30-40	10-20	5-10
Lift of particle, mm	47	48	51	55	58	75	87

We see from the foregoing results that the dust particles fly between the electrode rods and are lifted to a height of 5-8 cm. These results are in agreement with calculations [see formula (X.7)].

If the field is nonuniform in the transverse direction, a force F_x directed tangentially to the dust-laden surface will act on the detached particles as well as the force F_z.

Remembering the obvious relationships

$$F_x = m\frac{d^2x}{dt^2} \tag{X.8}$$

and

$$\frac{d^2x}{dt^2} = \frac{d}{dt}\left(\frac{dx}{dt}\right) = \frac{d}{dz}\left(\frac{dx}{dz}\cdot\frac{dz}{dt}\right)\frac{dz}{dt} = v_z^2\frac{d^2x}{dz^2}$$

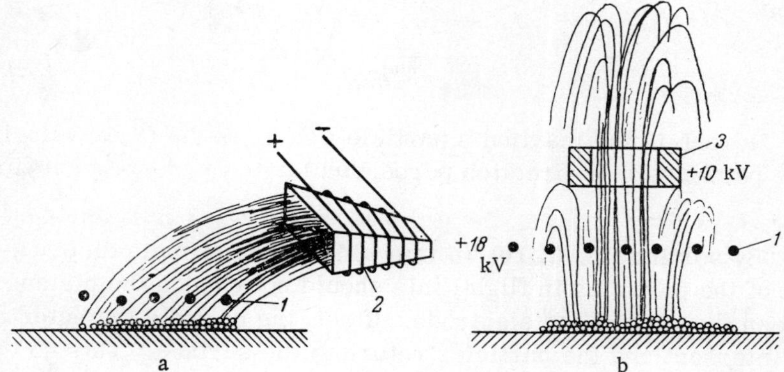

Fig. X.12. Flight trajectories of particles in the field created by means of a solenoid (a) and an electric current (b). 1) Rods of detaching electrode; 2) solenoid; 3) supplementary electrode.

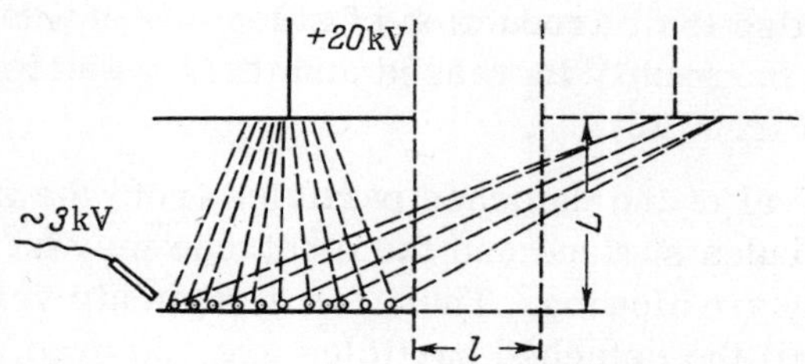

Fig. X.13. Combined action of ac and dc electric fields in removing attached particles.

we find

$$\frac{d^2x}{dz^2} = \frac{1}{v_z^2} \cdot \frac{d^2x}{dt^2} = \frac{1}{v_z^2} \cdot \frac{F_x}{m} \tag{X.9}$$

Integrating (X.9), we obtain

$$x = \frac{1}{mv_z^2} \int_0^L dz \int_0^L F_x \cdot dz = \frac{1}{mv_z^2} \int_0^L (L - z) F_x \cdot dz$$

Let us denote

$$\int_0^L (L - z) F_x \cdot dz = A$$

so that

$$x = \frac{A}{mv_z^2} \tag{X.10}$$

Here L is the maximum particle lift; v_z is the flight velocity of the particles in a direction perpendicular to the dust-laden surface.

By solving Eq. (X.10) we may obtain the optimum displacement of the particles in flight; this should not exceed the distance between the rods of the electrode. If not, the particle trajectories may intersect and the particles return to the surface. This no doubt explains the earlier-found optimum distance between the rods of the detaching electrode (see §49). In order to increase the lift of the particles and vary their flight pattern, a supplementary electrode may be used (Fig. X.12). The use of the supplementary electrode (which carries a potential lower than that of the detaching electrode) and also the introduction of solenoids enables the lift of the particles to be greatly increased and their trajectories to be varied at will (Fig. X.12).

The removal of the detached particles from the space between the dust-laden surface and the electrode may be accelerated and improved by air blowing. Thus, for a mean air velocity of 2.8 m/sec almost all the detached particles are removed.

A more universal method of removing particles is that based on the simultaneous action of dc and ac electric fields (Fig. X.13). Here the detached particles are not only lifted to a height L, but also displaced through a distance l relative to the dust-laden surface. The best trapping of the detached particles under the combined action of ac and dc fields is achieved when l = 30-50 cm and L = 20-40 cm.

Chapter XI

Adhesion Processes Under Conditions of Agricultural Production

§ 52. Adhesion of Soils to the Working Parts of Agricultural Machines

Under service conditions soil tends to stick to the working surfaces of agricultural machinery plowshares, for example) and this greatly impedes the operation of the machine, since the motive power supplied is used in overcoming the adhesive forces and friction between the soil and the working surface as well as the ordinary resistance of the soil.

The particles in the layer of soil sticking to the working surface interact strongly with each other, forming a continuous solid mass, so that there is no great difference between the areas of nominal and true contact separating the attached layer from the surface, such as occurs in the adhesion of a dust layer (see §2). Under these conditions we may consider that the value of the load F_N [see Eq. (I.24)] equals the adhesive force of the sticking layer (F_l). Allowing for the adhesion of the soil, the force overcoming the friction between the layer of soil and the working surface may be expressed by the following equation [468]:

$$F_{fr} = \mu(F_N + F_l) = \mu F_N + kS \tag{XI.1}$$

where F_{fr} is the frictional force, μ is the coefficient of friction, F_N is the normal force with which the attached layer acts on the surface, F_l is the adhesive force of the adhering layer, k is the cohesive (tenacity) factor ($k = F_s \mu$), and F_s is the sticking force referred to unit surface.

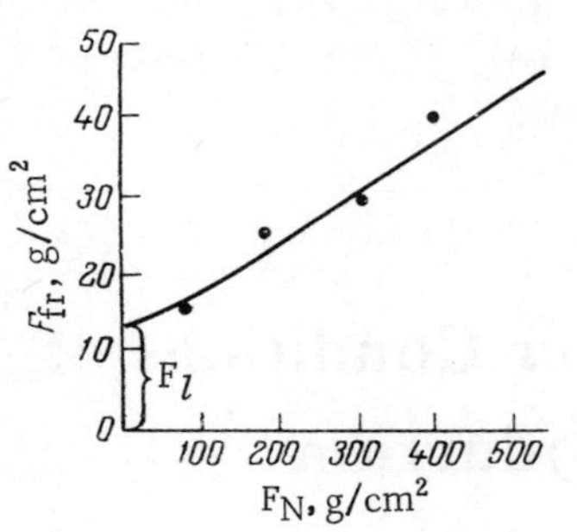

Fig. XI.1. Frictional force as a function of the normal load, referred to 1 cm^2.

The sum of the coefficients of friction and tenacity gives the shear index (φ):

$$\varphi = \mu + k$$

In the absence of sticking, i.e., when $F_s = 0$, we have $\varphi = \mu$. If, however, the force of adhesion greatly exceeds the normal force, i.e., $F_l \gg F_N$, then in accordance with (XI.1), the frictional force F_{fr} is given by the equation:

$$F_{fr} = F_N \cdot \mu S \qquad \text{(XI.2)}$$

Hence in order to calculate the adhesive forces between the soil and the working surface of agricultural machines we must know the specific sticking force and the coefficient of friction, as well as the areas of the surfaces in question (S).

The value of F_s may be determined experimentally. For this purpose we may use a tribometer to determine the relation between the shear force (which is equal and opposite to the frictional force) and the value of the normal load (in the absence of which the shear stress is simply employed in overcoming the forces of adhesion) (Fig. XI.1).

The sticking force, referred to 1 cm^2 of surface, depends on the properties of the soil and may vary by a factor of 150 for soils of different kinds.

Kachinskii [469] distinguished five classes of soils by reference to their sticking forces (referred to 1 cm^2 of area), subject to capillary saturation:

	g/cm^2		g/cm^2
Extremely tenacious	15	Weakly tenacious	0.5-2
Strongly tenacious	5-10	Loose or crumbling	0.1-0.5
Medium tenacious	2-5		

The coefficient of friction is determined by the dimensions of the soil aggregates, the moisture in the soil, the properties of

Table XI.1. Coefficient of Friction of Polished Steel Surfaces as a Function of the Size of Soil Aggregates and the Humidity of Argillaceous Soil

Limiting size of soil aggregates, mm	Type of steel	Coefficient of friction for soil humidity			
		10%	15%	20%	22%
1	St. 3	0.28	0.36	0.46	0.52
	27SG	0.22	0.29	0.32	0.46
	38KhSA	0.21	0.27	0.33	0.46
2	27SG	0.35	0.28	0.43	0.45
	38KhSA	0.25	0.32	0.40	0.36
3	St. 3	0.28	0.32	0.45	0.41
	27SG	–	0.22	0.28	0.39
	38KhSA	0.19	0.23	0.31	0.40

Table XI.2. Coefficient of Friction of Steel Surfaces for Different Humidities of Argillaceous Soils

Humidity of soil, %	Coefficient of friction				
	St. 3	27SG steel	38KhSA steel	chromium-plated steel	painted steel
20	0.47	0.55	0.57	0.48	0.52
25	0.63	0.56	0.65	0.50	0.79
30	0.70	0.56	0.72	0.54	1.05
35	0.74	0.55	0.74	0.58	–
40	0.76	0.52	0.73	0.57	–
45	0.73	–	0.73	0.55	–

the working surfaces of the agricultural machines, and also the presence or absence of layers of water between the soil and the surface.

Let us consider the factors determining the value of μ in more detail, noting incidentally that the absolute value of the coefficient of friction depends on the velocity of the touching surfaces (the working parts of the agricultural machines) [470, 471]; the coefficient of friction varies particularly rapidly for a slipping velocity of up to 2 m/sec.

It is hard to draw any conclusions regarding the relation between F_l and the size of the soil aggregates from the results of

Table XI.1, since the particle-size distribution of the soil was not specified in [472]. However, there is a tendency (Table XI.1) for the coefficient of friction and hence the value of F_l to increase with increasing humidity of the soil.

The ability of the working surface to become wetted with water also affects the value of the coefficient of friction. Since hydrophobic surfaces hold contaminating particles relatively poorly (see §27), the operation of the machines should be improved by hydrophobization. The hydrophobization of steel is effected by the addition of chromium and nickel, by chromium plating the surface, and also by treatment with silico-organic compounds [472]. Chromium—nickel steels also have a high value of hardness and suffer less severe wear than ordinary steel parts.

Table XI.2 shows the coefficients of friction of various steel surfaces as functions of soil humidity.

We see from the data presented in Table XI.2 that the hydrophobic 27SG steel and also steel of the chromium-plated type have the lowest coefficients of friction, and hence the lowest soil adhesion. A coating of oil paint makes the steel surface hydrophilic, and this increases the adhesion. These laws principally hold for soils with a 20-45% moisture content.

Materials kept in a methyltrichlorsilane atmosphere have still lower coefficients of friction. However, the incorrect application of chlorsilanes may lead to rapid abrasion and to the loss of the hydrophobic properties of the surface.

In order to prevent the adhesion of extremely tenacious soils such as those of the Hawaiian Islands, the moldboards of the plows are covered with a thin plastic film of Teflon or Teflon derivatives [473]. Polyethylene, polypropylene, polyvinyl chloride, Orlon, etc., may be used instead of Teflon. In order to make the surface abrasion-resistant, glass wool, glass fiber, soot, or graphite are added to the Teflon. Thus, Teflon mixed with 25% glass fiber gives a several times higher wear resistance to the surface, and the service life is doubled. Of all plastics tested for this purpose, the best so far may be regarded as the polyfluorethylene group (polytetrafluorethylene) and high-pressure polyethylenes. On using such plastic coatings there is hardly any sticking of the soil to the working surface.

Also used as a protective coating is enamel treated with a 20% HCl solution and boiled in paraffin. A layer of such an enamel 0.3-1 mm thick makes the underlying surface hydrophobic [474].

Bredun studied the adhesion of soil to a steel plowshare and also to plowshares coated with films of Capron, P-68 polyamide, polyfluorethylene, HD polyethylene, polyvinylbutyral, and polymethylacrylate [475]. The least sticking of soils occurred for polyfluorethylene and the greatest wear resistance for polymethylacrylate.

The amount of moisture in the soil has a considerable effect on the sticking powers. With increasing moisture content the adhesive force associated with the extra tackiness of the soil becomes greater (see Tables XI.1 and XI.2). For soils of the black earth type with more than 70% moisture, the strong adhesion of the soil to the metal surface results in detachment of the autohesive type when plowing, so that friction between metal and soil is replaced by friction between soil and soil [476]. According to data obtained from other sources [477], detachment of the autohesive type occurs at a humidity of 80-85% for well-structured clay and loam soils, and at 95% for light soils.

With increasing moisture content, the plowing speed at which the plowing resistance is lowest becomes greater. For example, in soddy podzol soils with a soil humidity of 60% the plowing speed is 1.06 m/sec, and for a humidity of 70% it is 1.46 m/sec.

In the presence of a continuous water film such as that created in plows with a "water channel" between the surface and the soil, friction and adhesion are reduced. However, the construction of such a "water channel," which essentially involves passing water from a container fixed to the plow into the space between the plowshare and the moldboard through a series of apertures, is cumbersome, and the machine requires constant replenishment of the water, which in practice is not worth the trouble.

This problem may be solved more efficiently by the use of electro-osmosis. If a steady electric current is fed to a plow sunk into the soil in such a way that the share and moldboard become the cathode, then under the influence of electro-osmosis the water in the soil will move toward the metal parts and wet them. A water film is thus formed between the surface of the plow and the soil particles, and this reduces adhesion and friction.

Baibakov determined the effect of electro-osmosis on the adhesion of soils [121]:

Sticking force (referred to 1 cm^2), g:	Without electro-osmosis	With electro-osmosis
to a polished steel surface	5.47/7.95*	1.95/2.80
to a chromium-plated surface . . .	5.80/7.07	1.53/3.68

These results were obtained for a current density of 2 mA per cm^2. Under these conditions electro-osmosis reduces the specific sticking force by a factor of two to three.

It is pointless to raise the current density to 5 mA/cm^2, as the soil then sticks harder again. In this case the increased adhesion is apparently due to the formation of Al- and Fe-base gels in the course of electrolysis.

The effect of electro-osmosis differs for different soils; its influence is felt at a particular soil humidity: for western pre-Caucasian black earth at a humidity of over 25%, for soddy podzol above 30%. The most favorable case is that of black earth containing clay and colloid particles; the transfer of moisture to the cathode depends on the motion of these.

§ 53. Adhesion in Treating Plants and Seeds with Pesticides

Adhesion of Pesticides to Plants. In dusting plants with pesticides, only about 10% of the powder preparations settle on the plant leaves. The greater part of the pesticide (about 90%) is wasted. This is not only because the area of the leaves is much smaller than the area dusted, but also because not all the particles falling on the leaves remain attached to the latter; if the leaves are dry, the particles are blown or shaken off, and if rain is falling the powder is washed away.

The efficient use of aerosol pesticides may be improved by increasing the adhesive force (see §35) between the particles and the leaves, for example, by giving a charge to the aerosol particles [478-481] (see §12). The charge on the particles must, of course, be made unipolar, since otherwise the particles might stick to-

*Numerator relates to a soil humidity of 25%; denominator to 27%.

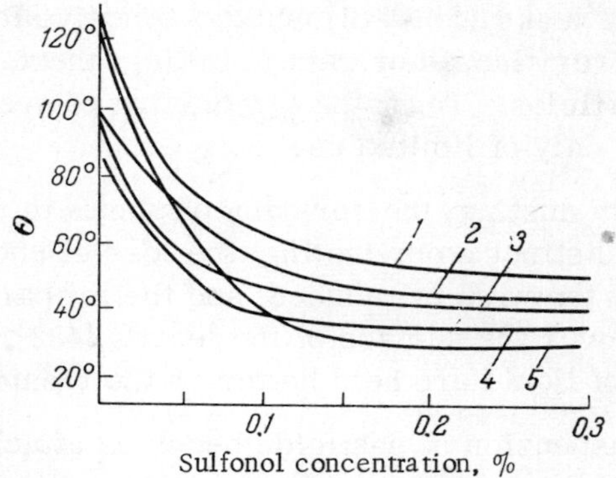

Fig. XI.2. Wetting of various surfaces by Sulfonol of various concentrations. 1) Backs of cabbage aphids; 2) wings of leaf beetles; 3) beet leaves; 4) wax plate; 5) wheat leaves.

gether in the field; hence, frictional charging must not be used. Aerosols charged in a unipolar manner may be obtained, for example, in the field of a corona discharge on passing the particles along a grounded metal tube containing an axial conductor emitting corona [478].

The number of particles of the same electrical polarity settling at the instant of time t is given by the equation

$$n = n_0 \cdot t - Wc \tag{XI.3}$$

where n_0 is the output of the aerosol source in particles/sec, t is the time in sec, W is the volume filled by the aerosol jet in m^3, and c is the particle concentration in particles/m^3.

The adhesion of a charged aerosol to the leaves of plants is 5-6 times that of an uncharged aerosol [478]. In dusting plants and defoliating cotton with a charged aerosol, the amount of the preparation used is halved as a result of the adhesive forces. In addition to the quantitative improvement, the charging of the particles ensures an even deposit over the whole surface of the plant leaves, including the underside [479].

The increased adhesion of the charged particles is only felt at the instant at which these come into contact with the surface.

Later the charge leaks (see § 12) and the adhesive forces diminish. When the air is very humid or rain is falling, there is no point in charging the particles. Thus, the dry dusting of preparations in an electric field is only of limited use.

Apart from dusting, the spraying of plants is widely employed. In spraying, as distinct from dusting, the loss of chemical poisons carried away by the wind is reduced, and the preparation is distributed evenly over the surface of the leaves [482]. In addition to this, the drops of liquid are held better on the treated surfaces.

When a suspension of pesticide becomes attached to a surface, the following processes take place: the wetting of the original surface, the partial evaporation of the liquid components, and the adhesion of the solid particles. Thus the adhering particles are more resistant to the effects of wind and rain.

Wetting may be characterized by contact angles; the values of these are given in Fig. XI.2 for several types of leaf and insect bodies as functions of Sulfonol concentration. The wettability of a wax surface is presented for comparison.

The shells of insects and the surfaces of plant leaves having greasy wax deposits, thick long fibers, and scales are poorly wetted (for example, the cabbage aphid, the caterpillar of the cabbage butterfly, beet-weevil bugs, cabbage and wheat leaves). Soft insect coats and smooth leaf surfaces free from wax deposits wet better (butterflies, beet, lemons, beans). The addition of Sulfonol increases the wettability of all insect shells and plant leaves. For surfaces with fibers and scales (wheat leaves, curve 5) this effect is particularly significant. In combating insects covered with a wax deposit (cabbage aphid, scale insects, nematode worms, etc.) [483], in addition to Sulfonol we may use wetting agents of the DB, OP-7, and OP-10 types in the form of 0.1% solutions, which increase the adhesion of anabasine and nicotine bases, colloidal sulfur, etc.

Surface-active substances raising the spreading and wetting capacities of pesticide liquids may at the same time reduce the adhesive qualities of the particles, particularly under the influence of atmospheric deposits. With increasing concentration of the SAS the adhesion usually diminishes. It is therefore important to introduce the optimum quantity of SAS into the spraying compositions [484].

Apart from wetting agents, the original preparation, which may be either a powder or a suspension, may be supplemented by additives increasing the adhesion by virtue of their tackiness. Such additives are called stickers. Thus, stearic acid, calcium stearate, paraffin, and similar compounds are added to DDT. Various oils may be used for this purpose as well as the polymers or copolymers of natural or synthetic elastomers, particularly butylene or isobutylene (with a molecular weight of 200-300 thousand) and salts of Li, Zn, Ca, Al, Cd [485] and fatty acids.

With the same intention, pesticides are also dispered in an ammoniacal solution of casein or carboxymethylcellulose (CMC) (40-60% with respect to the weight of the preparation) [486]. In addition to CMC one may use gelatin, animal glue, and also synthetic resins (polyvinylacetate, etc.) [487]. Gelatin, glue, dextrin, methyl cellulose, etc., may also be added to DDT suspensions. The addition of aminostearates or the preparations LOVO 190 and LOVO 192 to suspensions of Sevin and DDT improves the adhesion and reduces the loss of insecticide during heavy rain in tropical conditions [488]. On adding stickers to a Sevin suspension used in the treatment of cherry trees, the amount of preparation adhering increased by 35% [489].

The same component may be used as both wetting agent and sticker. Such preparations include polyethyleneglycolnonylphenol and the sodium salts of dinaphthyl methanedisulfo-acid [490]. The functions of wetting agent and sticker are fulfilled by liquid calcium sulfide for the insecticides DDT and hexachloran. Even severe rain cannot wash this preparation away [491]. The addition of 0.1% of the wetting agent Roplex to 50% Sevin prevents the preparation from being washed away from plant leaves for 12 days. However, even under these conditions the pesticides are washed away more rapidly from the tops of the leaves than from underneath [492].

The distribution of the preparation over the surface of the treated leaf depends on the delivery rate of the pesticide. Thus, in spraying a cherry orchard with a 0.8% suspension of Sevin and a delivery rate of 565 liters/ha, all the liquid falling on the plant leaves remains on them. On doubling the rate, some of the suspension runs away, the density of the deposit on the lower part of the leaves being 1.3 times that on the tops. The addition of the SAS "Plaik" reduces the expenditure of insecticide as a result of the increased adhesion [493].

Other methods are also available for increasing the adhesion of preparations.

When using insecticidal smoke pots, the liquid insecticide solidifies in flight and sticks to the plant leaves in the heated state (see §35). The crystals formed on cooling are firmly held on the treated surfaces [494].

Adhesion when Cleaning Seeds and Grain. When seeds are dusted for the purpose of disinfecting them, the disinfectant powder sticks to the surface of the seeds. Widely used disinfectants include, for example, the gamma isomer of hexachlorcyclohexane and mercury-organic insecticides [495]. In order to hold the disinfectant more firmly on the surface of the grain, dry dusting is replaced by semi-dry dusting with the use of additives giving the disinfectant tacky properties. Such additives may include polyvinylacetate, polyacrylamide, carboxymethylcellulose, and also silicate glue, wood pitch, and starch. However, whereas additives of the former class are fairly expensive, those of the second class are not always effective, since they tend to make the seeds themselves stick together [496].

The use of the gamma isomer of hexachlorcyclohexane with added tetramethylthiuramdisulfide (TMTD), OP-7 wetting agent, a concentrate of sulfite-alcohol residues (SAR), and other fillers was proposed by Bezuglii and colleagues [497].

In order to free clover and alfalfa seed from the seeds of certain weed grasses (particularly dodder), powders with magnetic properties are used. On mixing with the seeds, the magnetic powder sticks to the weed grass (since these have the rougher surfaces), and on applying a magnetic field these seeds are removed.

In order to prepare the magnetic powder we may use a mixture of ferrous and ferric oxide with 20% of mineral oil, promoting the binding of the composition and increasing its stickiness. The efficiency of the seed sorting increases on introducing ground talc into the mixture. It was found in this way [498] that the sorting efficiency obtained on removing weed seed from clover and alfalfa (i.e., the fraction of weed seeds originally present in the mixture which are extracted in the sorting process) was 96.5% when a mixture of mixed iron oxide and chalk was employed (powder used, 43.2 kg or 34.6 kg mixed iron oxide to 1 ton of seed) and 99.3% for a mixture of mixed iron oxide, talc, and mineral oil (amount of powder used, 18.2 kg or 7.4 kg of mixed iron oxide to 1 ton of seed). Thus, the total amount of powder used, and particularly the expensive mixed iron oxide, was considerably reduced in the second case, while the sorting efficiency was increased. The mineral oil, however, tended to make the powder stick to the inside of the bunkers used for sorting, and this was very undesirable, since it reduced the efficiency of the seed-sorting equipment.

§ 54. Erosion of Soils

The erosion of soils means the removal of soil particles and the transportation of these under the influence of wind and rain. Erosion by air currents is more correctly called deflation, but it is more frequently called wind erosion. The erosion of soils is a phenomenon responsible for a great deal of damage to agriculture, since it not only changes the structure and composition of soils but also involves the removal of all the fruitful topsoil by wind and flood.

Of course the erosion resistance of soils is determined by the properties of the soil itself, and particularly the autohesion of the soil particles, which is affected by the particle size, the humidity of the soil, the tackiness, etc. In addition to this, the erosion of soils is largely determined by kinematic conditions. Hence, the fight against erosion must be carried on by methods adapted to particular cases.

In this monograph, which is principally devoted to adhesive and autohesive phenomena, erosion will only be considered as the process of disrupting the autohesive forces acting between soil particles. Questions of combating erosion will also be considered from the point of view of possible means of increasing these forces.

Water and air erosion take place when the autohesive forces acting between the particles are overcome by forces arising from water and air flows. In this sense the two processes are identical. However, as we now know (see Chapters III and IV), adhesion in air and water may have very different characteristics.

Let us first consider processes associated with water erosion.

The first stage consists of the breakup of the soil structure under the mechanical action of water drops and the removal of the cementing bases, humus, and colloidal particles.

The second stage is the washing away or surface-layer particles, leading to the formation of soil caverns or cavities [499].

Water erosion under the influence of rain becomes possible when the kinetic energy of the drops and the flow of rain are capable of overcoming the forces of autohesion and the weight of the particles, as in the case of the washing of a surface (see §40).

The kinetic energy of the flow, referred to unit surface of the soil, may be expressed by the following semi-empirical formula [500]:

$$E_1 = \frac{43.5\, k_{sf} \cdot H l I_0 \cdot i^{0.15}}{k_\Delta} \tag{XI.4}$$

where k_{sf} is the coefficient of surface flow; H is the height of the flowing layer of water in m; l is the distance from the initial point of the flow to the section under consideration in m; I_0 is the intensity of the precipitation (rainfall) in m/sec; i is the angle of the slope (hillside); and k is the roughness coefficient of the surface of the slope.

The kinetic energy of the flow equals

$$E = \frac{43.5\, k_{sf} \cdot W l\, I_0 \cdot i^{0.15}}{k_\Delta} \tag{XI.5}$$

where W is the volume of water in the flow between the two sections.

Formulas (XI.4) and (XI.5) do not, however, consider the effect of the size of the raindrops. With increasing drop size the kinetic energy of the drops rises, and hence so does their disruptive effect on the soil.

A descriptive approach is conventionally adopted in the majority of papers on erosion. An attempt to break with this tradition was made by Zvonkov when studying water erosion [501]. Substituting the expressions for the forces in the equation of motion of the particles

$$m \frac{dv}{dt} = F_d - R \tag{XI.6}$$

where F_d (F_f or F_{lift}) is the force acting on the particle from a flow of velocity v; R is the force resisting the motion of the soil particles (R_1 in slipping before flying off, R_2 after flying off), Zvonkov obtained

$$m\frac{dv}{dt}=\frac{\pi}{4}d^2\left\{\tan\alpha\, c_x\cdot\frac{\rho_0}{2}v^2-(\mu+\lambda_c)\cdot[0.66(\rho-\rho_0)d+p_h+p_a]K_s\right\} \qquad \text{(XI.7)}$$

where α is the angle between the tangent to the trajectory of the particle motion and the horizontal plane; c_x, μ, and λ_c are the coefficients of resistance, friction, and cohesion (tenacity) of the particles; ρ and ρ_0 are the densities of the particle and the medium (air or water); p_h and p_a are the hydrostatic and atmospheric pressures; and K_s is the coefficient of stability (or protection) of the soil surface.

The coefficient K_s takes the autohesive forces of the particles into account. If we take this as unity for sand, the values for other soils will be as follows.

I. Soils without plant cover:

	K_s
Uncombined loose sand	1.0
Sandy loam	2.0-3.0
Loam	3.0-4.0
Argillaceous loess	8.0

II. Soils with plant cover:

Grass trampled by cattle	1.5- 2.5
Beet	5.0-10.0
Maize	7.0-15.0
Grain crops	200-400
Perennial grasses	300-500
Sparse woodland	300-500
Thick woodland with grass.	1000.0

It should be remembered that the values of K_s are approximate. The larger the coefficient of stability of the soil K_s, the higher is the erosion resistance.

On the basis of formula (XI.7), Zvonkov [501] calculated the critical velocities characterizing the main stages of erosion:

$v_{K,1}$, the first, at $F_f = R_1$, when rolling or sliding of the particles takes place (but not actual removal from the original place)*;

$v_{K,2}$, the second, at $F_{lift} = R_1$ (F_{lift} is the lifting force), when the particles stop running and start flying);

$v_{K,3}$, the third, at $F_{lift} = R_2$, when flight ceases and the particles are slowed down;

$v_{K,4}$, the fourth, at $F_f = R_2$, when the retardation ceases and the particle stops.

Earlier (see §§37, 38) we considered the conditions necessary for the detachment of particles by a flow of water, but made no distinction between the velocity determining the rolling or sliding of the particles along the bottom (first critical velocity) and the velocity of particle flight (second critical velocity). By the "dragging" or "detaching" velocity of the particles, we meant the velocity at which the lying or adhering particles were first set in

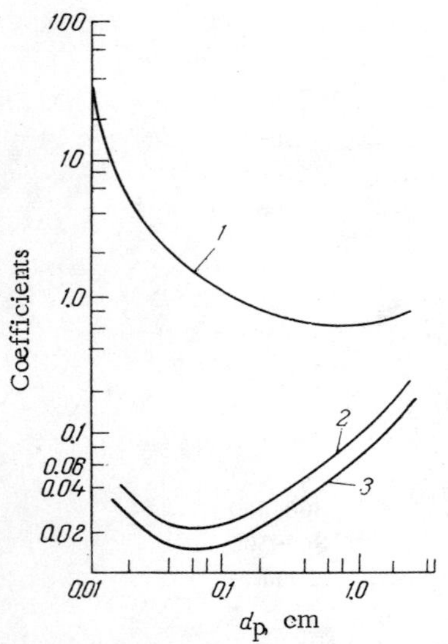

Fig. XI.3. Coefficients c_x (curve 1), λ_c (curve 2), and μ (curve 3) as functions of the dimensions of the soil particles for a 15-cm depth of flow.

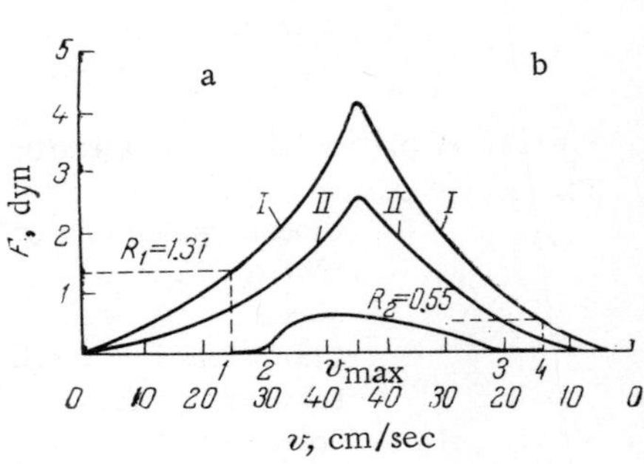

Fig. XI.4. Frontal force F_f (curve I) and lifting force F_{lift} (curve II) as functions of the velocity of the water flow. Points 1-4 give the critical velocities: $v_{k,1}$ = 23.9; $v_{k,2}$ = 29.8; $v_{k,3}$ = 19.2; $v_{k,4}$ = 13.9 cm/sec, respectively.

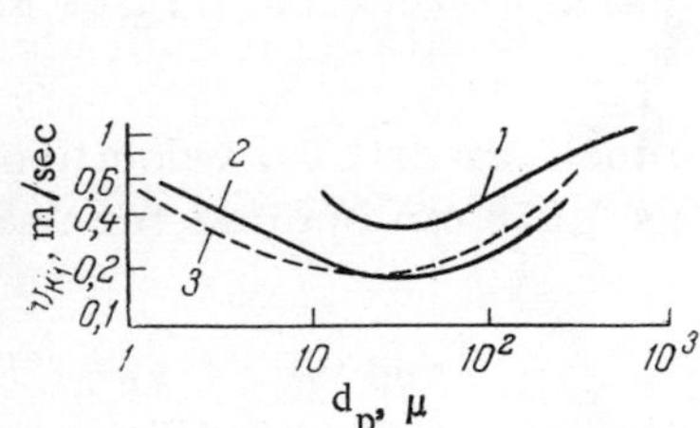

Fig. XI.5. First critical velocity as a function of the size of the soil particles according to: 1) Bagnold; 2) Sundborg; 3) Zvonkova (theoretical for sand).

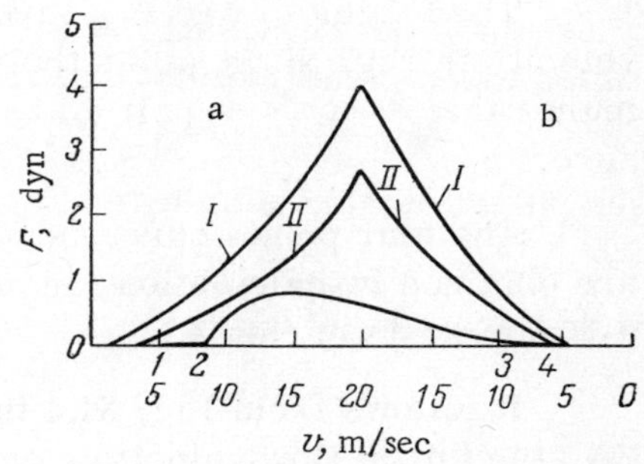

Fig. XI.6. Values of the forces F_f (curve I) and F_{lift} (curve II) as functions of the velocity of an air flow. Points 1-4 give the critical velocities: $v_{k,1}$ = 5.9; $v_{k,2}$ = 8.7; $v_{k,3}$ = 5.7; $v_{k,4}$ = 4.5 m/sec, respectively.

motion, and not the velocity determining the character of the motion (slipping, rolling, or flying).

The first and second critical velocities determine the amount of eroded material and the third and fourth the amount of accumulated material. For a sandy soil, the first critical velocity $v_{K,1}$ equals 18-22 cm/sec for a particle diameter of 0.015-0.033 cm. In this case the ratio of the second critical velocity to the first equals:

$$\frac{v_{K,2}}{v_{K,1}} = \sqrt{\frac{1}{\tan \alpha_1}} \tag{XI.8}$$

where α_1 is the angle between the vectors of the forces F_f and F_{lift} The ratio of the fourth critical velocity to the first equals:

$$\frac{v_{K,4}}{v_{K,1}} = \sqrt{\frac{\mu}{\mu + \lambda_c}} \tag{XI.9}$$

Knowing $v_{K,4}$, we may use (XI.8) and (XI.9) to calculate $v_{K,1}$ and $v_{K,2}$. The necessary values of the coefficients μ and λ_c may be found from the relation between the coefficients c_x, μ, and λ_c and the dimensions of spherical soil particles (Fig. XI.3).

*The frontal pressure of the flow (F_f) on the particles (see § 31) determines the motion of the particles in a direction parallel to the flow-velocity vector, and for a plane surface determines the rolling or slipping of the particles.

The forces F_f and F_{lift} are shown as functions of the flow velocity in Fig. XI.4: (a) as the velocity rises from zero to a maximum value ($0 < v < v_{max}$); (b) as the velocity falls from v_{max} to zero.

The four points corresponding to the four critical velocities are obtained by calculation for particles 0.058 cm in diameter and a 15-cm depth of flow.

It follows from Fig. XI.4 that $v_{K,1} > v_{K,4}$ and $v_{K,2} < v_{K,3}$, i.e., the initial flow velocities causing surface motion (rolling, sliding) and flight, respectively, are greater than the flow velocities for which the moving particles stop. This is because, when particles moving along the surface are retarded, they are not in contact with the ground but with a water layer on the surface of the latter, i.e., liquid adhesion occurs, whereas, at the moment of detachment the force of adhesion associated with solid contact has to be overcome (see §19). From the magnitude of $v_{K,1}$ and $v_{K,4}$ we may determine R_1 and R_2 (Fig. XI.4) by erecting perpendiculars from the points 1 and 4 to meet the curve 1.

Figure XI.5 presents the experimental values of the critical velocities $v_{K,1}$ as function of d_p according to Zvonkov [501], Sundborg [502], and Bagnold [503]. The minimum value of $v_{K,1}$ is approximately 0.2 m/sec for soil particles 100-400 μ in diameter. For removing particles with diameters of under 100 μ attached to the surface and also particles with diameters of over 400 μ (the weight of which increases in proportion to d_p^3) lying on the surface, the value of the velocity $v_{K,1}$ rises, i.e., we have the same situation as in the case of river flows (see §38).

The computing formulas for wind erosion are similar to the formulas proposed for water erosion. However, we must remember that the density of water is 819 times as great as that of air, while the adhesive forces in water are much smaller than in air (see Chapters III and IV).

Zvonkov extended the concept of the four critical flow velocities to the process of wind erosion. Figure XI.6 shows the way in which the forces F_f and F_{lift} vary with the velocity of an air flow for soil particles 0.058 cm in diameter, with $K_s = 1.0$, and zero slope relative to the horizontal. The general character of the relationships obtained and the relation between the critical velocities

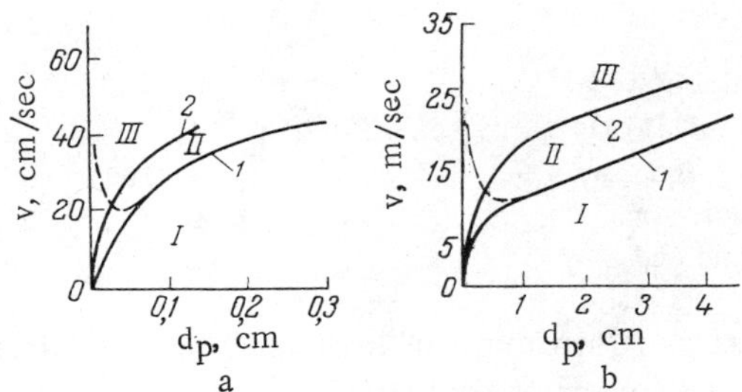

Fig. XI.7. Critical velocities $v_{k,1}$ (curves 1) and $v_{k,2}$ (curves 2) for water (a) and wind (b) erosion as functions of particle diameter. Zones I-III represent the following conditions: I) absence of erosion and onset of accumulation; II) motion of particles along the surface, i.e., rolling or slipping; III) suspended soil particles.

are the same as in the case of a flow of water (Fig. XI.4). However, owing to the low viscosity of air and the existence of rough places on the surface of the ground, we can hardly expect rolling or sliding of the particles in the section 1-2 (Fig. XI.6); it is more likely that in this range of velocities (from $v_{K,1}$ to $v_{K,2}$) the particles will move spasmodically [6]. The value of F_f corresponding to air and water erosion, respectively, reaches the same value for velocities differing by two orders of magnitude. This is supported by the results shown in Figs. XI.4 and XI.6.

The first and second critical velocities of the water and air flows as shown as functions of particle diameter in Fig. XI.7. As we should expect, the first critical velocity for wind erosion exceeds the velocity for erosion by a flow of water, owing to the difference between the adhesive forces in air and water. The broken line shows the change in the value of v_K when the force of interaction between the particles exceeds their weight (see Fig. XI.5 and §§31, 37, 38). In this case the first critical velocity produces the detachment of the adhering particles. The value of this velocity may exceed $v_{K,2}$, i.e., the velocity required to produce actual flight of the attached particles, while the zone corresponding to the motion of particles along the surface (zone II) diminishes or vanishes altogether.

By way of example, let us consider the numerical values of $v_{K,1}$ for soils consisting of relatively small particles:

	d_p, μ	$v_{c,1}$, cm/sec
Colloidal clay	5	130
Noncolloidal clay	5	76
Silt and fine clay	50-100	26
Light dusty sand	130	27
Fine sand and clay	400	15

The erosion resistance of soil may be classified not only in terms of the coefficient of stability (protection) (see page 379), but also by means of other indices. Thus another suitable index for the erosion resistance of soils containing sand particles is the ratio of the total amount of sand to the total amount of silt and silt fractions. This ratio equals 3.6 and 2.9, respectively, in noneroded sandy-loam black-earth and dark-chestnut sandy-loam soils, while in the eroded soils it reaches 6.6 and 5.2, i.e., it doubles. In places of severe and repeated erosion the ratio rises to 9. Soils containing a large quantity of sand and relatively few of the type of particles which give the quality of cohesion to the ground are more subject to erosion [504].

Another possible index of erosion resistance in black-earth soils is the humus content, which principally comprises particles of sizes between 0.001 to 1 mm. The humus gives the soil the quality of cohesion, and the more there is of it the better will the soil resist erosion. However, in water erosion the humus is washed away by the flow of water. This may be seen by comparing the fractional composition [505] of carbonate black earth before and after washing away by water* to a depth of not more than 10 mm:

d_p, mm	>1	1-0.1	0.05-0.001	<0.001
Composition of soil, %:				
before washing . . .	2.6	67.3	8.1	22.0
after washing	2.1	65.1	6.4	26.4

We see from the results presented that in water erosion the proportion of particles with diameters from 0.001 to 1 mm, i.e., the humus particles, is reduced.

*The measurements were made after treating a soil sample placed in a zinc-plated box lying at an angle of 10° with water (5 ml/min · cm^2 for 3 h).

The fact that on reducing the humus content the property of cohesion is reduced was verified by reference to the adhesion of soil to a metal surface. For this purpose a metal disc 8 cm in diameter was placed on the soil and then removed. On withdrawing a metal plate of this kind from the original unwashed soil, a force of 87.0 g had to be applied; for the soil remaining after water treatment the figure was 57.5 g, and for the soil carried away by the water, 100.0 g.

The change in the composition of the soil due to water erosion subsequently reduces the resistance of the soil to wind erosion. Thus an air jet traveling at 10 m/sec only carried 5.2% of carbonate black earth to a distance of 40-60 cm. If the same soil were first subjected to water erosion, then under the same conditions the air jet would carry away 28.0% of the black earth [505]

The resistance of soil to wind erosion may be increased by raising the humidity, so increasing the cohesion of the soil aggregates, i.e., increasing the autohesion as a result of capillary forces. In addition to this, the resistance to both wind and water erosion may be increased if the soil is worked in the "nonterracing" rather than the "terracing" manner; in this case, the stubble, which tends to fix the soil particles, is left in position. Thus in the Pavlodarsk station there was very little carrying away of the fine soil in dust storms when nonterracing working was applied; in the case of terracing the loss was 55-94 tons per hectare. The amount of humus in the dust carried to a distance of 20 m was 2% (i.e., almost all the humus was carried away) [506].

Since the erosion of soil largely depends on the force of the wind, erosion may be reduced by reducing the wind velocity. This is done by planting forests (wind-breaks) and also by increasing the roughness of the soil by raising the proportion of lumps in the top layer. In this way the wind velocity may be reduced to a value below $v_{K,1}$.

The erosion resistance of the soil may be increased by raising the cohesion of the soil aggregates; this may be done by treating the soil with binding substances [507]. Such substances include cellulose and its derivatives (lignin, humic acid), bitumen, peat cement, and various structure-forming substances from plant residues. The adhesive substances envelop the soil particles. The layer so formed, a few microns thick, imparts the quality of tackiness to the soil and thus increases the interactions between the soil particles [508]. Adhesive substances also include polymers, particularly the derivatives of polyacrylic acids, which have become known as "crylia" [509]. An area containing polyacrylamide additives constituting 0.1% of the weight of the soil withstood a 4-h onslaught from a water jet delivering 10 liters/sec at a velocity of 1.1-2 m/sec. The average thickness of the layer washed away from the areas with the reagents added was no greater than 1 mm, whereas, in the case of control samples (without reagents) it reached 35 mm [510].

The coefficient of stability (protection) of soil having an emulsion film [501] reaches 1000, i.e., it exceeds that associated with soils lacking plant cover by two or three orders (see page 379).

Bitumen emulsions and pastes may fix light soils such as sandy loam and sand [511]. Heated coal dust gives cohesion to sandy, silty-clay, and clay soils [512].

In order to fix fine silt in the soil one uses a product containing SAS and also anticorrosion and bacteriocidal substances (for example, triethanolamine phosphate and the fluoro-derivatives of ethanolamine and hexane are added to a solution of an organic polymer, with a concentration of 0.2-18%). On spreading this product, a film binding the silt into aggregates is formed so that the silt is not washed away by water [513].

§55. Caking and Adhesion of Powdered Fertilizers

Effect of Autohesion on Caking. When kept for a long time in a state of rest, powders have a tendency to "cake," i.e., become denser as a result of the redistribution of the fine grains among the larger ones, so that the area of contact between the particles increases and the forces of autohesion become stronger, as well as other factors. This phenomenon is exacerbated when the humidity of the surrounding air is increased. The same occurs with powdered fertilizers when these are stored for a long time.

The degree of caking of powdered fertilizers might be classified by reference to the autohesive force (see §1). However, in view of the fact that in time fertilizers cake into a solid mass, the experimental determination of the forces of autohesion between the

Table XI.3. Autohesive Forces of Certain Fertilizers as Functions of Humidity and Preliminary Compaction

Fertilizer	Fertilizer humidity, %	F_{aut} (referred to 1 cm^2), dyn: without compaction	after compaction under the following loads* in dyn/cm^2: 14	33	60	85
Powdered superphosphate	9.8	32	37	50	63	78
	13.0	38	42	52	65	80
	15.3	43	47	57	70	80
Granulated superphosphate	2.4	45	55	55	55	55
	8.8	50	55	55	55	55

* Load established for 20 min.

particles is very difficult. Hence, in practice, caking is determined by relative methods [514]: for example, by reference to the depth of penetration of a needle of specified size in the product, grinding in a ball mill, or the rupture resistance of a column (cylinder) of the product subjected to an axial load (σ_r).

If our criterion is the value of σ_r, all fertilizers may be divided [514] into seven categories of caking: slight ($\sigma_r < 1$ kg/cm^2), insignificant ($1 \leq \sigma_r \leq 2$), below average (according to the classification of [514] the product "cakes slightly" under these conditions), medium or average, considerable, severe (σ_r = 10–15 kg/cm^2) and very severe. Thus slight caking ($\sigma < 1$ kg/cm^2) characterizes powdered superphosphate, ammophos derived from Kara-Tau phosphorites, nitrophos (a nitrogen—potassium—phosphorus fertilizer), and manure mixture (based on superphosphate); "insignificant" caking characterizes granulated ammonium nitrate, urea, and diammophos; "severe" caking characterizes powdered ammonium nitrate and potassium chloride.

For some fertilizers the autohesive force has been determined (using the slope method) and compared with σ_r (see page 388).

We see from the data presented that there is not always a true correlation between the autohesive force and the rupture resistance, which is affected not only by the autohesive force but also

	F_{aut}, dyn/cm²	σ_r, kg/cm²
Urea	50	1.1
Potassium chloride	45	10-15
Powdered ammonium nitrate	25	10-15
Superphosphate:		
powdered	20	up to 1
granulated	0	0

by the packing of the particles, the density of the powdered mass, and other factors, including humidity as well as the preliminary compacting of the fertilizer (Table XI.3) [515].

For powdered superphosphate the autohesive forces increase with increasing humidity of the product and degree of preliminary compaction.

Combating Caking by Reducing the Autohesion. The caking of fertilizers may be prevented or greatly reduced if we lower the autohesive forces by changing the humidity of the product, the size of the particles, the interparticle contact area, and the properties of the particle surfaces.

Simple superphosphate obtained from the Kara-Tau phosphorites cakes slightly as a result of the hygroscopic magnesium phosphate which it contains. This kind of superphosphate is subjected to ammoniation. As the amount of MgO in the original product diminishes, the caking propensity does likewise [516].

Potassium chloride obtained by the chemical processing of silvinite ores cakes until a monolithic mass has been formed. The reason for the caking property is the very hygroscopic nature of the material [517]. Any attempt to combat caking by drying fails in its purpose, since the impurities (sodium chloride and magnesian salts) bind the water and prevent its evaporation. In addition to this some of the product is lost on drying.

The humidity of the product is not the only cause of caking. One particular type of material may cake to different extents for the same humidity if the particle-size distribution is different [518]. Thus the caking propensity of potassium chloride increases sharply if the diameter of the particles in the powdered product is smaller than 100 μ, i.e., for the same range of sizes as those for which the adhesive and autohesive forces exceed the weight of the particles (see §§16, 17). At the same time the product will not cake at all [517] if the particle size is greater than 700 μ. In this and similar cases it is more efficient to use methods based on changing the shape and size of the particles.

The value of the true contact area, which is determined by both the shape and size of the particles, also has an effect on the caking process. Thus for fused spherical particles obtained by the rapid cooling of potassium chloride no caking occurs [515]. The addition of calcium or magnesium nitrates promotes the crystallization of ammonium nitrate in the form of dendrites. This structure gives a "peaked" form to the surface, reducing the area of contact and hence the caking of the product (see §14) [519]. The addition of products obtained by the neutralization of the nitrate extract of Kara-Tau phosphorites to ammonium nitrate imparts hardness to the ammonium nitrate scales and prevents the formation of fine particles. This also reduces the caking properties [520].

An efficient method of combating caking is the enlargement of the particles; this is effected by the granulation of powdered fertilizers. Granulation not only increases the size of the particles, but also gives them a specific shape, so that their surfaces have the smallest possible chance of undergoing adhesion or autohesion. The value of the autohesive force is thus also made very stable, being unaffected by moisture and the preliminary compaction of the product (Table XI.7). The optimum size of the granules is 1-2 mm [515].

The caking of fertilizers may also be reduced by introducing a filler. The filler particles stick to the surface of the fertilizer particles. As a result of this, the surface of the fertilizer particles, which has a considerable capacity for adhesion and autohesion, is screened, and the true interparticle contact area is reduced. Direct contact of the fertilizer particles with each other is replaced by contact through the layer of attached filler particles. Such fillers may include highly dispersed powders [521] of kaolin, opoka (gaize), and diatomite (up to 20% of the weight of the fertilizer), and also talc and insoluble calcium salts. Talc may also be used for powdering granulated fertilizers [522]. Granulated water-soluble fertilizers (ammonium nitrate, superphosphate, ammonium phosphate, urea, and potassium chloride) are powdered with Attapulgit clay previously activated by heating at 316-538°C for 1 h. In this case the clay loses its colloidal properties and its adsorption capacity increases. Clay with particle sizes of under 15 μ is added to the extent of between 0.5 and 3% of the weight of the fertilizers [523].

In order to prevent the caking of fertilizers, powdered anhydrous $MgCO_3$, $MgCl_2$, and $CaCl_2$ may be added in pure form or mixed with mineral fillers (kieselguhr, koalin, talc, or $CaCO_3$). These substances are taken in quantities of between 0.1 and 5% of the weight of the fertilizers [524]. It is possible to carry out a step-by-step powdering of ammonium nitrate: first with fine limestone (particle size under 30 μ) at 80-100°C, calculated at 3-10% of the weight of the product, with subsequent cooling, and then additional powdering (for storage under very unfavorable conditions) with ground limestone, kieselguhr, and talc [525] to the extent of 0.5-1% of the weight of the product.

Substances reducing the autohesion of fertilizer particles may be used in the form of suspensions. After the evaporation of the liquid the powder is fixed firmly to the particles of fertilizer (see §26). Such an emulsion may be prepared from a concentrated solution of strongly hygroscopic salts or salt mixtures (for example, calcium and magnesium chlorides and nitrates) by adding nonionogenic SAS and gelatinous carriers such as $Mg(OH)_2$, etc. For example, a suspension for spraying granulated fertilizer is prepared by mixing 1.25 kg of kieselguhr with 1 kg of a 40% aqueous solution of $CaCl_2$. The mixture prepared is sufficient for treating 100 kg of fertilizer; a strongly adhering layer is thus formed on the granules, completely preventing caking [526].

However, special treatment is required for each different type of fertilizer. Thus the spraying of potassium chloride particles with a 1% solution of $MgCl_2$ and $CaCl_2$ gave no positive results. In this case it was better to use a 0.01% solution of the acetates of primary, secondary, and tertiary amines ("armaks") with 16-19 carbon atoms in the radical [527].

Hydrophobizing properties may be imparted to fertilizer particles not only by modifying the surface of the final product but also in the course of fertilizer production. This method is used in the production of potassium chloride. The method is based on adding a 1% solution of an amine hydrochloride (of the fatty or aromatic series with 16-20 carbon atoms in the chain) to the KCl pulp before filtration (centrifuging).

Of all fertilizers the one caking most severely is ammonium nitrate. The reasons for this are various. According to some research workers [528], the cementing action is due to fine particles of $Cu(NO_3)_2$ and $CaCO_3$ in the reacting mixture which stick to the surface of the product granules; according to others [529-531], the caking is due to the fact that on solidifying (at 32.3°C the ammonium nitrate undergoes a modifying transformation as a result of which the bulk weight of the product is reduced by 2.5-3%), the proportion of fine particles increases, and the granules lose strength. The caking of the product may in this case be prevented by introducing calcium and magnesium nitrates to the extent of 1.5-5% of the weight of the product [529]. An analogous result is obtained on adding a small quantity of products obtained from the nitrate decomposition of phosphates (containing 1-2% phosphorus, referred to P_2O_5) [530] or barium compounds [531] to the ammonium nitrate.

On fusing ammonium nitrate and limestone, a calcium—ammonium nitrate is obtained. The caking of this nitrate (NH_4NO_3 to $CaCO_3$ ratio 60 : 40) falls by a factor of 2-3 (in relative caking indices) as compared with the pure ammonium nitrate [532]. The introduction of SAS reduces the surface tension of the saturated aqueous solution of ammonium nitrate, which may also prevent caking. Spraying fertilizers with a Sulfonol solution leads to analogous results [533].

The possibility of combating caking by a combination of autohesion-reducing methods also arises. A popular method is to modify the particle surface and simultaneously reduce autohesion by the addition of suitable powders to the fertilizers. Thus the introduction of a small quantity of polyacrylamide to a manure mixture reduces the hygroscopic capacity (i.e., reduces the capillary component) and also modifies the surface of the granules (reduces the molecular component), giving the product a free-running property and preventing caking [534].

On reducing the caking of ammonium nitrate by "paraffining" either with pure paraffin or a mixture of this with other substances (for example, vaseline, colophony, bitumen, resin, or asphalt), i.e., by hydrophobizing the surface of the granules, the tackiness of the particles is increased at the same time, which is undesirable. In order to avoid this, the paraffined granulated product is powdered with finely ground, water-insoluble inorganic additives (kaolin, ash, apatite flour, limestone, etc.) to the extent of 2–3% of the weight of the product [535].

Ammonium nitrate granules should also be sprayed with a mixture of alkylized aromatic hydrocarbons having several aliphatic side chains (waste matter from polyolefine production), simultaneously powdering with kieselguhr [536]. Coating KCl granules with organic compounds to the extent of 0.025% of the weight of the KCl (especially Aermoflos No. 12) makes the product free-running [537].

Ammonium nitrate granules and fertilizers based on these are coated with a layer consisting of sodium phosphates (trimetaphosphate and polyphosphate) and salts of alkylaryl sulfo-acids. These reagents may be added either both together to the ammonium nitrate melt or separately: the sodium phosphates to the melt and the sodium salts of the alkylaryl sulfo-acids to the granules. The granules with added polyphosphates are then powdered with diatomite [538].

Adhesion of Powdered Fertilizers. Powdered fertilizers may stick to the surfaces of the vessels containing them. In this case, the adhesion can only be estimated indirectly in terms of the coefficient of friction.

Below we present the coefficients of friction of several fertilizers as a function of the properties of the material [515] (Table XI.4).

We see from the data presented in Table XI.4 that granulation reduces the friction, and hence the adhesion of the fertilizer particles. The lowest coefficient of friction occurs for polyethylene.

Table XI.4. Coefficients of Friction of Several Fertilizers on Various Surfaces

Fertilizer	Coefficient of friction		
	on a painted surface	on a steel surface	on polyethylene
Potassium chloride.	0.51	0.50	0.35
Urea.	0.64	0.56	0.45
Ammonium nitrate	0.66	0.60	0.49
Superphosphate:			
powdered.	0.71	0.70	0.60
granulated	0.55	0.53	0.43

The humidity of the product affects the adhesion, and hence the coefficient of friction. Thus the coefficient of friction of powdered superphosphate increases as the humidity of the product rises to 20%; this is evidently due to the increasing autohesive forces. Further raising the humidity of the product (above 20%) reduces the coefficient of friction, owing to the formation of a layer of water between the adhering particles and the surface.

In the processes considered in this chapter adhesion and autohesion are of particular importance. However, these effects are often masked, and it is not always easy to see how they operate, as in the case of other processes (see Chapters III, IV, and VI). We cannot, for example, trace the changes in adhesion and autohesion as functions of the particle size, the properties and state of the surface, the properties of the medium, etc.

References

1. B. V. Deryagin and N. A. Krotova, Adhesion, Izd. Akad. Nauk SSSR (1949).
2. S. S. Voyutskii, Autohesion and Adhesion of High Polymers, Gostekhizdat (1960).
3. D. Douglas, Adhesion, London (1961).
4. N. A. De Bruyne and J. R. Houwink, Klebtechnik, Die Adhäsion in Theorie und Praxis, Stuttgart (1957).
5. Adhesion [Russian translation], N. De Bruyne and R. Houwink (eds.), IL (1954). [English edition: Adhesion and Adhesives, Van Nostrand, Princeton, New Jersey.]
6. N. A. Fuks, Mechanics of Aerosols, Izd. Akad. Nauk SSSR (1956).
7. N. A. Fuks, Advances in the Mechanics of Aerosols, Izd. Akad. Nauk SSSR (1961).
8. A. Buzagh, Ann. Univ. Sci. Budapest., Sect. Chim., 1:32 (1959).
9. Short Chemical Encyclopedia, Vol. 1, Izd. "Sovetskaya entsiklopediya" (1961).
10. A. D. Zimon and Yu. P. Petunin, Lakokrasochnye Materialy i ikh Primenenie, No. 2: 63 (1963).
11. A. D. Zimon, Lakokrasochnye Materialy i ikh Primenenie, No. 2:40 (1963).
12. G. I. Fuks, V. M. Klychnikov, and E. V. Tsyganova, Dokl. Akad. Nauk SSSR, 65(3):307 (1959).
13. B. V. Deryagin and A. D. Zimon, Kolloidn. Zh., 23(5):544 (1961).
14. A. Buzagh, Koll.-Z., 47:370 (1929); 51:105, 230 (1950); 52:46 (1930); 53:294 (1930); 79:1956 (1937).
15. G. I. Fuks, Scientific Work of Members of the D. I. Mendeleev All-Union Chemical Society, No. 3 (1949), p. 33.
16. M. C. Kordecki and C. Orr, Am. Med. Assoc. Arch. Ind. Health, 1(1) (1960).
17. A. N. Dinnik, Impact and Compression of Elastic Bodies, Selected Works, Vol. 1, Izd. Akad. Nauk UkrSSR, Kiev (1952).
18. E. Cremer, F. Conrad, and Th. Kraus, Angew Chem., 64(1):10 (1952).
19. W. Batel, Chim.-Ingr.-Tech., 31(5):343 (1959).
20. F. Patat and W. Schmid, Chem. Ind. Techn., 32(1):8 (1960).
21. A. D. Zimon, Kolloidn. Zh., 24(4):459 (1962).
22. Yu. I. Chicherin, Yu. V. Abrosimov, and L. T. Bespalova, Khim. Prom., No. 5:393 (1963).
23. A. Kozhemyakin, Avtomob. Transp., No. 10:22 (1961).
24. G. Thouzeau and T. Taylor, Res. Rept. Safety Mines Res. Establ., No. 197:28 (1961).
25. V. Mackrle and S. Mackrle, Adhese ve Filtračnim Loži, Praha (1959).
26. V. Mackrle, L'Etude du Phénoméne d'Adhérence, Colmatage dans le Milieu Poreux, Prague (1961).
27. M. Joly, J. Chim. Phys., 59(3):249 (1962).
28. E. F. Kurgaev, Kolloidn. Zh., 19(1): 72 (1957).

29. R. Korn, Chem. Zvesti, 16(9): 645 (1962).
30. G. I. Fuks and N. I. Nikolaeva, Dokl. Akad. Nauk SSSR, 153(2): 398 (1963).
31. E. D. Yakhnin and A. B. Taubman, Dokl. Akad. Nauk SSSR, 155(1): 179 (1964).
32. F. Bowden and D. Tabor, The Friction and Lubrication of Solids, Oxford (1950).
33. G. I. Fuks, in collection: Research on Surface Forces, Izd. "Nauka" (1964), p. 176.
34. B. V. Deryagin and V. P. Lazarev, Kolloidn. Zh., 1(4): 293 (1935).
35. B. V. Deryagin, Zh. Fiz. Khim., 5(9): 1165 (1934).
36. V. P. Lazarev and B. V. Deryagin, Transactions of the Second All-Union Conference on Friction and Wear in Machines, Vol. 3, Izd. Akad. Nauk SSSR (1949), p. 106.
37. M. P. Volarovich, D. M. Tolstoi, and A. K. Pukhlyanko, Kolloidn. Zh., 3(7): 621 (1937).
38. A. S. Akhmatov, Research on Surface Forces, Izd. Akad. Nauk SSSR (1961), p. 93.
39. A. S. Akhmatov, Molecular Physics of Boundary Friction, Fizmatgiz (1963).
40. B. V. Deryagin, Yu. P. Toporov, and M. D. Futran, Transactions of the Third All-Union Conference on Friction and Wear in Machines, Vol. 2, Izd. Akad. Nauk SSSR (1960), p. 152.
41. E. A. Moelwyn-Hughes, Physical Chemistry [Russian translation], IL (1962). [English edition: Cambridge University Press.]
42. O. K. Davtyan, Quantum Chemistry, Izd. "Vysshaya Shkola" (1962).
43. R. S. Bradley, Phil. Mag., 13(86): 853 (1932); Trans. Faraday Soc., 32(8): 1088 (1936).
44. H. C. Hamaker, Physica, 4(10): 1058 (1937).
45. A. D. Zimon and B. V. Deryagin, Kolloidn. Zh., 25(2): 159 (1963).
46. G. I. Fuks and N. I. Kaverina, Kolloidn. Zh., 21(6): 718 (1959).
47. J. H. De Boer, Trans. Faraday Soc., 32(1): 10 (1936).
48. P. L. Howe, D. P. Benton, and J. E. Puddington, Can. J. Chem., 33(9): 1375 (1955).
49. J. T. Overbeek and E. J. Verwey, Theory of the Stability of Lyophobic Colloids, New York (1948).
50. H. Casimir and D. Polder, Phys. Rev., 73(1): 36 (1948).
51. J. S. Courtney-Pratt, Proc. Roy. Soc., A202: 505 (1950).
52. B. V. Deryagin and I. I. Abrikosova, Soviet Phys. JETP, 4(1): 2 (1957).
53. I. I. Abrikosova, Research on Forces of Molecular Attraction Between Solids, Master's Thesis, Inst. Fiz. Khim., Akad. Nauk SSSR (1955).
54. M. M. Dubinin, Izv. Akad. Nauk SSSR, No. 10: 1739 (1960).
55. E. M. Lifshits, Dokl. Akad. Nauk SSSR, 97(4): 643 (1954); Zh. Éksperim. i Teor. Fiz., 29(1): 7 (1955).
56. B. V. Deryagin, I. I. Abrikosova, and E. M. Lifshits, Usp. Fiz. Nauk, 64(3): 493 (1958).
57. B. V. Deryagin, Zh. Fiz. Khim., 6(10): 1306 (1935).
58. V. V. Demchenko, Zh. Fiz. Khim., 36(7): 1544 (1962).
59. G. I. Fuks and E. V. Tsyganova, Collections of Papers on Oil Technology, No. 1, Gostoptekhizdat (1968), p. 266.
60. I. N. Putilova, Handbook for Colloidal Chemistry Practice, Izd. "Vysshaya Shkola" (1961).

61. H. Pecht, Chem.-Ingr.-Tech., 33(10): 691 (1961).
62. C. N. Davies, M. Aylward, and D. Leacey, Arch. Ind. Hyg. Occupat. Med., 4:354 (1951).
63. J. W. Beams, 43rd Annual Technical Proceedings of the American Electroplater's Society, Vol. 1 (1956); J. W. Beams, J. B. Breazeale, and W. L. Bart, Phys. Rev., 100(6):1657 (1955).
64. G. Böhme, H. Krupp, H. Rabenhorst, and G. Sandstede, Trans. Inst. Chem. Engrs. (London), 40(5): 252 (1962).
65. H. Krupp, G. Sandstede, and K. Schramm, Chem.-Ingr.-Tech., 32(2): 99 (1960).
66. B. B. Morgan, Monthly Bull. Brit. Coal Utilis. Res. Assoc., 25(4): 125 (1961).
67. A. D. Zimon and T. S. Volkova, Kolloidn. Zh., 27(4): 529 (1965).
68. S. Moses, Ind. Eng. Chem., 41:2339 (1949).
69. R. I. Larsen, Am. Ind. Hyg. Assoc. J., 19(4): 265 (1958).
70. A. D. Zimon, N. I. Dovnar, G. A. Belkina, and G. V. Nozdrina, in collection: Research on Surface Forces, Izd. "Nauka" (1964), p. 330.
71. V. N. Glazanov, On the Electrostatic Enrichment of Coal Fragments, Ugletekhizdat (1950).
72. B. V. Deryagin, Yu. P. Toporov, and I. N. Aleinikova, Kolloidn. Zh., 26(3): 394 (1964).
73. G. Bohme, W. Kling, H. Krupp, H. Lange, and G. Sandstede, Z. Angew. Phys., 16(6): 486 (1964).
74. T. M. Lowes, Fuel Soc. J., Univ. Sheffield, 16: 35 (1965).
75. B. V. Deryagin, Yu. P. Toporov, I. N. Tomfel'd, I. N. Aleinikova, and B. N. Parfanovich, Lakokrasochnye Materialy i ikh Primenenie, No. 4: 62 (1964).
76. A. D. Zimon and N. G. Klishin, Kolloidn. Zh., 27(5): 685 (1965).
77. J. Burson, K. Linoya, and C. Orr, Nature, 200 (4904):360 (1963).
78. S. A. Tauraitene, N. M. Gal'vidis, K. P. Strazdas, and A. S. Tauraitis, Zh. Nauchn. i Prikl. Fotogr. i Kinematogr., 8(6): 267 (1963).
79. W. Stone, Phil. Mag., 9(58): 610 (1930).
80. M. Corn, J. Air Pollution Control Assoc., 11(11: 523 (1961); 11(12): 566, 584 (1961).
81. J. McFarlane and D. Tabor, Proc. Roy. Soc., A220(1069): 224 (1950).
82. D. Beischer, Kolloid-Z., 89(2): 214 (1939).
83. Zh. Tekenev, Interaction of Dust Particles from an Air Flow with the Surface of an Obstacle, Master's Thesis, Tashkent (1962).
84. Zh. Tekenov and B. Mazitov, in collection: Some Questions of Applied Physics, Izd. Akad. Nauk UzSSR, Tashkent (1961), p. 11.
85. G. Tomlinson, Phil. Mag., 6: 695 (1928).
86. A. D. Malkina and B. V. Deryagin, Kolloidn. Zh., 12(6): 431 (1950).
87. T. N. Voropaeva, B. V. Deryagin, and E. N. Kabanov, in collection: Research on Surface Forces, Izd. Akad. Nauk SSSR (1961), p. 143.
88. G. I. Fuks, Transactions of the Third All-Union Conference on Colloid Chemistry, Izd. Akad. Nauk SSSR (1956), p. 301.
89. J. T. Overbeek and M. J. Sparnay, J. Colloid. Sci., 7: 343 (1952); Faraday Society General Discussion on Coagulation and Flocculation, 18: 12 (1954).
90. A. D. Zimon, Lakokrasochnye Materialy i ikh Primenenie, No. 6: 57 (1961).

91. H. Witzmann, Staub, 22(1): 2 (1962).
92. A. D. Zimon and V. I. Kovrigo, Kolloidn. Zh., 28(5): 656 (1966).
93. H. Arnold, Phil. Mag., 22(6): 755 (1911).
94. H. Liebster, Ann. Physik, 82(4): 541 (1927).
95. J. Schmiedel, Phys. Z., 29: 593 (1928).
96. W. B. Dandliker, J. Am. Chem. Soc., 72: 5110 (1950).
97. P. H. Scott, G. C. Clark, and C. M. Sliepcevich, J. Phys. Chem., 59: 849 (1955).
98. P. H. Scott and S. W. Churchill, J. Phys. Chem., 62: 1300 (1958).
99. A. Westgren, Ann. Physik, 52(4): 308 (1917).
100. H. S. Allen, Phil. Mag., 50(5): 323 (1900).
101. J. Zeleny and L. W. McKeehan, Phys. Z., 11: 78 (1910).
102. E. J. Baier, Am. Ind. Hyg. Assoc. J., 22(4): 307 (1961).
103. V. N. Favstova and I. N. Vlodavets, Kolloidn. Zh., 17(6): 456 (1955).
104. Yu. O. Ran'ko, Kolloidn. Zh., 10(1): 42 (1948).
105. V. S. Nikitin and V. I. Usadkov, Author's Certificate, USSR 115491, Feb. 29, 1958.
106. J. W. Phillips, D. Y. Thomas, and W. H. Walton, U.S. Patent 2948469, August 9, 1960.
107. Zh. Tekenov, Dokl. Akad. Nauk UzbSSR, No. 2: 17 (1962).
108. V. I. Klassen, V. A. Nevskaya, and N. S. Vlasova, Ugol', No. 7: 41 (1961).
109. Tovohiko and Norio, Japan Scientific Laboratory Journal, 34(11) (1958); cited in Staub, 19(9): 341 (1959).
110. H. P. Dibbs, J. Appl. Chem., 10(9): 372 (1960).
111. P. Connor and W. H. Hardwick, Ind. Chemist., 36: 427 (1960); Intern. J. Air Pollution, 3(1-3): 427 (1960).
112. R. Challande, Comptes Rend., 237: 35 (1953).
113. H. Born and K Zimmer, Naturwiss., 28: 447 (1940); Die Gasmaske, 12: 25 (1940).
114. A. Siemaszko and R. Plejewski, Kernenergie, 6(10): 561 (1963).
115. N. S. Smirnov, in collection: Aerosols and Their Use, Izd. Minsel'khoza SSSR (1959), p. 72.
116. M. A. Konstantinova-Shlezinger (ed.), Luminescence Analysis, Fizmatgiz (1961).
117. N. F. Dergachev, Izv. Vses. Politekhn. Inst., No. 6: 7 (1949).
118. G. I. Fuks, Study and Classification of the Effectiveness of Gas Flame Treatment, NIIChASPROM (1963).
119. V. V. Karasev and G. I. Izmailova, Zh. Tekhn. Fiz., 24(5): 871 (1954).
120. I. I. Zhukov, Colloid Chemistry, Izd. LGU (1949).
121. A. Z. Baibakov, Bulletin of Scientific and Technical Information on Agricultural Physics, No. 3, 4, Agrofizicheskii NII (1957).
122. V. I. Lygin, N. V. Kovaleva, N. N. Kavtaradze, and A. V. Kiselev, Kolloidn. Zh., 22(3): 334 (1960).
123. A. V. Kiselev, A. Ya. Korolev, R. S. Petrova, and K. D. Shcherbakova, Kolloidn. Zh., 22(6): 671 (1960).
124. I. Yu. Babkin, A. V. Kiselev, and A. Ya. Korolev, Dokl. Akad. Nauk SSSR, 136(2): 373 (1961).
125. L. M. Vinogradova and A. Ya. Korolev, Fifth All-Union Conference on Colloid Chemistry, Summaries of Contributions, Izd. Akad. Nauk SSSR (1962), p. 82.

126. A. Ya. Korolev, in collection: Glues and Glue Technology, Oborongiz (1962), p. 35.
127. A. Ya. Korolev, P. V. Davydov, and A. M. Vinogradova, in collection: Adhesion of Polymers, Izd. Akad. Nauk SSSR (1963), p. 3; L. M. Vinogradova, A. Ya. Korolev, P. V. Davydov, and R. V. Kuchenkova, Ibid., p. 137.
128. D. W. Jordan, Brit. J. Appl. Phys., Suppl., No. 3:194 (1954).
129. K. A. Andrianov, B. V. Deryagin, N. N. Zakhavaeva, M. V. Sobolevskii, and M. V. Talaev, Zh. Prikl. Khim., 32(12):2682 (1959).
130. M. G. Voronkov and B. N. Dolgov, Priroda, No. 5:22 (1954).
131. N. A. Krotova and L. P. Morozova, in collection: Research on Surface Forces, Izd. Akad. Nauk SSSR (1961), p. 83; Kolloidn. Zh., 24(4):473 (1962).
132. A. V. Kiselev, N. V. Kovaleva, A. Ya. Korolev, and K. D. Shcherbakova, Dokl. Akad. Nauk SSSR, 124(3):617 (1959).
133. Yu. M. Luzhnov, in collection: Research on Surface Forces, Izd. "Nauka" (1964), p. 357.
134. J. McFarlane and D. Tabor, Seventh International Congr. Appl. Mech., London 4:335 (1948).
135. G. I. Fuks and L. V. Timofeeva, in collection: Clocks and Clock Mechanisms, No. 5 (1961), NIIChASPROM, p. 57.
136. M. S. Sharlovskaya, Izv. Sibirsk. Otd. Akad. Nauk SSSR, No. 5:42 (1962).
137. W. R. Harper, Proc. Roy. Soc., A231(1186):388 (1955).
138. E. F. Shpol'skii, Atomic Physics, Tekhteoretizdat (1951).
139. W. C. Dunlap, Introduction to the Physics of Semiconductors [Russian translation], IL (1959). [English edition: Introduction to Semiconductors, Wiley, New York (1957).]
140. M. M. Bredov and I. Z. Kshemyanskaya, Zh. Tekhn. Fiz., 27(5):921 (1957).
141. V. P. Smilga, in collection: Research on Surface Forces, Izd. Akad. Nauk SSSR (1961), p. 76.
142. N. A. Krotova, L. P. Morozova, et al., All-Union Conference on Colloid Chemistry, Summaries of Contributions, Izd. Akad. Nauk SSSR (1962), p. 84.
143. E. M. Balabanov, Physical Bases of the Electrical Separation of Materials by Means of a Corona Discharge, Master's Thesis, Mosk. Gos. Univ. (1945).
144. E. M. Balabanov, Dispersed Systems in the Electric Field of a Corona Discharge, Doctoral Dissertation, P. N. Lebedev Physics Institute (1953).
145. W. B. Kunkel, J. Appl. Phys., 21(8):820 (1950).
146. W. Jutzi, Beitrag zur Kenntnis der Elektrischen Eigenschaften Grobkolloider Aerosole, Dusseldorf (1954).
147. E. Bodenstedt, Z. Angew. Phys., 6(7):297 (1954).
148. V. S. Bogdanov, Izv. Akad. Nauk SSSR, Otd. Khim. Nauk, No. 8:1520 (1961).
149. A. D. Zimon, Kolloidn. Zh., 25(3):317 (1963).
150. B. V. Deryagin, Transactions of the Third All-Union Conference on Colloid Chemistry, Izd. Akad. Nauk SSSR (1956), p. 301.
151. B. V. Deryagin, Report presented at the Symposium on Powders in Industry held in the Royal Institution, London, September, 1960.
152. H. Reumuth, Staub, 22(8):301 (1962).
153. R. Meldau, Handbuch der Staubtechnik, Vol. 1, Dusseldorf (1956).

154. Nature, 192(4801): 400 (1961) (a review).
155. D. Kothari, Nuclear Explosions [Russian translation], IL (1958). [English edition: Nuclear Explosions and Their Effect, W. S. Heinman, New York.]
156. Chemistry, 31(2):41 (1958) (a review).
157. V. Nishivaki, Atomic Sci., No. 4: 279 (1955).
158. K. Spurnyi, Ch. Iekh, B. Sedlachek, and O. Shtorkh, Aerosols, Atomizdat (1964).
159. A. D. Zimon and T. S. Volkova, Kolloidn. Zh., 27(3): 365 (1965).
160. P. I. Bryant, Trans. Ninth Nat. Vacuum Sympos. Am. Vacuum Soc., Los Angeles, Calif., 1962, New York (1962), p. 311.
161. H. Meerbach, Staub, 22(7): 260 (1962).
162. V. N. Uzhov, Combating Dust in Industry, Goskhimizdat (1962).
163. V. I. Bekoryukov and I. A. Karol', in collection: Questions of Nuclear Meteorology, Gosatomizdat (1962), pp. 221, 249.
164. G. I. Fuks, Effect of the Composition of the Boundary Layers on Coagulation and Frictional Interactions and the Improvement of Lubricants, Abstract, Inst. Fiz. Khim. Akad. Nauk SSSR (1965).
165. O. Brandt, Koll.-Z., 85(1): 24 (1938).
166. W. Fett, Atmospheric Dust [Russian translation], IL (1961).
167. D. Fiat, M. Folmann, and U. Garbatski, Proc. Roy. Soc., A260(1302): 409 (1961).
168. E. R. Mitchell and G. K. Lee, Can. Mining and Metallurg. Bull., 55(607): 768 (1962).
169. G. I. Fuks and V. M. Klychnikov, Transactions of the All-Union Scientific-Research Institute of Fertilizers, Agricultural Technology, and Soil Science (VIUAA), No. 28 (1948), p. 215; (1949), pp. 186, 215.
170. A. D. Zimon and B. V. Deryagin, Fifth All-Union Conference on Colloid Chemistry, Summaries of Contributions, Izd. Akad. Nauk SSSR (1962), p. 37.
171. P. A. Rebinder and A. A. Trapeznikov, Dokl. Akad. Nauk SSSR, 18(7): 421; 18(3): 185; 18(7): 425 (1938).
172. B. V. Deryagin and V. V. Karasev, Dokl. Akad. Nauk SSSR, 62(6): 761 (1948); 101(2): 289 (1955).
173. B. V. Deryagin and E. V. Obukhov, Kolloidn. Zh., 1:385 (1935).
174. B. V. Deryagin and M. M. Kusakov, Izv. Akad. Nauk SSSR, Ser. Khim., No. 5: 741 (1936).
175. M. M. Kusakov and A. S. Tatievskaya, Dokl. Akad. Nauk SSSR, 82(4): 333 (1952).
176. B. V. Deryagin, Transactions of the Third All-Union Conference on Colloid Chemistry, Izd. AN SSSR (1956), p. 225.
177. B. V. Deryagin, Kolloidn. Zh., 17(3): 207 (1955).
178. B. V. Deryagin and A. S. Tatievskaya, Dokl. Akad. Nauk SSSR, 89(6): 1041 (1953).
179. A. D. Sheludko, Certain Properties of Thin Layers of Liquid, Doctoral Dissertation, Inst. Fiz. Khim. Akad. Nauk SSSR (1961).
180. V. S. Istomina, Kolloidn. Zh., 12(4): 279 (1950).
181. Ya. I. Frenkel', Kinetic Theory of Liquids, Izd. Akad. Nauk SSSR (1945).
182. A. D. Sheludko, Dokl. Akad. Nauk SSSR, 123(6): 1074 (1958).
183. B. V. Deryagin and M. M. Kusakov, Zh. Fiz. Khim., 26(10): 1536 (1952).

184. B. V. Deryagin and A. S. Tatievskaya, Kolloidn. Zh., 15(6):416 (1953).
185. G. A. Korchinskii, Kolloidn. Zh., 19(3):307 (1957).
186. S. V. Nerpin, Hydromechanics of Thin Layers, Doctoral Dissertation, Inst. Fiz. Khim. Akad. Nauk SSSR (1956).
187. B. V. Deryagin and L. D. Landau, Zh. Eksperim. i Teor. Fiz., 2:802 (1941); 15:662 (1945).
188. B. V. Deryagin and V. G. Levich, Dokl. Akad. Nauk SSSR, 98(6):985 (1954); B. V. Deryagin, Kolloidn. Zh., 16(6):425 (1954).
189. B. V. Deryagin, Izv. Akad. Nauk SSSR, Ser. Khim., No. 1:1153 (1937).
190. A. Buzagh and Z. Szabo, Koll.-Z., 83:139 (1938).
191. A. Buzagh and K. Dux, Koll.-Z., 84:279 (1938).
192. A. Buzagh, Koll.-Z., 85:329 (1938).
193. A. Buzagh and J. Zimmerman, Koll.-Z., 84:16 (1938).
194. A. Buzagh, Acta Chim. Hung., 1:181 (1950).
195. A. Buzagh, Koll.-Z., 125:14 (1951).
196. G. I. Fuks, Kolloidn. Zh., 12(3):228 (1950).
197. A. Taylor and F. Wood, Trans. Faraday Soc., 53(4):523 (1956).
198. S. A. Nikitina and A. B. Taubmann, Dokl. Akad. Nauk SSSR, 116(1):113 (1957).
199. J. L. Rabate, Peintures, Pigments, Vernis, 32(10):852 (1956); J. L. Rabate, Effect of the Contamination of the Atmosphere in Various Regions on Accelerating the Aging of Paint Coatings [Russian translation from the French], GlavNIIproekt pri Gosplane SSSR, VNIIZh, Leningrad (1960).
200. W. Burkhart, Automob. Rev., 35(11):32 (1960).
201. Lacquers, Paints, and Auxiliary Materials, Parts I and II, Standartgiz (1963).
202. I. I. Khain, Zh. Prikl. Khim., 33(7):1526 (1960).
203. S. N. Kazarnovskii and G. M. Skalina, Transactions of the All-Union Scientific-Research Institute of Railroad Transport, No. 208:25 (1961).
204. A. Ya. Korolev, Perchlorvinyl Enamels, Gostekhizdat (1948).
205. G. Champetier and G. Rabate, Chemistry of Lacquers, Paints, and Pigments, Vol. I, Vol. II [Russian translation], Goskhimizdat (1960), (1962).
206. G. Kittel', Cellulose Lacquers, Goskhimizdat (1957).
207. B. M. Malyshev, Vest. Mosk. Gos. Univ., Ser. Fiz.-Mat. i Est. Nauk, No. 3:3 (1952).
208. T. Gillespie, J. Colloid Sci., 10(3):266 (1955).
209. T. Gillespie and E. Rideal, J. Colloid Sci., 10(3):281 (1955).
210. S. V. Yakubovich, Testing Painting Material and Paint Coatings, Goskhimizdat (1952).
211. B. V. Lyubimov, Special Paint Coatings in Engineering, Mashgiz (1959).
212. M. Richter, Wiss. Fortschr., 7(11):390 (1957).
213. W. F. Libby, Atomics, 9(7):236, 245 (1958); Bull. Atomic Scientists, 14(1):27 (1958).
214. C. H. Junge, Unschau, 60(1):8 (1960); H. Flohn, Wasser, Luft, und Betrieb, 2(7):170 (1958).
215. E. Spode, Technik, 12(10):696 (1957); C. Junge, Artificial Stimulation of Rain, London (1957), p. 3.
216. D. Doran, J. Roy. Aeron. Soc., 66(621):565 (1962).

217. F. Wilckens, Soldat u Techn., 2(6):292 (1959).
218. E. K. Fedorov (ed.), Meteorology and Atomic Power [Russian translation], IL(1959).
219. R. J. Eichelberger and J. W. Cechring, ARS Journal, 32(10):1583 (1956).
220. A. B. Taubmann and S. A. Nikitina, Dokl. Akad. Nauk SSSR, 110(4):600; 110(5):816 (1956).
221. P. I. Ermilov, Izv. Vysshikh Uchebn. Zavedenii, Khim. i Khim. Tekhnol., 2(1):134 (1959).
222. P. E. Tucker and T. W. Watson, Paint Manufact., 28(1): 12 (1958).
223. H. Reumuth, Seifen-Oele-Fette-Wachse, 84:811 (1958); 85:5 (1959).
224. A. D. Zimon, Kolloidn. Zh., 27(2):193 (1965).
225. N. L. Cross and R. G. Picknett, Mechanism Corros. Fuel Impurities, Butterworths, London (1963), p. 383.
226. S. N.Kazarnovskii, Transactions of the All-Union Scientific-Research Institute of Railroad Transport, No. 208:159 (1961).
227. R. Smith-Johansen, U. S. Patent 2617269.
228. N. P. Myagkov, Colloid-Chemical and Electrochemical Properties of Nonmetallic Coatings and Their Protective Action, Master's Thesis, Rizhsk. Politekhn. Inst. (1961).
229. T. Steudel, Glasers Ann., 87(2):37 (1963).
230. K. A. Andrianov, D. Ya. Zhinkin, and A. F. Moiseev, Khim. Prom., No. 2:106 (1959).
231. Silicone Enamels, Izd. Akad. Nauk SSSR, Exhibition of Achievements of the National Economy of the USSR (1964).
232. K. I. Karasev and Z. V. Kurochkina, Stroit. Materialy, No. 3:32 (1962).
233. R. N. Thomas, Paint Ind. Mag., 76(3):10 (1961).
234. A. V. Sokolov, Lakokrasochnye Materialy i ikh Premenenie, No. 3:79 (1960).
235. Ya. L. Raskin and R. M. Livshits, Lakokrasochnye Materialy i ikh Primenenie, No. 5:33 (1960).
236. E. V. Iskra, Koks i Khim., No. 3:47 (1957).
237. T. A. Ermolaeva, Lakokrasochnye Materialy i ikh Primenenie, No. 5:46 (1961).
238. M. A. Shtern, Khim. Nauka i Promy., 4(5): 641 (1959).
239. A. A. Trapeznikov and M. A. Chupeev, Lakokrasochnye Materialy i ikh Primenenie, No. 5:17 (1962).
240. W. Krauss, Farbe Lack, 66(10): 575 (1960).
241. T. R. Santelli, U.S. Patent 2982672, May 2, 1961.
242. E. S. Gurevich and A. M. Frost, Lakokrasochnye Materialy i ikh Primenenie, No. 3:11 (1961).
243. K. I. Karasev, Stroit. Materialy, No. 9:24 (1961).
244. Structural Materials, Parts, and Equipment [Russian translation], No. 5:30 (1961); No. 4:4; No. 7:5; No. 7:19 (1962).
245. D. S. Yakubovich, Lakokrasochnye Materialy i ikh Primenenie, No. 2: 73 (1963).
246. E. I. Gambardella, Lakokrasochnye Materialy i ikh Primenenie, No. 6:27 (1960).
247. R. Prakash, Paint India, 11(8): 25 (1961).
248. A. D. Yakovlev, Obtaining Coatings from Powdered Resins in the Suspended State, Leningrad Otd. Obshch. po Raspr. Polit. i Nauchn. Znanii RSFSR, Seriya "Sinteticheskie Materialy," No. 1 (1961).

249. Method of Painting with Dry Painting Materials (Abstract), Inform. Byull. o Zarubezhn. Khim. Prom. NIIPEkhim., No. 3:38 (1960).
250. Minami Etsu and Okada Itaka, Japanese Patent 15536, September 6, 1961; Journal of Abstracts: Chemistry (1963), 11T189P.
251. A. A. Sinegub-Lavrenko and R. V. Katseva, Plastmassy, No. 1:37 (1963).
252. P. Gueriaux, French Patent 1197122, November 27, 1959.
253. D. R. Davis, Plastics Technol., 8(6):37 (1962).
254. P. A. Polonik, Electrification of Chemical Fibers and Ways of Overcoming It, Gizlegprom (1959).
255. K. Karasev and G. Lopatin, Sel'skoe Stroitel'stvo, No. 11:13 (1962).
256. Technical Bulletin Armour Chemical Division, L 111 3 (1956).
257. E. R. Frederick, Chem. Eng., 68:13, 107 (1961).
258. A. V. Shubnikov, I. S. Zheludev, V. P. Konstantinova, and I. M. Sel'vestrova, Study of Piezoelectric Textures, Izd. Akad. Nauk SSSR (1955).
259. G. A. Lushcheikin, V. E. Gul', and B. A. Dogadkin, Kolloidn. Zh., 25(3):334 (1963).
260. P. I. Vasserman, Ya. M. Kolotyrkin, V. V. Chebotarevskii, and A. A. Feoktistova, Kolloidn. Zh., 21(4):392 (1959).
261. V. N. Lofitskii and B. F. Rel'tov, Combating the Adhesion of Soil to the Bodies of Dump Trucks and the Scoops of Excavators, Vses. Nauchn.-Issled. Inst. Gidrotekhniki im. B. E. Vedeneeva (1953).
262. V. P. Bardina, L. A. Suprun, and P. S. Shcherbakov, Lakokrasochnye Materialy i ikh Primenenie, No. 4:38 (1961).
263. G. Nottes and H. Sasse, West German Patent 1093168, May 10, 1961.
264. East German Patent 626379.
265. I. Ya. Klinov, V. K. Khnyazev, and N. P. Aleksakhin, Lakokrasochnye Materialy i ikh Primenenie, No. 3:46 (1963).
266. H. Dreiheller and E. Graul, Atompraxis, 4:177 (1958).
267. A. D. Zimon, Kolloidn. Zh., 28(3):394 (1966).
268. N. N. Neprimerov, Experimental Study of Some Aspects of the Extraction of Paraffin-Containing Petroleum Products, Izd. Kazan. Univ. (1958).
269. P. P. Galonskii, Combating Paraffin in Oil Productions, Gostoptekhizdat (1955).
270. S. F. Lyushin, V. A. Rasskazov, D. M. Sheikh-Ali, R. R. Insanova, and E. P. Lipkov, Combating Paraffin Deposits in the Production of Petroleum, Gostekhizdat (1961).
271. V. A. Rasskazov, A. A. Gonik, and S. F. Lyushin, Preventing the Deposition of Paraffin in Oil Production by Using Paint Coatings, Ufa (1962).
272. R. J. Cole and F. W. Lassen, Oil and Gas. J., 58(9) (1960).
273. C. F. Parks, Oil Gas J., 58(4) (1960).
274. R. A. Abdullin, Lakokrasochnye Materialy i ikh Primenenie, No. 4:41 (1963).
275. V. F. Nezhevenko, V. M. Grigor'ev, and V. I. Kudentsov, Use of Paint Coatings for Protecting Oil Equipment from Paraffin Deposits and Corrosion, Byull. Kuibyshev. Covnarkhoza, No. 7, Kuibyshev (1962), p. 28.
276. H. Anders, Farbe Lack, 66(7):393 (1960).
277. I. T. Matov and B. B. Sushkov, Morsk. Flot, No. 8:4 (1959).
278. Sudostroenie, No. 10:74 (1961).

279. V. A. Bershtein, I. N. Kasheev, E. Sh. Ryt, and S. T. Tsodikova, Sudostroenie, No. 4:41 (1962).
280. Corrosion, Prevent. Control, 9(7):29 (1962).
281. H. Benninghoff, Farbe Lack, 66(10):583 (1960).
282. E. S. Gurevich and A. M. Frost, Lakokrasochnye Materialy i ikh Primenenie, No. 2:42 (1962).
283. V. G. Levich, Physicochemical Hydrodynamics, Fizmatgiz (1959).
284. G. Schlichting, Theory of the Boundary Layer [Russian translation], IL (1956). [English edition: Boundary Layer Theory, Pergamon, New York.]
285. V. N. Goncharov, Uniform Turbulent Flow, Gosenergoizdat (1951).
286. S. N. Syrkin, Conditions of Pneumatic Transport along Tubes, Tsent. Kotlotrub. Inst. (1935).
287. H. Rumpf, Chem.-Ingr.-Tech., Nos. 5-6:317 (1953).
288. V. I. Kravets and I. A. Ryzhenko, Izv. Vysshikh. Uchebn. Zavedenii, Gorn. Zh., No. 4:63 (1962).
289. F. V. Tideswell and R. Hattersley, Twenty-Second Annual Report of the Safety in Mines Research Board, London (1943).
290. I. P. Petrov, Izv. Vysshikh. Uchebn. Zavedenii, Gorn. Zh., No. 4:92 (1964).
291. K. V. Kochnev, V. V. D'yakov, and V. I. Kovalev, Collection of Papers on Silicosis, No. III, Ural. Filial Akad. Nauk, Sverdlovsk (1961), p. 119.
292. V. I. Kovalev, Collection of Scientific Papers of the All-Union Central Trade-Union Council Institutes, No. 5, Profizdat (1961).
293. A. D. Klimanov, in collection: Scientific Papers on Mining Questions, No. 16, Moscow Mining Institute (1956), p. 101.
294. S. Ya. Kheifits and A. S. Burchakov, Reducing the Dustiness of the Air in the Long Working Faces of Coal Mines, Ugletekhizdat (1958).
295. E. V. Donat and V. V. D'yakov, in collection: Industrial Ventilation, Vses. Nauchn.-Issled. Inst. Okhrany Truda, Sverdlovsk (1957), p. 91.
296. D. B. Bekirbaev, G. S. Grodel', et al., Combating Coal and Rock Dust in Mines, Gosgortekhizdat (1959).
297. D. A. Haal, College Canadian, 191(4927):129 (1955).
298. N. A. Kil'chevskii, Theory of the Collision of Solid Bodies, Gostekhizdat (1949).
299. L. A. Galin, Contact Problems of the Theory of Elasticity, Tekhteoretizdat (1953).
300. M. A. Gol'dshtik and A.K. Leont'ev, Inzh.-Fiz. Zh., 3(11):83 (1960).
301. W. Ranz, G. Talandis, and B. Gutterman, A. Ch. E. I. J., 6:124 (1960).
302. L. M. Levin, Research on the Physics of Coarsely Dispersed Aerosols, Izd. Akad. Nauk SSSR (1961).
303. I. A. Ryzhenko, Izv. Vysshikh. Uchebn. Zavedenii, Gorn. Zh., No. 11:121 (1960).
304. I. A. Ryzhenko, A Study of the Parameters of the Ventilating Draughts Characterizing the Carrying Away of Dust from Mine Workings, Master's Thesis, Institut Teploenergetiki Akad. Nauk UkrSSR, Kiev (1960).
305. V. N. Voronin, Combating Silicosis, Vol. 1, Izd. Akad. Nauk SSSR (1953), p. 47.
306. F. Herning, Gas-Wasserfach, No. 9:127 (1951).
307. I. A. Ryzhenko and A. I. Shcherbina, Dokl. Akad. Nauk UkrSSR, No. 2:197 (1961).

308. Ya. D. Averbukh, On Retaining Dust Particles on the Surfaces on Which They Have Settled under Conditions of Turbulent Flow, Master's Thesis, Ural Industrial. Institute im. S. M. Kirov, Sverdlovsk (1946).
309. V. N. Voronin, L. D. Voronina, and A. D. Bagrinovskii, Handbook on the Design and Practice of Dust-Preventing Air Conditioning in Metallurgical Mines, Gosgortekhizdat (1960).
310. S. S. Yankovskii, Dust Trapping and Purification of Gases in Nonferrous Metallurgy, Collection of Scientific Transactions of the Gintsvetmet, No. 20, Metallurgizdat (1963), p. 30.
311. Ch. M. Dzhuvarli and M. M. Dzhafarova, Izv. Akad. Nauk AzSSR, Ser. Fiz.-Mat. i Tekhn. Nauk, No. 2:65 (1961).
312. F. Albrecht, Phys. Z., No. 32:48 (1931).
313. W. Sell, Gebiete Ing. Forsch., No. 347:1 (1931).
314. W. E. Ranz and J. B. Wong, Ind. Eng. Chem., 44:1371 (1952); J. Appl. Phys., 26(2) (1955).
315. G. Asset and D. Pury, Arch. Ind. Hyg., 9(4):273 (1954).
316. L. V. Radushkevich and V. A. Kolganov, Izv. Akad. Nauk SSSR, Otd. Khim. Nauk, No. 1:23 (1962).
317. A. G. Amelin and M. I. Belyakov, in collection: Aerosols and Their Use, Izd. Minsel'khoza SSSR (1959), p. 53.
318. V. I. Ignat'ev and N. I Zverev, Teploenergetika, No. 3:36 (1958).
319. V. V. D'yakov, A. I. Korzon, and Yu. A. Chudov, Raising the Efficiency of the Dust-Reducing Air Conditioning in the Ural Mines, Allowing for the Natural Self-Purification of the Mine Air, Vses. Nauchn.-Issled. Inst. Okhrany Truda, Sverdlovsk (1963).
320. J. Rosinski, C. Nagamoto, and A. Ungar, Kolloid-Z., 164(1):26 (1956).
321. H. Reumuth, Glasers Ann., No. 9:365 (1962).
322. S. S. Grigoryan, Prikl. Mat. i Mekh., 22(5):577 (1958).
323. L. Loeb, Static Electrification, Gosenergoizdat (1963).
324. W. G. Kunkel, J. Appl. Phys., 21(8):833 (1950).
325. W. A. Runge, Phil. Mag., 24:853 (1912); 25:481 (1913).
326. D. A. Thomas, Brit. J. Appl. Phys., Suppl., 2:55 (1953).
327. R. Walther and W. Franke, Braunkohle, 28:789 (1929).
328. Chao Tse-san, Vestn. Mosk. Gos. Univ., Ser. 4, Geologiya, No. 2:61 (1963).
329. Chao Tse-san, Questions of Engineering Geology and Soil Science, Izd. Mosk. Gos. Univ. (1963), p. 232.
330. M. A. Velikanov, Dynamics of River Flows, Vol. 1, Gostekhizdat (1954).
331. V. N. Goncharov, Dynamics of River Flows, Didrometioizdat (1962).
332. S. Kh. Abal'yants, Tr. Sredneaz. Nauchn.-Issled. Inst. Irrigatsii, No. 96:3 (1958).
333. Tou Kuo-jeng, Displacement of Deposits and Stability of the Bottom of Water Channels, Doctoral Dissertation, Leningrad Institute of Water Transportation Engineers (1960).
334. B. A. Shulyak, Modeling Phenomena in the Atmosphere and Hydrosphere, Izd. Akad. Nauk SSSR (1962), p. 151.
335. U. Kh. Tvenkhofel, Formation of Deposits, ONTI (1936).

336. V. S. Knoroz, Izv. Vses. Nauchn.-Issled. Inst. Gidrotekhniki im. B. E. Vedeneeva, 59:62 (1958); Gidrotekhnicheskoe Stroitel'stvo, No. 8:21 (1953).
337. L. S. Klyachko, Theory and Practice of Dust-Removing Ventillation, Vses. Nauchn.-Issled. Inst. Okhrany Truda VTsSPS (Leningrad), Book 5 (1952), p. 60.
338. E. A. Zamarin, Transporting Power and Permissible Rates of Flow in Channels (Canals), Gosstroiizdat (1951).
339. G. N. Roer, Gidrotekhn. Stroit., No. 8:9 (1948).
340. I. K. Nikitin, Turbulent River Flows and Processes in the Region near the Bottom, Izd. Akad. Nauk UkrSSR, Kiev (1963).
341. P. D. Lobasov, Collection of Papers from the All-Union Scientific-Research Institute of Hydrotechnology and Sanitation Technology, No. 4 (1953), p. 80.
342. Ts. E. Mirtskhulava, Transactions of the All-Union Scientific-Technical Conference on Hydrant Systems and River Processes, Vol. 2, Tbilisi (1960), p. 30.
343. D. M. Mints, Nauchn. Tr. Akad. Kom. Khoz. im. K. D. Pamfilova, No. 4-5:16 (1949); No. 2-3:3 (1959).
344. Yu. M. Shekhtman, Filtration of Low-Concentration Suspensions, Izd. Akad. Nauk SSSR (1961).
345. D. M. Mints, Theoretical Fundamentals of the Technology of Water Purification, Stroiizdat (1964).
346. D. M. Mints and P. B. Krishtul, Nauchn. Tr. Akad. Kom. Khoz. im. K. D. Pamfilova, No. 1:21 (1960).
347. D. M. Mints and V. P. Krishtul, Zh. Prikl. Khim., 33(2):304 (1960).
348. Yu. I. Veitser and L. N. Paskutskaya, Nauchn.-Tr. Akad. Kom. Khoz. im. K.D. Pamfilova, No. 30:69 (1964).
349. E. A. Baranov, Nauchn. Tr. Akad. Kom. Khoz. im. K. D. Pamfilova, No. 1:73 (1960).
350. Yu. L. Sturlis, Mechanization of the Washing of Automobiles, Avtotransizdat (1961).
351. I. P. Plekhanov, Use of Chemical Substances for Washing Automobiles, Avtotransizdat (1956).
352. K. Lidl, Omnibus Rev., 12(5):122 (1961).
353. A. M. Chaika and N. F. Tsyapko, Hydraulic Coal Cutting in Underground Workings, Gosgortekhizdat (1960).
354. M. I. Khmel'nik, Izv. Akad. Nauk SSSR, Otd. Tekhn. Nauk; Mekhan. i Mashinostr., No. 4:167 (1961).
355. D. C. Enkins and J. D. Booker, Intern. J. Air Pollution, 3(1-3):97 (1960).
356. W. Kling, E. Grotte, and H. Mahl, Melliand Tetiler., 35(11):1252 (1954).
357. H. Reumuth, Glasers Ann., 85(7):221 (1961); 86(3):71 (1962).
358. G. Cerizzell, Com. Vehicles, 38(6):36 (1964).
359. W. Rehnert, Deut. Eisenbahntech., 9(7):341 (1961).
360. S. P. Zhebrovskii, Electric Filters, Gosenergoizdat (1950).
361. H. E. Rose and A. J. Wood, An Introduction to Electrostatic Precipitation in Theory and Practice, London (1956).
362. G. W. Penney and E. H. Klingler, Trans. Am. Inst. Elec. Engr., 81(1):200 (1962).
363. A. Duffield and P. Grotenhuis, J. Inst. Metals, 87:33 (1958-59).

364. H. Brandt, Staub, 21(9):392 (1961).
365. G. W. Penney, Arch. Environ. Health, 4(3):301 (1962).
366. I. Sh. Plotinskii, Dust Trapping and Purification of Gases in Nonferrous Metallurgy, Collection of Scientific Papers of the Gintsvetmet, No. 17, Metallurgizdat (1959), p. 67.
367. H. J. White, Industrial Electrostatic Precipitation, Portland State College, London (1963).
368. A. Winkel and A. Schutz, Staub, 22(9):343 (1962).
369. G. Eishold, Arch. Eisenhuttenw., 32(4):221 (1961).
370. I. L. Peisakhov and P. I. Voskresenokii, Dust Trapping in Nonferrous Metallurgy, Collection of Scientific Papers of the Gintsvetmet, No. 9, Metallurgizdat (1955), p. 23.
371. G. M. Gordon and N. P. Ozolit, Dust Trapping and the Purification of Gases in Nonferrous Metallurgy, Collection of Scientific Papers of the Gintsvetmet, No. 17, Metallurgizdat (1959), p. 93.
372. A. M. Arslanova, Collection of Information on Dust Trapping in Nonferrous Metallurgy, Metallurgizdat (1957), p. 160.
373. S. G. Libin, Collection of Information on Dust Trapping in Nonferrous Metallurgy, Metallurgizdat (1957), p. 177.
374. R. Flossman and A. Schutz, Staub, 23(10):443 (1963).
375. W. Jessnitz, Chem. Ztg., 87(19):707 (1963).
376. Iron Steel Engr., 39(5):154 (1962).
377. I. Sh. Plotinskii, Collection of Information on Dust Trapping in Nonferrous Metallurgy, Metallurgizdat (1957), p. 114.
378. G. M. Gordon, Collection of Information on Dust Trapping in Nonferrous Metallurgy, Metallurgizdat (1957), p. 147.
379. G. S. Proshkin, Collection of Information on Dust Trapping in Nonferrous Metallurgy, Metallurgizdat (1957), p. 186.
380. H. Brandt, Staub, 23(8):378 (1963).
381. O. H. Dieter, Staub, 22(9):360 (1962).
382. H. J. Lowe and D. H. Lucas, Brit. J. Appl. Phys., Vol. 24, Suppl. No. 2, p. 40 (1953).
383. M. A. Kornev and E. V. Kirsh, Vestn. Tekhn. i Ekon. Inform. NIITEKhim, No. 11:25 (1962).
384. V. Krystofek, Czechoslovak Patent 102566, February 15, 1962.
385. P. Burnier, Rev. Polytech., No. 1177:37 (1962).
386. J. Orna and J. Nowotny, Czechoslovak Patent 102830, March 15, 1962.
387. J. Wiemer, West German Patent 975017, July 6, 1961.
388. M. Drbohlav, J. Bajger, and J. Uherek, Czechoslovak Patent 98323, January 15, 1961.
389. N. N. Eliseev, Collection of Information on Dust Trapping in Nonferrous Metallurgy, Metallurgizdat (1957), p. 251.
390. S. S. Yankovskii, Dust Trapping and Purification of Gases in Nonferrous Metallurgy, Collection of Scientific Papers of the Gintsvetmet, No. 17, Metallurgizdat (1959), p. 104.

391. V. P. Savraev, Collection of Scientific Papers of the All-Union Mining and Metallurgical Scientific-Research Institute of Metals, No. 6, Metallurgizdat (1960), p. 187.
392. E. S. Zhakabatirov and G. I. V'yugov, Combating Gas and Dust in the Mining Industry, Scientific Papers of the Karelofinsk Scientific-Research Coal Institute, No. 7, Gosgortekhizdat (1963).
393. I. V. Piskarev, Filter Cloths, Preparation and Use, Izd. Akad. Nauk SSSR (1963).
394. J. L. Englesberg, Ind. Eng. Chem., 54(11): 50 (1962).
395. A. Marzocchi, F. Lachut, and W. Willis, J. Air Pollution Control Assoc., 12(1): 38 (1962).
396. G. M. Gordon and B. Z. Karpov, Collection of Scientific Papers of the Gintsvetmet, No. 9, Metallurgizdat (1955), p. 53; G. M. Gordon, N. P. Ozolit, B. V. Petrov, and V. I. Sasin, Dust Trapping and the Purification of Gases in Nonferrous Metallurgy, Collection of Papers of the Gintsvetmet, No. 20, Metallurgizdat (1963), p. 82.
397. H.Ochs, Kalte, 17(10): 562 (1964).
398. T. Gillespie, Internat. J. Air Pollution, 3(1-3): 44 (1960).
399. T. I. Babanina and L. I. Filippenko, Collection of Scientific Papers of the Institute of Mining, Krivorozhsk. Filial Akad. Nauk UkrSSR, No. 1 (1962), p. 123.
400. E. Butterworth, Manuf. Chemist, 35(1): 66; 35(2): 65 (1964).
401. B. V. Petrov and B. S. Gudin, Dust Trapping and the Purification of Gases in Nonferrous Metallurgy, Collection of Scientific Papers of the Gintsvetmet, Metallurgizdat (1963), p. 99.
402. R. Dennis and L. Silverman, Am. Soc. of Heating, Refrigeration, and Air Conditioning Engineers Journal, 4(3): 43 (1962).
403. S. M. Skinner, R. L. Savagen, and J. E. Rutzler, J. Appl. Phys., 24(4): 438 (1953).
404. P. A. Pozdnikov and B. V. Mal'tsev, Tsvetn. Metal., No. 8: 28 (1959).
405. W. O. Vedder and W. F. Gibby, U. S. Patent, October 19, 1959.
406. L. D. Kropp and A. Sh. Bronshtein, Use of Battery Cyclones (Dust Catchers), Izd. "Energiya" (1964).
407. M. K. Gafurbaev, Gigiena i Sanit., No. 8: 102 (1961).
408. O. Ya. Mogilevskaya, Gigiena i Sanit., No. 4: 30 (1960).
409. V. A. Evgrafov, Tr. Leningr. Inst. Vodn. Transp., No. 42: 41 (1963).
410. O. Schmid, West German Patent 1125408, October 4, 1962.
411. L. Silverman and D. M Anderson, U. S. Patent 2992700, July 18, 1961.
412. K. Slobodan, Koncar-Djurdjevoc, and Vaclay Raytor, French Patent 1297089, May 14, 1962.
413. G. Sander, West German Patent 1117980, May 30, 1962.
414. A. V. Ryzhov, Author's Certificate USSR 151558, October 31, 1962; G. Sandstrom, Swedish Patent 180239, August 7, 1962.
415. C. Allander, Swedish Patent 179922, July 3, 1962.
416. Filter-Zyklon, Maschinenmarkt, 68(58): 38 (1962).
417. F. Schaub and J. Hibbel, West German Patent 1090642, June 6, 1961.
418. M. T. Berkovich and Ya. Z. Bukhman, Industrial Dust, Sverdlovsk (1960).
419. W. Gutowski, Energetyka Przemyslowa, 10(12): 435 (1962).

420. M. T. Berkovich, Collection of Papers on Silicosis, No. 1, Sverdlovsk (1956), p. 41.
421. I. N. Plaksin and N. F. Olofinskii, Collection on the Enrichment of Minerals, Izd. Akad. Nauk SSSR (1960), p. 124.
422. N. F. Olofinskii, Electrical Methods of Beneficiation, Gosgortekhizdat (1962).
423. I. N Plaksin, N. F. Olofinskii, and V. A. Novikova, Scientific Transactions of the A. A. Skochinskii Institute of Mining, No. 19 (1963), p. 3.
424. R. Brison and O. Tangel, Mining Eng., 12(8):913 (1960); E. Occella, Atti e Rass. Tecn., 15(1):23 (1961).
425. A. Wesner, Mining World, 23(5):40 (1961).
426. A. Z. Yurovskii, V. D. Gorshko, V. I. Korshunov, and I. D. Remesnikov, in collection: Technological Progress in the Enrichment of Coals, Gosgortekhizdat (1963), p. 104.
427. V. M. Fridkin, Physical Basis of the Electrophotographic Process, Izd. "Énergiya" (1966).
428. R. M. Schaffert, Electrophotography, New York (1965).
429. A. B. Dravin, Study of the Process Underlying the Production of Electrophotographic Images, Master's Thesis, Moscow Polygraph. Inst. (1962).
430. I. I. Zhilevich, in collection: Electrophotography and Magnetography, Vilna (1965), p. 166.
431. G. F. Daletskii, Uch. Zap. Saratovsk. Gos. Univ., Vyp. Khim., 21:161 (1949).
432. A. B. Dravin and Zh. V. Kirillova, Preparation of Offset Prints by the Electrophotographic Method, Izd. VNIIPP (1962); collection: Electrophotography and Magnetography, Vilna (1965), p. 135.
433. I. I Zhilevich and E. L. Nemirovskii, Electrophotography, Iskusstvo (1961).
434. D. H. Dessauer, G. R. Mott, and H. Bogdonov, in collection: Problems in Electrography [Russian translation], IL (1960), p. 44.
435. G. R. Mott and E. H. Lehmann, US PATENT 3008826, November 14, 1961.
436. I. V. Anfilov, Study of the Development Process in Electrophotography, Master's Thesis, Moscow (1962).
437. K. A. Metcalfe and R. J. Wright, J. Oil and Color Chem. Assoc., 39(11):845 (1956).
438. H. E. Copley, US Patent 2484782, October 11, 1949; in collection: Problems in Electrography [Russian translation], IL (1960), p. 11.
439. K. A. Metcalfe and R. D. Wright, in collection: Problems in Electrography [Russian translation], IL (1960), p. 33.
440. A. D. Zimon and G. V. Nozdrina, in collection: Technology of Obtaining Images and Materials Used in Electrophotography, Vses. Nauchn.-Issled. Inst. Poligr. Prom. (1964), p. 59.
441. H. E. Crumrine and C. L. Huber, U. S. Patent 3009402, November 21, 1961.
442. S. V. Arshvila and I. V. Chekavichus, Preparation of Offset Prints by the Electrophotography Method, VNIIPP (1962), p. 16.
443. G. R. Mott, The Prenrose Annual, 50:133 (1956).
444. G. A. Vinogradov and L. M. Komarova, Poroshkovaya Met., No. 1(7):27 (1962).
445. Yu. V. Naidich, I. A. Lavrinenko, and V. N. Eremenko, Poroshkovaya Met., No. 1(19):5 (1964).

446. W. Susse, Staub, 20(12): 429 (1960).
447. N. A. Ermush, Tr. Inst. Lesokhoz. Probl. i Khim. Drevesiny Akad. Nauk LatvSSR, 21: 55 (1960).
448. J. Mikes, Hungarian Patent 147530, October 1, 1960.
449. I. S. Gorbunov, I. G. Shatalova, V. I. Likhtman, N. V. Mikhailov, and P. A. Rebinder, Poroshkovaya Met., No. 6: 10 (1961).
450. A. Pokorny, Strojirenska Vyroba, 10(6): 296 (1962).
451. G. I. Aksenov, Fundamentals of Powder Metallurgy, Kuibyshev. Knizhnoe Izd. (1962).
452. O. Telle and R. Erdmenger, West German Patent 945086, June 28, 1956.
453. P. S. Titov and E. Ya. Besidovskii, Izv. Vysshikh. Uchebn. Zavedenii Tsvetn. Met., No. 2: 162 (1966); E. Ya. Besidovskii, Study and Development of a Method of Depositing Protective and Decorative Coatings by Means of Metal Powders, Master's Thesis, Moscow Institute of Steels and Alloys (1966).
454. N. V. Kuznetsov, Working Processes and Problems in Perfecting the Convective Surfaces of Boiler Installations, Gosenergoizdat (1958).
455. I. P. Epik, Effect of the Mineral Part of Shales on the Conditions of Operation of a Boiler Plant, Estgosizdat (1961).
456. A. M. Gurvich and R. S. Prasolov, Teploenergetika, No. 7: 80 (1960).
457. J. Crossley, Inst. Fuel, 25(145): 9 (1952).
458. N. N. Lebedev and I. P. Skal'skaya, Zh. Tekhn. Fiz., 32: 375 (1960).
459. A. M. Gurvich, A. G. Blokh, and A. I. Nosovitskii, Teploenergetika, No. 2: 3 (1955).
460. R. S. Prasolov, External Deposits on the Heating Surfaces of Boiler Plants, I. I. Polzunov Central Scientific-Research Boiler-Turbine Institute (IKTI), Review No. 3, Mashgiz (1958); Mass and Heat Transfer in Boiler Systems, Izd. "Energiya" (1964).
461. A. V. Sinyavskii, Effect of the Introduction of Lime into the Gas Ducts of Boilers on the Structure and Composition of the Ash Deposits Formed on the Combustion of Fuel Oil, Ordzhonikidze (1957).
462. H. Norda, Mitt. Ver. Grosskesselbesitzer, No. 80: 364 (1962).
463. Oil Engine and Gas Turbine, 23(229): 280 (1955).
464. G. I. Skanavi, Physics of Dielectrics, Fizmatgiz (1958).
465. S. M. Skinner, R. L. Savagen, and J. E. Rutzler, J. Appl. Phys., 24(4) (1953).
466. P. Burnier, Rev. Polytechn., No. 1177: 37 (1962).
467. N. A. Kaptsov, Electrical Phenomena in Gases in Vacuum, Tekhteoretizdat (1950).
468. M. Kh. Pugilevskii, Collection of the Scientific-Research Papers of the All-Union Scientific-Research Institute of Agricultural Machine Building, No. 2, Theory, Construction, and Production of Agricultural Machines (1936).
469. N. A. Kachinskii, Mechanical and Microaggregate Composition of Soils; Study Methods, Izd. Akad. Nauk SSSR (1958).
470. M. U. Nickols, Agr. Eng., 12(8): 321 (1931); F. A. Kummer and M.U. Nickols, Agr. Eng., 19(2): 13 (1938).
471. A. N. Kosarchuk, Agrophysical Estimation of the Quality of Tilling (Plowing) at High Speeds, Master's Thesis, Agrofiz. Nauchn.-Issled. Insg. Leningrad (1960).

472. A. Z. Baibakov, Physical Interaction of Soil with a Metal Surface on Working, [in Russian] Master's Thesis, Agrofiz. Nauchn.-Issled. Inst. Leningrad (1963).
473. Express Information of Agricultural Machinery, No. 42:767 (1962); Sel'skoe Khoz. Za Rubezhom (Rastenievodstvo), No. 2:78 (1963).
474. Plastic Coatings for Plowshares, Agricultural Technology, No. 5, Sel'khozgiz (1962), p. 35.
475. M. I. Bredun, Vestn. Sel'skokhoz. Nauki, No. 7:71 (1963).
476. A. F. Vadyunina, Scientific Records of the M. V. Lomonosov Moscow State University, Soil Science, No. 44 (1940).
477. N. A. Kachinskii, Structure of Soil, Izd. Mosk. Gos. Univ. (1963).
478. A. V. Kitaev, Vestn. S.-Kh. Nauki, No. 9:127 (1957).
479. V. N. Livshits and V. M. Moiseev, Electrical Phenomena in Aerosols and Their Application, Izd. "Energiya" (1965).
480. M. Pauthenet, in collection: Use of the Force of an Electric Field in Industry and Agriculture [Russian translation], VNIIEM (1964), p. 230.
481. A. V. Kitaev, Possibility of Using Electrostatic Forces for Depositing Chemicals to Protect Plants, Mastera's Thesis, Moscow (1958).
482. J. Walker, Agr. Eng. Res., 6(2) (1961).
483. A. A. Smirnova, Tr. Vses.Inst. Zashchity Rast., No. 4:69 (1952); Dokl. Vses. Akad. Sel'skokhoz. Nauk im. V. I. Lenina, No. 2:19 (1952).
484. S. F. Bezuglyi, L. Z. Osityanskaya, and B. A. Akimov, Khim. Nauka i Promy., 4(5):636 (1959).
485. Soc. des Ets. Abel, French Patent 1286891, January 29, 1962.
486. V. Dejmek, Czechoslovak Patent 104296, July 15, 1962
487. S. F. Bezuglyi, Zh. Veses. Khim. Obshchestva im. D. I. Mendeleeva, 5(3): 298 (1960).
488. Q. A. Geering and J. N. Lloyd, Mededel. Landbouwhogeschool Opzoekingsstat. Staat Gent., 26(3):1471 (1961).
489. D. Pielou and K. Williams, Proc. Entomol. Soc. Brit. Columbia, 59(6):18 (1962).
490. Hamada Iosikadzu and Hirota Koki, Japanese Patent 14250, September 28, 1969; cited in Journal of Abstracts: Chemistry (1963), 19N290P.
491. Use of Aerosols in Agriculture (collection) [Russian translation], IL (1955), p. 74.
492. D. Pielou and K. Williams, Can. J. Plant Sci., 42(2):371 (1962).
493. K. Williams and D. Pielou, Can. Entomologist, 94(8):874 (1962).
494. E. De Ong, C. Kenneth, and L. Fancher, J. Econ. Entomol., 43(4):542 (1960); in collection: Use of Aerosols in Agriculture IL 1955), p. 112.
495. K. Ya. Kalashnikov, Zashchita Rastenii ot Vreditelei i Boleznei, No. 8:24 (1962).
496. B. I. Rukavishnikov, Zashchita Rast. ot Vreditelei i Boleznei, No. 2:54 (1957).
497. S. F. Bezuglyi, A. F. Litvinova, K. A. Gar, and E. I. Andreeva, Author's Certificate 157582; Byull, Izobret. i Tovarnykh Znakov, No. 1884 (1963).
498. V. A. Ocheretyanyi, Selektsia and Sadovodstvo, No. 8:63 (1952).
499. C. Giovagnotti, Erosione Suolo Italia, 2:23 (1962); cited in the abstracts of the journal Pochvovedenie, 6:57, 90 (1964).
500. V. V. Slastikhin, Vestn. Sel'skokhoz. Nauki, No. 11:149 (1958).

501. V. V. Zvonkov, Wind and Water Erosion, Izd. Akad. Nauk SSSR (1962).
502. A. Sundborg, The River Klar-elven, Study of Fluvial Processes, Geografika Annaler., Stockholm, 38(2-3) (1956).
503. R. A. Bagnold, The Physics of Sand and Desert Dunes, London (1954).
504. T. F. Yakubov, Pochvovedenie, No. 5:41 (1957).
505. A. N. Kiselev, Pochvovedenie, No. 9:840 (1952).
506. K.I. Andrianova, Zemledelie, No. 11:43 (1960).
507. A. P. Bocharov, Estimating the Wind Resistance of the Surface of Eroded Soils, Sel'khozizdat (1963).
508. P. V. Vershinin, Soil Structure and Conditions of Its Formation, Izd. Akad. Nauk SSSR (1958).
509. S. A. Modina, Pochvovedenie, No. 3:62 (1962).
510. V. I. Bobchenko, Pochvovedenie, No. 10:71 (1961).
511. L. N. Yastrebova, Avtomob. Dorogi, No. 8:10 (1960).
512. K. P. Mashin, Avtomob. Dorogi, No. 6:15 (1960).
513. Soc. P. R. L. Belgocodex, Belgian Patent 551502, December 18, 1959.
514. N. E. Pestov, Physicochemical Properties of Granula and Powdered Chemical Products, Izd. Akad. Nauk SSSR (1947).
515. A. M. Shcherbakov, M. L. Kruglyakov, and M. F. Burmistrova, Khim. v Sel'skom Khoz., 3(9):17 (1965).
516. M. N. Nabiev and A. A. Vishnyakova, Dokl. Akad. Nauk Uz.SSR, No. 2:25 (1957).
517. R. A. Dymshits, Zh. Vses. Khim. Obshchestva im. D. I. Mendeleeva, 7(5):550 (1962).
518. S. Pawlikowski and S. Szymonik, Chemik, 16(1):14 (1963); E. Douglass and F. Schold, Fertilizer and Feeding Stuffs J., No. 37:634 (1951).
519. A. M. Dubovitskii, F. G. Margolis, and T. V. Glazova, Khim. Prom., No. 2:93 (1954).
520. N. I. Krylova and N. K. Nabiev, Izv. Akad. Nauk Uz.SSR, No. 10:29 (1957).
521. M. G. Makvelyan, G. O. Grigoryan, et al., Author's Certificate 129648, July 1, 1960.
522. H. Pecht, Chem.-Ingr.-Tech., 34(8):564 (1962).
523. Minerals Chemicals Philipp Corp. and H. Smith, US Patent 3041159, June 26, 1962.
525. O. Balz, S. Mohr, and W. Wagner, West German Patent 1062713, January 14, 1960.
524. Soc. An. du Blanc Omya, French Patent 1214828, April 12, 1960.
526. Chemische Fabrik Kalk G. m.b.H., F. Brandt, and A. Dolbe, West German Patent 1070198, January 13, 1958.
527. J. Ordonneau, Ann. Mines, 139:35, 63 (1956).
528. Sin Ben-don and N. S. Torocheshnikov, Zh. Neorgan. Khim., 2(4):896 (1957).
529. A. L. Shneerson, V. A. Klevke, and M. A. Miniovich, Zh. Prikl. Khim., 29(5): 682 (1956).
530. N. I. Krylova, A Study of the Mechanisms Responsible for Stopping the Caking of Ammonium Nitrate, Master's Thesis, Akad. Nauk Uz.SSR, Inst. Khimii, Tashkent (1958).

531. T. V. Zabolotskii and K. I. Shvetsova, Khim. Prom., No. 6:11 (1953).
532. Ya. I. Kil'man and V. A. Klevke, Tr. Nauchn.-Issled. i Proektn. Inst. Azotnoi Prom., No. 6:239 (1956).
533. E. Jankowiak, Z. Waligora, and H. Zajonz, Przemysl. Chem., 42(3):140 (1963).
534. N. V. Babenko, Khim. v Sel'skom. Khoz., No. 2:19 (1964).
535. N. P. Kurin, Izv. Tomsk. Politekhn. Inst. im. S. M. Kirova, 71:25 (1952); Khim. Prom., No. 5:129 (1953).
536. Scholven-Chemie A. G., G. Meyer, and H. Thaden, West German Patent 1137049, September 1, 1960.
537. K. Seshadri, B. Chouhari, and M. Oza, Res. and Ind., 8(6):164 (1963).
538. Phillips Petroleum Co. and V. Vives, U.S. Patent 3034848, May 15, 1962.

Index